Road Congestion Pricing in Europe

Road Congestion Pricing in Europe

Implications for the United States

Edited by

Harry W. Richardson

The James Irvine Chair of Urban and Regional Planning, School of Policy, Planning and Development and Professor of Economics, University of Southern California, USA

and

Chang-Hee Christine Bae

Associate Professor of Urban Design and Planning, University of Washington, Seattle, USA

Edward Elgar

Cheltenham, UK • Northampton, MA, USA

Published by
Edward Elgar Publishing Limited
Glensanda House
Montpellier Parade
Cheltenham
Glos GL50 1UA
UK

Edward Elgar Publishing, Inc.
William Pratt House
9 Dewey Court
Northampton
Massachusetts 01060
USA

A catalogue record for this book
is available from the British Library

ISBN 978 1 84720 380 9

Printed and bound in Great Britain by MPG Books Ltd, Bodmin, Cornwall

Contents

Contributors

Chang-Hee Christine Bae Department of Urban Design and Planning, University of Washington, Seattle, WA, USA.

David Banister Transport Studies Unit, Oxford University Centre for the Environment, Oxford, UK.

Alon Bassok Department of Urban Design and Planning, University of Washington, Seattle, WA, USA.

Michael G.H. Bell Centre for Transport Studies, Imperial College London, UK.

Terence Bendixson Independent Transport Commission, University of Southampton, UK.

Karin Brundell-Freij Department of Technology and Society, University of Lund, Sweden.

Sungbin Cho ImageCat Inc., Long Beach, California, USA.

Marcial Echenique Department of Architecture and Martin Centre, Cambridge University, UK.

Jonas Eliasson SEKTEC, Stockholm, Sweden.

Stephen Glaister Centre for Transport Studies, Imperial College London, UK.

Peter Gordon School of Policy, Planning and Development, University of Southern California, Los Angeles, CA, USA.

Daniel J. Graham Centre for Transport Studies, Imperial College London, UK.

Anthony J. Hargreaves Department of Architecture and Martin Centre, Cambridge University, UK.

Kenny Ho Department of Economics, University of Birmingham, UK.

Muriel Beser Hugosson SEKTEC, Stockholm, Sweden.

Stephen Ison Transport Studies Group, Department of Civil & Building Engineering, Loughborough University, UK.

David King Department of Urban Planning, University of California at Los Angeles, CA, USA.

A. Koh Institute of Transport Studies, University of Leeds, UK.

Pierre Kopp Economics Research Center at the Sorbonne, University of Paris I, France.

Bumsoo Lee School of Policy, Planning and Development, University of Southern California, Los Angeles, CA, USA.

Shin Lee School of City and Regional Planning, Cardiff University, UK.

David Maddison Department of Economics, University of Birmingham, UK.

Michael Manville Department of Urban Planning, University of California at Los Angeles, CA, USA.

Bryan Matthews Institute of Transport Studies, University of Leeds, UK.

Anthony D. May Institute of Transport Studies, University of Leeds, UK.

Batool Menaz Institute of Transport Studies, University of Leeds, UK.

James E. Moore II Department of Industrial and Systems Engineering, Viterbi School of Engineering, University of Southern California, Los Angeles, CA, USA.

Chris Nash Institute of Transport Studies, University of Leeds, UK.

Qisheng Pan Urban Planning and Environmental Policy, School of Public Affairs, Texas Southern University, Houston, USA.

Rémy Prud'homme Institute of Urbanism for Paris and University of Paris XII, France.

Martin G. Richards Consultant and Author, Coldharbour, Surrey, UK.

Harry W. Richardson School of Policy, Planning and Development, University of Southern California, Los Angeles, CA, USA.

Tom Rye Transport Research Institute, Napier University, Edinburgh, UK.

Georgina Santos Transport Studies Unit, Oxford University, UK.

S.P. Shepherd Institute of Transport Studies, University of Leeds, UK.

Donald Shoup Department of Urban Planning, University of California at Los Angeles, CA, USA.

A. Sumalee Institute of Transport Studies, University of Leeds, UK.

Muanmas Wichiensin Centre for Transport Studies, Imperial College London, UK.

Preface

Most of the chapters in this book were first presented at the Moeller Centre at Churchill College, Cambridge University, on 6–7 May 2006. A few were commissioned separately. The focus of the workshop was road congestion pricing, inspired by the London Congestion Charging Scheme introduced in February 2003, although the scope of the workshop was much wider, looking at Western Europe (especially the Stockholm Trial) and the United States. A motivating idea was whether the European experience had implications for implementing road pricing in the United States. Although there is a renewed interest in road pricing in the United States, in most cases the reduction in congestion objectives are secondary to the highway financing potential of road pricing schemes.

Harry W. Richardson
Chang-Hee Christine Bae

1. Introduction

Harry W. Richardson and Chang-Hee Christine Bae

This introduction has two purposes: to present the book's central theme, that is, the implications of London's Congestion Charging Scheme and the Stockholm Trial for the United States, and to summarize the key points of the contributing chapters.

The idea of pricing for the use of roads has been around for a long time, stretching back at least to the turnpike roads of the eighteenth century (much more common in the United Kingdom and other parts of Western Europe than in the United States), and more recently, the privilege under the Interstate Highway System for each State to designate one highway as a toll road (implemented more in Northeastern States, such as New York and New Jersey) and the plethora of toll bridges throughout the country. But all these examples were either to raise revenue or to recover construction costs not to decongest roads.

The idea of pricing as an instrument to tackle road congestion is based on literature in economic theory from the early 1960s in which economists such as Walters (1961), Vickrey (1963) and Johnson (1964) developed the standard road congestion analysis to demonstrate that the market equilibrium derived from unpriced roads results in excessive congestion. The key idea was obvious: drivers pay only for their own congestions not those of others. Setting the price of driving equal to the social marginal cost, however, would reduce traffic to its optimal level. There may still be some level of congestion, but it will be 'optimal'. The possibility that this could be translated into transportation policy, however, lagged for a long time primarily on the grounds of political feasibility: no politician subject to regular elections would risk the wrath of the voter-driver by implementing such a proposal.

The first breach in this view was made in a less democratic society (Singapore) where a successful downtown congestion pricing scheme has been in place since 1975. A later example, on a ring road in Trondheim, Norway, from 1986, was for revenue raising not congestion control, with a very moderate toll that was supposed to be abolished once the construction costs had been recovered.

This introduction deals with a much more recent and highly publicized scheme, the London Congestion Charging Scheme begun in 2003 (Leape, 2006; Richardson and Bae, 2006), and to a much lesser extent with the so-called Stockholm Trial of 2006. These are discussed in great detail in some of the later chapters; here we focus on an overview. However, the main purpose of this preliminary evaluation is to raise the question of whether the European experiences enhance the prospects of a wider application of congestion pricing in the United States. In that context, we should also take into account the beginnings of such applications, either implemented or in the planning stage, in several States, 23 according to the Government Accountability Office (GAO, 2006).

1 THE LONDON CONGESTION CHARGING SCHEME (LCCS)

Introduced in February 2003, the LCCS was the brainchild of the first elected mayor of London, Ken Livingstone (formerly known as 'Red Ken' during the Thatcher era when he was Leader of the Greater London Council). The scheme is an area licensing scheme in which charges are imposed when vehicles cross a boundary into or out of the zone. The zone is part of Central London, specifically the Borough of Westminster (see Figure 8.1 in Santos, Chapter 8 of this book) although an approximate doubling of its size to include the Borough of Kensington and Chelsea took place in February 2007 (Figure 8.2, also in Santos, Chapter 8). The initial zone was small, covering only 8.4 square miles, or about 1.3 per cent of the total area of Greater London; the extended zone is almost double the original.

The charge applies between 7:00 am and 6:30 pm, Monday to Friday (excluding public holidays such as the quite frequent Bank Holiday Mondays). The fee was initially £5 per day but was raised to £8 (almost $16) in July 2005. There are about 165,000 violations per month with a rising scale of penalties up to £150 after a month. Currently, more than a third of permit purchases are made through retail outlets (typically newsagents and tobacconists). Exemptions include buses, taxis, police cars, emergency service vehicles, disabled drivers and certain alternative fuel vehicles. Residents of Westminster and the adjacent borough (Kensington and Chelsea) receive a 90 percent discount after paying an annual registration fee of £10. The scheme is enforced via automatic number plate recognition (ANPR), based upon camera sites located at each entry and exit point and also within the zone, reinforced with a manual check.

The western extension has two free corridors (one of them North-to-South along the original zone boundary), and the hours have been shortened, with the charge ceasing at 6:00 pm rather than 6:30 pm (in both the extension and the original zone). However, it does not have the same impact as the original zone because two-thirds of the traffic entering the zone is not subject to charging. Nevertheless, trip reductions are more than 10 percent with a similar increase in trip speeds. On the other hand, traffic has increased by about 2 percent in the original zone with free travel into Kensington and Chelsea.

Livingstone had three major objectives in introducing the charge: raising revenue; promoting public transit; and, of course, reducing congestion (although this was the only goal claimed by Transport for London: TfL). Attaining the first objective fell far short of expectations. The original goal was a *net* revenue of £130 million, but only £50 million were raised in the first year (despite 110,000 participants per day). The two explanations for the shortfall were much higher administrative and operating costs than anticipated (mainly because of technological glitches) and a sharper reduction in vehicles entering the zone than predicted. The raising of the charge has a problematic impact on revenues because of the price elasticity effect, although the extension of the boundary should have a positive impact on revenues. Nevertheless, the target *net* revenue of up to £100 million remains problematic.

Meeting the second objective (promoting public transit) was very successful. TfL added hundreds of buses, and of the 60,000 reductions in the number of drivers entering the zone, about one-half shifted to taking the bus (a much smaller number shifted to the Underground (that is, the subway) where capacity was not increased). Part of the reason was a dramatic decline in bus traffic delays by 60 percent.

The overachievement of the third objective was perhaps the most surprising. Total traffic entering the zone declined by 18 percent (the drop in charged vehicles was somewhat greater, but was offset by an increase in exempt vehicles), cars declined by 33 percent, and traffic circulating within the zone dropped by 15 percent. These reductions were also associated with a 21.5 percent increase in trip speeds. In addition to the shift to buses and the underground, about one-quarter of the vehicles moved around the zone, with some negative effects such as more congestion (resulting from a 4 percent per annum increase in vehicle miles traveled (VMT) in the inner ring roads, somewhat mitigated by improvements in transportation management) and more air pollution although there was significantly less pollution within the zone (see Ho and Maddison, Chapter 10, and Banister, Chapter 9). With respect to air pollution, there is a proposal being discussed to impose differential tolls based on emission rates (this would represent an important step in internalizing two kinds of transportation externalities, both congestion and pollution). Other consequences of the traffic reduction included: (i) shifts to the less important modes (carpools, motorcycles, bicycles and walking); (ii) diversion of traffic to uncharged hours; and (iii) trip reductions. These last effects explain a reduction of about 12,000 cars entering the zone (or a fifth of the total).

A key question is how the LCCS stands up in a cost–benefit analysis. An early internal analysis by TfL was based not on optimality, which requires marginal analysis, but on a comparison of total costs and benefits. The study estimated annual benefits of £180 million and annual costs of £130 million. The major item on the benefits side of the ledger was time savings while the major cost item was the operating costs of the scheme. Santos and Fraser (2006) undertook a more sophisticated cost–benefit analysis that consistently showed benefit–cost ratios below unity.

Santos (Chapter 8) has undertaken a detailed analysis of the price elasticities and marginal costs associated with the LCCS. Her estimates of the point elasticity for cars is quite high because the ubiquity of public transit facilitated shifts from cars to transit. The elasticity for taxis is very high (note that in this case taxis are exempt and taxi use increased substantially) because the charge encouraged a strong substitution between cars and taxis. The elasticities for both vans and trucks are low because there are few substitution modes, and routes (and to a lesser extent times) are difficult to change. The congestion charge elasticities are very low because the increase in the charge in 2005 had a negligible deterrent effect. Also, truck traffic increased when the charge went up, primarily because of a truck fleet discount scheme.

Santos also estimates the marginal congestion costs by mode and compares them with the actual congestion charge per mile driven. Her results show that an efficient scheme would charge individual modes a different rather than a uniform charge. This would be relatively easy to do from an administrative point of view but difficult politically because the truck lobby would strongly resist raising the charge for the grossly undercharged trucks. The data suggest that cars are overcharged, while trucks and (to a lesser extent) vans are severely undercharged. The increase in the charge to £8 aggravated the overcharging of cars, and brought the charge for vans closer to their marginal congestion cost while trucks remain significantly undercharged.

Interestingly, while the LCCS has received enthusiastic support abroad, largely because of its reductions in automobile use, European observers have, in general, been more critical. Santos and Fraser (2006) have criticized the scheme as being inefficient, arguing that the charge does not approximate the marginal social costs of congestion. Similarly, Prud'homme and Bocarejo (2005) claimed that the scheme is an economic loss in cost–benefit terms, offset somewhat by its political success. Also, research at Imperial College London suggested that congestion charging has had a significant effect on sales in some central London retail establishments because of the shift in shopping habits to suburban outlets (*Local Transport Today*, 2004).

2 THE STOCKHOLM TRIAL

Even more recent than the LCCS is the trial with congestion pricing in Stockholm that began with an expansion of public transport from August 2005 and continued with a congestion pricing experiment between January and July 2006 authorized by the Congestion Charges Act of 2004. In the General Election on 17 September 2006, the result of the vote was a modest 5 to 4 majority to make the experiment permanent (Eliasson et al., Chapter 15).

The main component of the public transit initiative was the purchase of new buses, the opening of new bus lines, and the establishment of new park-and-ride facilities with 13,800 parking spaces (these met an unfulfilled demand, regardless of the congestion charge). The toll scheme was a cordon operating between 6:30 am and 6:30 pm on weekdays. The most interesting feature of the Stockholm Trial is its variable time-of-day tolls. There is no toll on the major bypass road (Essingeleden) or on drivers from Lidingö Island in the East provided that they cross the cordon within 30 minutes (Stockholm is the only land connection for the island).

What were the main results of the six-month trial? The most important is a 22 percent decline in all auto passages (about 100,000), somewhat steeper in the afternoon/evening than in the morning peak, on all control points (there was also a 10 percent decline in truck passages). An expected consequence was faster travel times. Less expected was the very modest, almost negligible, increase in public transit use (see Chapter 15 and Hugosson and Eliasson, 2006 for detailed results).

Other results include: a decline in exhaust emissions in the inner city (by 14 percent in the city but much less (3 percent) in Stockholm County); a decline in accidents within the toll zone (by up to 10 percent); a detectable but insignificant decline in noise levels; no increase in carpooling; and a negligible impact on the regional economy. The technology worked well, as measured by the very few appeals against the charges, with almost none being accepted. As for cost–benefit analysis, the start-up costs were very high and clearly were not recouped in a six-month trial. However, Eliasson et al. argue that a permanent toll will generate a net present value of SEK 760 million, with a quite rapid 4-year payback period, largely based on travel time savings. On the other hand, the bus investments are definitely 'unprofitable' in terms of cost–benefit evaluation. Another point is that the short-term adaptation to a previously announced six-month trial may be very different from the long-term response. Certainly, public opinion changed over the trial period, shifting from a 55 percent to a 41 percent disapproval of the toll.

Prud'homme and Kopp (2006) have taken a more critical approach. While we could argue about the different numbers from alternative sources, Prud'homme and Kopp bring out some broader issues, although they do produce data that indicate that the toll was too high (leading to an 'excessive' reduction in auto traffic) and estimated that costs were three times the benefits, largely because of the expensive public transit component. On the other hand, they acknowledge that the scheme might generate net social benefits by about 2020, as operating costs stabilize or fall and the value of time increases.

Their concerns are largely political. Stockholm residents gain, while national taxpayers lose because a high proportion of the start-up costs were financed out of the national budget. The vote for the toll is difficult to interpret because it was fudged in the ballot with a vote for public transit service improvement. In addition, because different political parties adopted strongly divergent views about the toll, party loyalty might have overridden serious thought about the merits or demerits of the toll. Furthermore, the high start-up costs involved in setting up the trial were perceived as a 'sunk' cost skewing the vote in favor of approval.

They also spell out three conditions for success: (i) severity of congestion (very low in Stockholm compared with London, implying that perhaps the time savings benefits in London might be up to 10 times higher); (ii) low implementation costs (Stockholm's were more than half those of London, despite the difference in benefits); and (iii) cheap public transportation (but it is quite expensive in Stockholm). So, their conclusion is that the Stockholm toll is not cost effective.

3 RECENT DEVELOPMENTS

There have been recent developments that point to an expansion of road congestion pricing experiments in Europe. These have been partially but not wholly driven by the recent new air pollution regulations passed by the European Parliament in December 2007. These are estimated to produce health benefits of more than 40 billion euros by 2015 and achieve a 40 percent reduction in premature deaths over the period 2000–2020. Among other developments, this has helped to promote a new charge program in Milan ('Eco-Pass') started in January 2008, and the acceleration of new road pricing plans in several East European countries (e.g. Poland and Hungary). It has also resulted in a plan to increase the London charge for high-polluting vehicles (such as SUVs) to £25 sterling. Perhaps the most ambitious program in the European Union is the Valletta CVA (Controlled Vehicle Access) in Malta which began in May 2007 before the passage of the new European Union regulations. A plan was developed for Dublin in September 2007 to be initiated soon on the A50 highway with a variable toll of 2–3 euros. However, there have been developments outside the European Union too. Especially prominent is the Oslo Toll Ring automated system initiated in February 2008 (in fact, six Norwegian cities have schemes in operation, and the original goes back to Trondheim in 1986 when a toll was imposed to finance road construction).

4 IMPLICATIONS FOR THE UNITED STATES

There are several types of congestion pricing schemes: area licensing (such as London); a cordon (such as Stockholm); segment (corridor) tolling; joint HOT/HOV (high occupancy toll/high occupancy vehicle) lanes; and system-wide tolling. The choice among these depends upon many factors such as the spatial structure of the metropolitan area and/or its inner core and the objectives of the tolling policy. If the prime candidates for congestion pricing in the United States are the very large cities, area licensing or cordon schemes are ruled out in most cases because of spatial structure and the dynamics of metropolitan growth; in most cities there are too many access routes, and a downtown area is not the most attractive area for tolling and could easily accelerate the decentralization that city governments are so desperately trying to slow down. Two possible exceptions are New York City and San Francisco because of their topography. New York's toll bridges and tunnels (with very few free access routes into Manhattan) make Manhattan almost a cordon. However, a sensible pricing strategy for Manhattan probably requires tolls that are differentiated by route and time of day, and this demands at least a mini-system-wide (that is, cordon and internal area) scheme. Nevertheless, both San Francisco and New York City have discussed schemes very similar to that of London.

There was a major development in New York City in December 2006. The Manhattan Institute published a study (Schaller, 2006) that explored the possibility for a cordon pricing scheme in the two Manhattan central business districts (CBDs) (Midtown and Downtown), south of 60th St. The Manhattan Institute held a conference soon after. The immediate reactions were not positive. They included the costs to some low-income workers, especially those living in the peripheral parts of the outer boroughs, the difficulties of parking near subway stations, and subway capacity problems. Even more telling was the response regarding political acceptability. It was argued that no politician in New York would include road pricing as part of a campaign platform as Ken Livingstone did in London or submit the measure to a voter referendum as in Stockholm. Mayor Michael Bloomberg's initial reaction was that people would regard it as a commuter tax and the State legislature would veto the idea. However, in April 2007, he changed his mind and came out in favor of a \$8 entry toll south of 86th St. and for drivers within Manhattan itself a \$4 charge. It would require State approval, and some legislators objected on equity grounds. It would not be quite as revolutionary as it appears because the toll would be defrayed by bridge tolls, and most entry points into Manhattan are toll bridges. Also, many car owners living in Manhattan keep their cars in other boroughs for weekend and vacation use, and area tolls might reinforce that trend. However, given New York City's comprehensive transit system and permitting some relaxation of taxi medallion restrictions, road congestion pricing there is both technically feasible and economically viable. However, in July 2007 the State legislature refused to take up the proposal and an important Federal deadline for a possible \$500 million grant was missed. This meant that the implementation of a congestion pricing scheme for New York City became a dead issue, at least in the short run.

However, in August 2007 the story took yet another turn. The US Department of Transportation selected New York City as one of five beneficiaries (the others were San Francisco, Miami, Minneapolis and Seattle) of the new Urban Partners program, despite New York City's failure to make the application deadline. The grant was for \$354.5 million, although there were other components in addition to congestion pricing. There were still hurdles to be overcome: opposition from some politicians in the Outer Boroughs and approval needed from several bodies (the new 17-member New York City Traffic Congestion Mitigation Committee, the State Legislature and the City Council) by March 2008. Nevertheless, it appeared to be an impressive victory for Mayor Bloomberg.

However, the victory was temporary. Although New York City Council voted in favor of congestion pricing by a 30–20 vote on 31 March 2008, the State Assembley (under pressure from Democrats in the New York City Outer Boroughs and suburbs) refused to take the proposal to the floor for a vote on 7 April 2008. As a result, for the foreseeable future, the plan is dead and New York City lost Federal funds for the project.

The easiest approach is probably an expansion of HOT/HOV lanes given the existence of many HOV (or carpool) lanes in many metropolitan areas. Opening up these lanes to paying solo drivers, that is, converting them to HOT lanes, represents a very simple approach to expanding the scope of congestion pricing (Lee and Gordon, Chapter 17). A variant of the HOT/HOV concept is the existence of adjacent free lanes and HOT/HOV lanes as in the example of SR91 in Southern California (Richardson et al., Chapter 18), a 10-mile stretch in Orange County between Los Angeles County and Riverside County, formerly in private ownership and now back in the hands of the Orange County Transit District (OCTD).

Corridor projects are often stimulated as much by construction financing needs as by congestion pricing goals. For example, Washington State has been plagued by transportation revenue shortfalls in recent years, aggravated by stalemate in the legislature (somewhat relieved by a 9 cents per gallon increase in the state gasoline tax) and voters' mandated constraints on vehicle registration taxes. The result is recommendations for tolls not only on new roads but even on some existing roads to finance the construction of new roads and bridges (WSDOT, 2006; Bae et al., Chapter 16).

A comprehensive congestion pricing scheme would be system-wide in the sense of applying to the whole metropolitan region, even if in practice it was restricted to freeways and arterial roads. A few years ago this would have required massive investments in road sensors and other pricing infrastructure. Recent technological changes have made this approach easier but not necessarily cheap. On-board global position system (GPS) equipment can measure charges for any road and change the price at any time, thus dealing with the two bugbears – differentiation by route and by time of day (see Chapter 16). The main infrastructure investment would be the remote monitoring stations, equipment and personnel. There is obviously a cost associated with installing the equipment on the vehicle, not an overwhelming obstacle given the strides in (and falling costs of) onboard navigation technology, and with economies of scale the cost on new cars might be quite modest. The trickier problem is the transition: what do you do about the car fleet already on the road? Perhaps a fixed annual fee might be a workable if third-best solution.

Within the United States, there has been slow and somewhat unsteady progress towards road congestion pricing in San Francisco, Seattle, New York City, Minneapolis, Miami, Houston, Boston and Portland, among others. Table 1.1 summarizes the projects (operational and planned) as of 2000. A more recent GAO study expands the list to 23 states, 16 operational and seven in the planning stage (GAO, 2006), and new proposals are made almost month by month.

An interesting question, not necessarily with a clear-cut answer, is whether the European experience in general and London's in particular sheds any light on the scope for expanding congestion pricing in the United States.

First, London illustrates the importance of political leadership and political will. The LCCS was the personal priority of London's first elected

Table 1.1 Projects as of 2000

State	Location	Facility	Status
Arizona	Phoenix	All freeways	Study
California	Alameda Co.	1-680, I-880	Study
	Contra Costa	SR 4W	Study
	Los Angeles	Various	Post-study
	Orange Co.	SR 91 Express Lanes	Operational
	Orange Co.	SR 57	Study
	Riverside Co.	SR 91 extension	Study
	San Diego Co.	I-15	Operational
	Santa Cruz Co.	SR 1	Authorized
	Sonoma Co.	US 101	Post-study
Colorado	Denver	I-25	Study
Florida	Miami	I-95, SR 836	Study
	Orlando	I-4	Study
Maryland	Baltimore suburbs	Various	Study
Minnesota	Minneapolis	All freeways	Study
Oregon	Portland	Various	Study
Pennsylvania	Philadelphia	US 1	Study
Texas	Austin	I-35	Study
	Dallas	I-635	MIS
	Houston	I-10	Operational
	Houston	I-10 extension	MIS

Source: Poole, Robert W. and C. Kenneth Orski (2000), 'HOT lanes: a better way to attack urban highway congestion', *Regulation*, **23** (1).

mayor, Ken Livingstone, and without his enthusiasm for the scheme, it would never have been implemented (Rye and Ison, Chapter 14). There were other UK plans that failed because of a lack of political support, for example, Edinburgh and Cambridge. In Stockholm, the Green Party was a major champion in the national legislature. In the United States, the projects that have got under way have usually been based on technical rather than political criteria. They have succeeded because of their limited scale. Although there is support from think-tanks and interest groups (for example, the Reason Foundation), there has been no individual (or broad) political support for a major scheme, with the exception of a mechanism for highway financing, often on inter-urban roads. In a nation of motorists, this has been a very difficult cause to sell, and most politicians are risk averse.

Second, the success of London (if measured by reductions in automobile use) owes much to the availability and expansion of public transit (especially buses) that explain the huge shift of 33,000–42,000 from cars to transit. This would not happen in the United States. The modal share of cars in the United States of 89.5 percent compares with 64.4 percent in the UK. Even in New York City cars account for 37 percent of commutes compared with a pre-LCCS share in London of 9.9 percent (Lee, Chapter 11). It would be difficult to expand transit on the London scale because of its low financial priority. Even if it were possible, the modal shift would not occur on a similar proportionate scale (55 percent), perhaps because of a predisposition against riding on public transport but primarily because of its inability to compete with door-to-door travel times.

Third, the spatial structure of US metropolitan areas would dictate a different congestion pricing scheme from London and elicit different responses. As suggested above, tolled corridors might work. A downtown cordon or area licensing scheme would be inappropriate in most cases because of the small role of downtown in metropolitan region population and employment (an efficient downtown cordon scheme for a US city might present a difficult design problem; see May et al., Chapter 7). A freeway-only scheme might decongest the freeways because many US metropolitan areas have surface highway redundancy but all of them have some congested arterials that might benefit from congestion pricing. As a generalization, most US metropolitan areas are very dispersed and decentralized, usually with several suburban choke points, so dealing with the complications of a system-wide congestion scheme seems the most appropriate approach.

Fourth, in an important sense Stockholm offers more guidance than London. This is because of its time-of-day pricing which even in the limited Stockholm Trial showed different responses at different times of the day. In the United States, there are substantial efficiency opportunities for time-of-day pricing because of the time distribution of daily trips with long morning and afternoon peaks with a lesser mid-day hump, and especially because of the large number of peak non-work trips (especially in the afternoon) that might be tolled off into non-peak hours (Gordon and Richardson, 1989). Today's technology allows us to implement time-of-day pricing relatively easily. For example, the SR91 FasTrak system (Richardson et al., Chapter 18) allows infinite time-of-day adjustments (to the individual cent if required), but this has never been adopted by sticking to the time block approach with minimum 50-cent differentials on the ground that more sophisticated adjustments would be too complicated and confusing to drivers.

Fifth, the pressure for congestion pricing is a function of the severity of congestion. As pointed out above, London is much more congested than Stockholm. It is almost more congested than the average large American cities, with travel speeds approximately half of those in the United States (Giuliano and Narayan, 2003). Although commuting times have increased in US metropolitan areas since 1995, as shown in both the 2000 Census and the 2001 National Household Transportation Survey, their long-term stability means that there is much milder congestion than experienced in London. There have been dire predictions, but they have not happened yet. This may help to explain why transportation revenue generation has been a more powerful stimulant to tolls than congestion reduction.

Sixth, institutional constraints are more severe in the United States than in Europe. Municipal governments in the United States have a high degree of independence and any system-wide scheme would involve charging on roads that cross many jurisdictions. Obtaining agreement among all the parties is complex; one idea is to obtain political consensus by some kind of revenue-sharing mechanism in favor of cities rather than regional transit agencies or other uses (King et al., Chapter 19). In other cases, the political quagmires can be minimized by reliance on a single organization such as a regional transit agency. This is one of the reasons why the Edinburgh proposal failed.

Seventh, equity issues have not been foremost in Western Europe on the congestion pricing issue, despite widespread concern about 'social inclusion' in many other areas. Yet in the United States, they have been critical. Texas, for instance, is thinking about a credit-based tolling scheme that would return toll revenues in advance, even to non-motorists. There has also been much discussion about 'Lexus lanes', based on the idea that low-income drivers cannot afford to pay tolls. The best solution to this problem is to have adjacent free and toll lanes that offer a choice. However, the equity arguments do not provide an overwhelming objection (Richardson and Bae, 1998; Bae and Mayeres, 2005), in part because of a variety of revenue-redistribution mechanisms, in part because even moderate-income households place a value on travel time.

Finally, there are financial constraints. In both London and Stockholm, most of the funding came from the central government. In the United States, the Federal government has funded a few small-scale pilot projects, but is unlikely to fund congestion pricing schemes across an array of metropolitan areas. Private corporations may be able to finance both the capital and operating costs of a congestion scheme via bond issues repaid out of toll revenue, but the willingness to rely on the private sector varies from state to state. An expensive component in both London and Stockholm was investment in public transit. It would be difficult to implement this in the US context. Federal financing and sales tax revenues have been frequent sources, but the flow of funds is often sporadic. If GPS technology is adopted, another issue is: who would pay for the onboard equipment? The answer is probably the motorist via additional equipment cost at the time of purchase. The current cost is about $300; if the automobile manufacturers were to install the equipment in all new vehicles, economies of scale would reduce costs significantly, probably by more than half. What to do about the existing car fleet is a vexing problem. Many of these issues are difficult to resolve but are manageable.

The prospects for road congestion pricing in the United States remain problematic. However, the recent European experience suggests that political feasibility is no longer a major obstacle. Also, the most important lesson from Europe is that pricing does reduce automobile congestion by significant amounts.

5 CHAPTER SUMMARIES

The book is organized as follows. Bell and Wichiensin (Chapter 2) develop an inter-modal equilibrium model that links an urban road network subject to a congestion charge to a parallel urban transit market, with a view to finding the optimum congestion charge consistent with the commercial decisions of transit operators. A congestion charge is set to maximize social surplus. Travel behavior is assumed to conform to elastic-demand user equilibrium traffic assignment. The transit market is assumed to be either a profit-maximizing monopoly or a profit-maximizing duopoly competing non-cooperatively. The operator(s) set the fares to maximize profits and the supply of transit services is determined by the associated demand. The problem has been formulated as a bi-level program with the determination of the congestion charge on the upper level and the setting of transit fares on the lower level. In the case of non-cooperating operators, the Bertrand–Nash equilibrium fares are sought. The results of the model are analysed for an example reflecting the Edinburgh transit market. This reveals the importance of competition in the market for distributing the social surplus between providers and travelers.

The Independent Transport Commission (ITC), a land-use and transport think-tank linked to the University of Southampton, has for several years been studying the application of national schemes of variable road pricing to Great Britain. Bendixson (Chapter 3) discusses the implications of two scenarios developed for the ITC by Glaister and Graham (Chapter 4). In the first scenario, surplus revenue is redistributed in the form of reduced fuel duty to drivers on uncongested roads. In the second, an additional £16 billion is collected and transferred to the Exchequer and devoted to enhanced public services. The scenarios raise the key issue of how road pricing revenue is spent and who decides how it should be spent. Various possibilities are discussed. In one, the roads would be assigned to a set of regional public utilities supervised by a regulator. In others the level of road charges would be determined by an elected national roads board or by public referendum.

Glaister and Graham (Chapter 4) continue this discussion. For transport systems the issues of pricing, service quality, funding and investment in urban areas are inextricably interdependent. They argue that no policy can be set for any of these aspects of transport in isolation from any other. Transport planners and urban policy makers can choose to tolerate congestion, or build new capacity or introduce road user charging. These issues are explored and analysed in the context of London: Europe's most obviously resurgent city and the one with the most recent experience of road pricing in the form of the congestion charge. However, despite the evidence that in the centre, where it applies, the congestion charge has had broadly the effects that economic theory would predict, there is still a growing problem for the rest of London and the UK caused largely by the combined effects of rising real incomes and the improving fuel efficiency of cars which reduces the impact of fuel taxes. This suggests a growing pressure for a national system of road pricing. To date 'prices', in the form of fuel duty (over £0.50 out of each £0.80 for a litre of fuel), have been set on the basis of historical precedent or political expediency. The chapter sets out a regionally based model to analyse the implications of setting alternative levels of congestion charging and environmental taxes covering the whole country. This includes modeling the implications for other transport modes and the net changes accruing to drivers and the Exchequer. The sooner that user charging can be introduced the better. However, the difficulties are real and somewhat intractable. The most significant problem is not the technical feasibility of such a system but finding a sound method for administering the funds that the system would generate.

Hargreaves and Echenique (Chapter 5) test a congestion charging scheme for Cambridge using a MEPLAN land-use transport model combined with a SATURN traffic model. The scheme would include a daily toll for drivers crossing a cordon around the edge of the city and a lower charge for residents driving entirely within the cordon. The congestion charge would dramatically reduce the number of cars entering the city and improve traffic conditions. However, the charge would result in higher property prices as higher-income groups would displace lower socio-economic groups by outbidding them to move into the city in order to avoid paying the cordon charge. This would increase the cost of living and employers' production costs, and some employers would move out of the city, especially those in the retail and service sectors. The revenue raised and environmental benefits might be insufficient to compensate for the negative impacts on the local economy and the social equity. Cambridge Futures then tested this congestion charging scheme in combination with transportation investments. These would include expanding the public transit system and creating an orbital road outside the cordon linking the park-and-ride sites together and making it easier for through-traffic to bypass the city. This combination of road user charging with transportation improvements has a synergistic effect, making areas outside the city more accessible, and reducing average rents by facilitating more residential dispersal.

Unlike the technical approach of Hargreaves and Echenique, Richards (Chapter 6) adopts a more political interpretation. He points out that congestion charging has been on the policy agenda in the UK since the early 1960s, but it was not until the Blair government, elected in 1997, identified it as a way of 'breaking the logjam' of congested roads and introduced enabling legislation that it became a real prospect. With the policy firmly grasped by Livingstone, expelled from the Labour Party but elected the first mayor of London, the Blair government's interest cooled, except for a national Lorry Road User Charging (LRUC) system. This was intended to replace a part of existing fuel duties with a distance-based charge, and level the playing field with operators from other EU countries (where fuel duties are lower) competing in the UK domestic market.

With Livingstone's LCCS widely accepted as a success, and with a new transport secretary (Alistair Darling) the government renewed its interest, and commissioned a Steering Group to undertake a feasibility study for a national system of road pricing for all vehicles. Although the Steering Group's report was published in 2004, with a General Election ahead, little public progress was made until the government had been re-elected in May 2005. Although the citizens of Edinburgh had emphatically rejected a proposed congestion charging scheme in February 2005, and the government canceled the LRUC scheme in July 2005, Darling made it clear that he saw road user charging as an important policy option for reducing traffic congestion and increasing national efficiency. In November 2005, he announced that the government would be funding demand management studies in seven areas. However, he made it clear that he was not willing to introduce charges on the inter-urban highway network under the government's direct control, except where capacity is also increased.

The primary purpose of the chapter is to review the development of changes in policy since 1997 to identify the key policy drivers that might jumpstart extensions of road charging to cities outside London in the next decade or two. A major one might be using charges as a means of reducing the need for new highway investments.

May et al. (Chapter 7) address a much more technical problem, how to design optimal cordons, under the assumption that other UK cities may adopt congestion charging schemes in the future. A cordon will reduce congestion within the area but aggravate it outside because of an increase in bypass traffic. The net welfare impact will depend substantially on the specifics of cordon design. More complex schemes could generate higher benefits than the single cordon approach adopted in the LCCS.

The chapter discusses two major design options. The first is based on genetic algorithms (an artificial intelligence searching technique; see Chapter 7 for more details). This is tested on the Edinburgh road network. The second, short-cut approach combines a simpler model (selective link analysis) with judgment to improve cordon design without requiring an optimal solution. Tests suggest that this method can achieve more than 90 percent of the benefits of an optimal solution.

Santos (Chapter 8) describes and assesses the first three years of the LCCS and its impacts. In a very detailed analysis she computes elasticities for different types of vehicle to analyse whether the charges are efficient, and also estimated marginal congestion charge elasticities (that is, what happened to traffic when the charge increased). She also examines the costs and benefits of the zone and its extension to Kensington and Chelsea. Although the original LCCS had positive impacts overall, the proposed extension will result in economic and social losses.

The two chapters by Banister (Chapter 9) and Ho and Maddison (Chapter 10) focus on a relatively ignored problem, the environmental issues associated with congestion charging. One of the main benefits from the LCCS has been improvements in air quality and reductions in noise and accidents in the central area. These chapters outline the evidence from the existing scheme on the environment, and then discuss the environmental issues raised in the consultation on the western extension. The bottom line is that air quality improved within the charging zone (Banister) but deteriorated outside (Ho and Maddison), reflecting trip diversion. These environmental issues are then placed in the wider context of plans for a London-wide low emissions zone and other technological initiatives being taken in the transport sector to improve air quality. The conclusions argue for the inclusion of environmental costs and benefits in the evaluation of congestion charging, and for the possibility of making the charges directly related to the emissions from vehicles rather than the same charge being applied to all vehicles.

Lee (Chapter 11) considers the transfer of the LCCS to US cities. Her primary objectives are: (i) to identify and apply a framework of analysis for policy transfer; (ii) to argue that a successful transfer should involve incorporating key aspects of the overall planning process to overcome implementation barriers in the UK–US context; and (iii) to examine the key travel characteristics likely to affect the outcome of congestion charging in the US.

Nash et al. (Chapter 12) examine a very different problem, the European Commission's policy on pricing for inter-urban freight traffic. The analysis focuses on competition with rail and on how to avoid discriminatory pricing policies between one country and another. The principle of marginal social cost pricing is supported, although its application to both road and rail is subject to political constraints. The directive on inter-urban road goods vehicle charging links tolls to equating *average* revenue and *average* infrastructure cost, but with variations in differentials in environmental, accident and congestion costs. The chapter examines experiences in Germany, Austria and Switzerland (although Switzerland is outside the EU).

Prud'homme and Kopp (Chapter 13) shift attention from London to Paris. Bertrand Delanoe, a leftist mayor elected at about the same time as Livingstone, also wanted to reduce traffic, partly to reduce air pollution, partly to reduce congestion, mostly because he (and, above all, his Green allies) hated cars. For ideological reasons, he did not want to introduce a congestion charge.

What he did, at least in the 2000–04 period, was to make the life of car and truck drivers more difficult. This was achieved primarily by reducing the amount of road space available to cars and trucks (by enlarging and isolating bus lanes, by creating bicycle lanes and by widening pavements), but also by stopping underground parking construction, by increasing street parking prices for non-Parisian residents, and by semi-closing some areas to traffic.

Interestingly enough for the analyst, the supply of public transport did not increase in the period considered, in part because RATP, the public transport company, is not under the direct control of the municipality (although a new tramway was introduced in 2006). This makes it possible to associate changes in motor vehicle traffic, which were of the same magnitude as in London, to changes in car-usage costs, particularly in additional time spent because of increased congestion or, more precisely, reduced speeds.

Available data show that bus speeds and bus patronage did not increase at all in the period considered. There was no modal shift at all from car to bus. Metro patronage did increase, but not faster (in fact, more slowly) than in the preceding period, so that it is unclear that there was a modal shift from car to metro. Public transport users gained nothing.

What is sure is that car and truck users are now spending more time traveling within Paris. The time lost because of the policy is estimated to be worth around €700 million. Worse, the policy slowed down the decline in motor vehicle-related pollution. At urban speeds, the elasticity of pollution emissions to speed appears to be high (if not precisely estimated), so that we have slightly fewer cars polluting much more, resulting in more pollution (relative to what would have happened in the absence of the policy) rather than less, the opposite of the policy objective. The only clear winners are bicycle users. However, they accounted for only 0.01 percent of passenger-km in Paris in 2000. They now account for 0.014 percent, a 40 percent increase. But even the most generous estimate of their gain is far from justifying the policy. By comparison, a congestion charge, in spite of all its limitations, is an efficient instrument for reducing traffic.

Rye and Ison (Chapter 14) consider the factors that have been key to the successful implementation of congestion charging in the relatively few schemes that exist in Europe and Asia, developing, in effect, a conceptual framework for implementation. They then look at a few US metropolitan areas, specifically those that applied for funding to the Federal Highway Administration's (FHWA's) Value Pricing Initiative. Conditions in these areas are compared with the conceptual framework to assess the likelihood of successful implementation and to suggest recommendations.

Eliasson et al. (Chapter 15) analyse the recent Stockholm Trial. The trial consisted of two parts: a congestion charging scheme in place between January and July 2006, and extensions in public transport between August 2005 and December 2006. Initially, the trial was meant to consist only of a congestion charging scheme. Later, it was decided that the charging scheme should be complemented by public transit extensions: several new bus lines, additional capacity on commuter trains and subways, and more park-and-ride facilities. Somewhat surprisingly, apart from the park-and-ride facilities of which there had been a severe shortage, the additional public transit opportunities had little impact. The trial produced more than a one-fifth decline in automobiles entering the cordon, a result very close to that achieved in London. Public opinion has come round to support the scheme, so it is being made permanent.

Bae et al. (Chapter 16) report on a federally-sponsored pilot project in the Seattle metropolitan area on road pricing (one of the FHWA's value pricing initiatives). This was one of several such experiments in the United States (for example, in Georgia, Iowa, Minnesota and Oregon), and it takes place against a backcloth of much more attention being paid to road pricing than some years ago. While it is true that the primary driver is the transportation funding problem, road pricing in urban areas would have substantial congestion-reduction effects. The most interesting aspect of the Seattle experiment was its use of GPS technology rather than the more standard transponder plus road sensors. Although the experiment was small scale, it offers opportunities to judge the feasibility of the GPS approach. If it works effectively, it may be more suitable for a system-wide approach (that is, freeways plus arterials) than other alternatives.

Economists agree that most auto-highway systems are mismanaged because congestion is the default rationing mechanism and this is inefficient. Modern toll collection technology and real-time speed-flow data-gathering technology make peak-load pricing feasible and attractive. To date, the effects of pricing on land use are not yet well understood. Lee and Gordon (Chapter 17) analyse the spatial impacts of various freeway pricing scenarios, including the conversion of HOV to HOT lanes. Spatial allocations occur via a congestible highway network of almost 90,000 links, including almost 5,000 freeway links.

Richardson et al. (Chapter 18) apply the Southern California Planning Model (SCPM) to an important prototype application, a 10-mile segment of California SR91. SCPM is an integrated model that estimates trip production densities (and employment and population) for over 3,000 spatial zones of the Greater Los Angeles area at the level of 47 economic sectors. The possible widening of this route via extra tolled or extra general-purpose lanes has been the subject of considerable controversy. A non-compete provision in the franchise awarded to the California Private Transportation Company (CPTC) had stood in the way of public agencies' efforts to provide additional capacity in the corridor. The approach sheds light on this controversy. The main finding is that, whereas congestion tolls are widely presumed to be efficient, the efficiency outcomes are complex when only a small part of the network is tolled. In sensitivity tests, the most plausible results, and the largest user benefits from adding a new tolled lane, are for the mid-range values of our various assumptions. This result is consistent with recent theoretical investigations of second-best pricing. Flows on congested, untolled, parallel routes benefit from the addition of untolled facilities. The discussion is extended to an examination of impacts throughout the Los Angeles network, including changes in destination choice by drivers and freight operators. Most research on road pricing (usually on corridors) has been of a partial equilibrium nature, and does not consider network effects.

In the final chapter, King et al. (Chapter 19) point out that the political feasibility of congestion pricing depends in part on how the toll revenue is used. They argue that congestion pricing on freeways will have the greatest chance of success if the revenue is distributed to cities, and particularly to cities through which the freeways pass. In contrast to a number of previous proposals, they further argue that cities are stronger claimants for the revenue than either individual drivers or regional authorities. Drawing on theory from behavioral economics, the idea is illustrated with data from several metropolitan areas. In Los Angeles, where potential congestion toll revenues are estimated to be almost $5 billion a year, the distribution plan could be both politically effective and highly progressive.

Overall, the book aims to contribute to the debate about road congestion pricing in the United States, primarily by drawing upon recent European experience. The theoretical rationale for pricing was made by several transportation and urban economists more than four decades ago, but now there are real-world case studies to buttress the argument.

REFERENCES

Bae C.-H.C. and I. Mayeres (2005), 'Transportation and equity', in K.P. Donaghy, S. Poppelreuter and G. Rudinger (eds), *Social Dimensions of Sustainable Transport: Transatlantic Perspectives*, Aldershot, UK: Ashgate.

Giuliano G. and D. Narayan (2003), 'Another look at travel patterns and urban form: the US and Great Britain', *Urban Studies*, **40**, 2295–312.

Gordon P. and H.W. Richardson (1989), 'Counting non-work trips: the missing link in transportation, land use and urban policy', *Urban Land*, **48**, 6–12.

Government Accountability Office (GAO) (2006), *Highway Finance: States' Expanding Use of Tolling Illustrates Diverse Challenges and Strategies*, Washington, DC: USGAO.

Hugosson M.B. and J. Eliasson (2006), *The Stockholm Congestion Charging System: An Overview of the Effects after Six Months*, Stockholm: Transek AB.

Johnson M.B. (1964), 'On the economics of road congestion', *Econometrica*, **32**, 137–50.

Leape J. (2006), 'The London congestion charge', *Journal of Economic Perspectives*, **20**(4), 157–76.

Local Transport Today (2004), 'Study says London congestion charge is harming retail sales', 22 April, p. 1.

Prud'homme R. and J.P. Bocarejo (2005), 'The London congestion charge: a tentative economic appraisal', *Transport Policy*, **12**, 279–87.

Prud'homme R. and P. Kopp (2006), 'The Stockholm toll: an economic evaluation', mimeo.

Richardson H.W. and C.-H.C. Bae (1998), 'The equity impacts of road congestion pricing', in K.J. Button and E.T. Verhoef (eds), *Road Pricing, Traffic Congestion and the Environment*, Cheltenham, UK and Lyme, USA: Edward Elgar, pp. 247–62.

Richardson H.W. and C.-H.C. Bae (2006), 'Congestion pricing: the implications of London's Congestion Charging Scheme (LCCS) and the Stockholm Trial for the United States', Paper presented at the ACSP Conference, Fort Worth, TX, October.

Santos G. and G. Fraser (2006), 'Road pricing: lessons from London', *Economic Policy*, **21**, 263–310.

Schaller B. (2006), *Battling Traffic: What New Yorkers Think About Road Pricing*, New York: Manhattan Institute.

Vickrey W.S. (1963), 'Pricing in urban and suburban transport', *American Economic Review*, **53**, 452–65.

Walters A.A. (1961), 'The theory and measurement of private and social cost of highway congestion', *Econometrica*, **29**, 676–99.

Washington State Department of Transportation (WSDOT) (2006), *Measures, Markers and Mileposts*, Olympia, WA: WSDOT.

PART I

UK applications

2. Profit-maximising transit in combination with a congestion charge: an inter-modal equilibrium model

Michael G.H. Bell and Muanmas Wichiensin

1 INTRODUCTION

Traffic congestion in urban areas is one of the most serious problems for both government and transport planners. Since a congestion charging scheme was first introduced in Singapore more than 30 years ago, big cities like Seoul and Tokyo have considered such schemes, with London implementing congestion charging in 2003 (see reviews in Gómez-Ibáñez and Small, 1994; May and Milne, 2000). From this evidence, many studies have been made for the auto mode network in order to determine the optimal amount of the congestion charge (see, for example, Arnott and Small, 1994; Liu and McDonald, 1999).

However, congestion charging affects not only car drivers but also the users of alternative modes as well as decisions about whether or not to travel. Hence, a model which allows for variable demand as well as mode choice is required. In particular, cities considering congestion charging will normally have at least two transit modes (bus and train). These services are often provided by the private sector in some regulated way.

In the UK, following bus privatisation in 1980, several studies have focused on the characteristics of the transit market. Some say that the market is contestable while others say the evidence is inconclusive or disputable (Gwilliam et al., 1985; Evans, 1991). Some say that the tendency is for operators to merge rather than for new entries into the market. The merger of operators has therefore been studied (Salant et al., 1983; Perry and Porter, 1985; Beesley, 1990; McAffee et al., 1992).

In the interests of simplicity we focus in this chapter on two transit markets; the first is one where a single agency/company runs both transit modes (the monopoly case) and the second is where each mode is run by an independent company (the duopoly case). We assume that the competition between the duopolists is strict, that is, there is no price communication, merger or collusion to maximise joint profits. In both cases, fares are set to maximise profits and capacity adapts to the meet the demand. Hence we adopt the Bertrand (simultaneous) game (Bertrand, 1883; Maskin, 1986) to set bus and train fares.

In modelling, we consider the congestion charge which maximises social surplus subject to a multi-modal network equilibrium with fares set by the operator(s) to maximise profits. The effect of car and bus flows on congestion and the impact of the resulting travel times on mode choice and demand is considered.

The model formulation is shown in the next section. The model uses the elastic-demand user equilibrium traffic assignment model to reflect travel behaviour in response to the prices for using the modes (the congestion charge and fares) which will be charged according to decisions taken by the government agency and the transit operator(s) respectively. Section 3 presents the results. We demonstrate the impact of profit-maximising transit on the optimal congestion charge. Also we show the effect of a congestion charge on social welfare, consumer surplus, operator profits and government revenues. Finally, conclusions and policy implications are summarised in Section 4.

2 THE MODEL

The transport system is assumed to consist of three types of agent (the government, the transit operators and the travellers). The government sets the congestion charge, the operator(s) set the bus and train fares, and the travellers choose whether to travel and if so by which mode. Route choice is not considered. The implicit underlying network has two links, one for road and the other for rail, connecting a single origin to a single destination. Bus and private transport share the road link and so experience the same congestion. It is assumed that both bus and train systems have sufficient capacity to carry any demand that might arise without necessitating changes to waiting times.

For clarity, the model formulation can be described as a bi-level programme (see Figure 2.1). The first level, representing the government, deals with social surplus maximisation by setting the congestion charge. The second level, representing the trip makers and transit operator(s), deals with congested network equilibrium mode choice and profit-maximising fare setting.

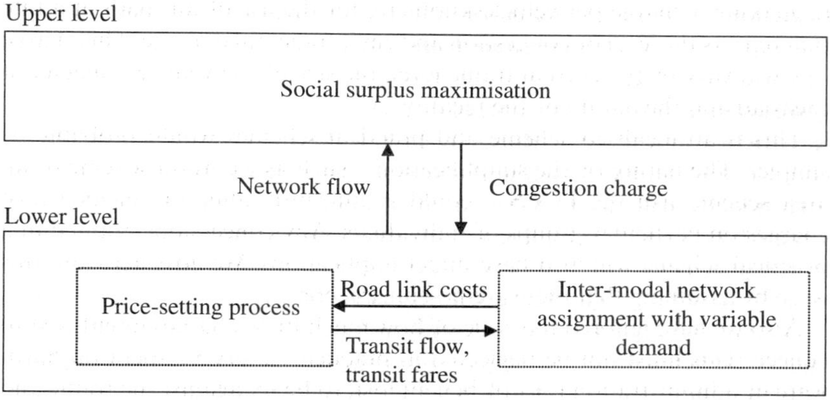

Figure 2.1 Framework of the model

The notation used in this chapter is defined as follows:

m	mode index (a = auto, $t1$ = bus, $t2$ = train);
v_m	trips by mode m;
$c_m(v_a, v_{t1})$	cost by mode m (a = auto, $t1$ = bus);
t_{0a}	auto free-flow travel time;
t_{0t1}	bus free-flow travel time;
t_{0t2}	train travel time;
p_m	fare for each mode (p_a = congestion charge or toll);
ω	value of time;
C	minimum generalised travel cost for the singular origin–destination (O-D) pair;
C^0	an observed minimum generalised travel cost for the singular O–D pair;
q	travel demand between origin and destination;
q^o	an observed demand between origin and destination corresponding to the observed minimum generalised travel cost;
θ	positive dispersion parameter related to auto and transit mode choice, that is, the parameter for log sum function (estimated from data);
μ	positive dispersion parameter related to bus and train mode choice, that is, the parameter for log sum function (estimated from data); and

γ sensitivity parameter for the travel demand function (also estimated from date).

We consider a simple example of a congested road and a railway between one origin and one destination. Demand is elastic. For the demand function, we adopt the exponential form which is widely used in urban transport models. Instead of a potential demand q^o occurring when travel cost is zero (Evans, 1992), we represent the demand function q as:

$$q = q^o \exp\left(-\frac{C - C^0}{\gamma}\right) \quad (q^o > 0,\ C^0 > 0,\ \gamma > 0) \tag{2.1}$$

where q^o is an observed demand corresponding to an observed minimum travel cost C^0. C^0 is an input which locates the demand curve. In the context of the numerical example presented later, this could be set equal to the average travel cost for Edinburgh corresponding to the total quantity of travel in Edinburgh q^o.

It is assumed that travellers minimise their generalised costs of travel with imperfect travel information. The generalised cost for the car mode is composed of the road link cost and the congestion charge. For the road link, travel cost is flow dependent.

Assuming that there is interaction between the bus and car modes, in-vehicle travel time on the road link is a non-negative, increasing function of link flow. The relationship between the flow and the travel time is represented by a travel time function (see equation (2.2)) adopted from the well-known Bureau of Public Roads (BPR). Travel time is converted into generalised cost to which is added the direct cost of travel (the congestion charge in the case of the car and the fare for the bus). Buses do not pay the congestion charge (see equation(2.3)).

$$c_a(v_a, v_{t1}) = p_a + \omega t_{0a}\left[1 + 0.15\left(\frac{v_a + bv_{t1}}{k}\right)^4\right] \tag{2.2}$$

$$c_{t1}(v_a, v_{t1}) = p_{t1} + \omega t_{0t1}\left[1 + 0.15\left(\frac{v_a + bv_{t1}}{k}\right)^4\right]. \tag{2.3}$$

In equations (2.2) and (2.3), k represents the notional road capacity measured in equivalent car trips and b converts trips by bus into equivalent trips by car. Note that $0 < b < 1$, to reflect that *ceteris paribus* a transfer of trips from car to bus would reduce congestion and therefore in-vehicle travel time. Out-of-vehicle travel time consists of waiting time and time for walking to and from the station. The waiting time element is assumed to be fixed, which means that public transport has enough capacity to absorb the increase in patronage. If instead waiting time were a function of demand as a result of service frequency adjustment, the public transport ridership in response to a congestion charge would be similar or larger than predicted by this model. Time for walking to the car, bus or train and from the car, bus or train to final destination is omitted for simplicity. Trains are assumed to run to a fixed schedule, so the generalised cost of travel by train is:

$$c_{t2} = p_{t2} + \omega t_{0t2}. \tag{2.4}$$

The expected perceived minimum cost of travel by public transport is:

$$c_t = -\frac{1}{\mu}\ln\left[\sum_{m\in(t1,t2)} \exp(-\mu c_m)\right], \tag{2.5}$$

while the expected perceived cost of travel assuming that the least-cost mode is chosen is:

$$C = -\frac{1}{\theta}\ln\left[\sum_{m\in(a,t)} \exp(-\theta c_m)\right]. \tag{2.6}$$

Flow conservation is assumed, hence:

$$v_a + v_t = q. \tag{2.7}$$

Mode choice is calculated by assuming that travellers minimise their travel costs subject to imperfect information. Trips by car are given by:

$$v_a = q\frac{1}{1 + e^{\theta(c_a - c_t)}} \tag{2.8}$$

and trips by bus are given by:

$$v_{t1} = v_t\frac{1}{1 + e^{\mu(c_{t1} - c_{t2})}}. \tag{2.9}$$

The governmental objective function is social surplus which consists of consumer surplus plus producer surplus. We maximise social surplus as it is a measure that can be used to evaluate the efficiency of a proposed policy (in this case, congestion charging) and the consequences associated with it. Consumer surplus, equal to γq, expresses the perceived benefits experienced by potential travellers (see equation (2.10)). This simple expression for consumer surplus comes from the integration of the exponential demand function (Evans, 1992). Finally, the producer surplus is operator profits. Government revenue from the congestion charge is assumed to be a transfer payment from drivers to the community at large and so does not affect welfare. The objective is therefore:

$$\max_{p_a}\left[\gamma q + \sum_{i\in 1,2}(p_{ti} - ac_{ti})q_{ti}\right], \tag{2.10}$$

where ac_{ti} is the average cost to operator *ti* of supplying a unit of service, which is assumed to be invariant with respect to demand.

The single operator in a monopoly market chooses train and bus fares to maximise profit. In the duopoly case, we adopt the Bertrand–Nash game (Bertrand, 1883; Nash, 1951) to set bus and train fares. The capacity (number of seats) provided by the operators does not constrain the demand. We assume that operators accept the equilibrium price solution (they do not collude, compensate each other or maximise their joint profits). In the situation where duopolists who are not willing to accept the solution and try to adjust their prices, there may be no unique solution and profits may reach zero (Lewis, 1948; Bishop, 1960; Maskin, 1986).

In our problem, the services offered by operators, although not identical, are relatively good substitutes. Travellers do not necessarily choose the service with a lower fare but would tend to do so, depending on their state of information about services. The elasticity of substitution between mode *t*1 with respect to the cost of mode *t*2 is equal to $\mu\, p_{t2}\, c_{t2}$, which is proportional to μ. Similarly, the elasticity of substitution between mode *t*2 with respect to the cost of mode *t*1 is equal to $\mu\, p_{t1}\, c_{t1}$, which is also proportional to μ. The elasticity of substitution may in practice be increased by providing better passenger information about the quality and cost of services.

Suppose that U_{t1}, U_{t2} are the profits of the two operators and that R_{t1}, R_{t2} are their best-response fare functions. The best-response fare function of an operator is defined in a way that gives this operator the best profit for any choice of fare by its competitor. According to Bertrand's concept of equilibrium, we obtain equations (2.11) and (2.12):

$$U_{t1}[R_{t1}(p_{t2}), p_{t2}] \geq U_{t1}(p_{t1}, p_{t2}) \tag{2.11}$$

$$U_{t2}[p_{t1}, R_{t2}(p_{t1})] \geq U_{t2}(p_{t1}, p_{t2}). \tag{2.12}$$

By the definition of the Bertrand equilibrium solution, the optimal fares p^*_{t1} and p^*_{t2} satisfy equations (2.13) and (2.14):

$$p^*_{t1} = R_{t1}(p^*_{t2}) \tag{2.13}$$

$$p^*_{t2} = R_{t2}(p^*_{t1}). \tag{2.14}$$

3 RESULTS

The multi-modal road network is reduced to two links, one representing the road network and the other the rail network, which connect a single origin to a single destination. The model is approximately calibrated to fit data for Edinburgh around 1999 to 2002 from the Statistical Bulletin Transport Series Trn/2004/4 (Scottish Executive, 2004). The link lengths (5 km) correspond to the average trip length in Edinburgh. Mode share for car, bus and train are 78, 16 and 6 per cent, respectively, in the absence of a congestion charge.

Travel time is available as an average for Scottish cities (no distinct value is given for Edinburgh). The data show that the most frequent duration for car journeys fell in the range of 11–20 minutes, for bus journey times in the range of 21–30 minutes and for rail journeys in the range of 31–40 minutes. We use a free-flow travel time of 15 minutes for cars, 25 minutes for buses and 35 minutes (25 + 10 minutes penalty representing greater access time) for train.

The value of time in 1998 prices for average work time is quoted as £9.23 per hour in the UK Department for Transport *Transport Economic Notes* (2001). By converting this to a value representative of Scotland assuming that the value of time varies in proportion to the wage rate, as was done in Glaister and Graham (2004), we obtained the value of £6.47 per hour used here.

Our results are hence reasonably representative of the situation in Edinburgh. We show the consequences of various levels of congestion charge for the elements of social surplus and the modal shares. The toll is varied in steps of £1. Figure 2.2 shows the changes in social surplus with the toll charge for a normal elasticity of substitution between rail and bus ($\mu = 0.6$) on the left and an enhanced elasticity of substitution on the right ($\mu = 1.2$). There is an optimal toll under both forms of transit market. However, the beneficial effect of competition in the market is less clear when travellers are more sensitive to fare differences between the transit modes.

Fare determination is shown in Figures 2.3 and 2.4. The contour lines indicate the profit for each operator, with the inner contours representing higher profit. At a Bertrand–Nash equilibrium, the tangents to the intersecting contour lines are horizontal and vertical, and therefore at right angles to each other. The best-response curve of each player runs through the contours where the tangents are horizontal, in the case of the bus operator, and vertical, in the case of the rail operator. The intersection of the best-response curves defines the Bertrand–Nash equilibrium. At the monopoly solution, the profit contours share the same tangent as there is only one operator in the market. As expected, the fares are higher in the monopoly market.

Figure 2.5 shows in more detail how the optimal bus and train fares change with the toll and the elasticity of substitution between train and bus. The optimal fares in the monopoly market are greater than those in the duopoly market. The optimal train fare is greater than the optimal bus fare

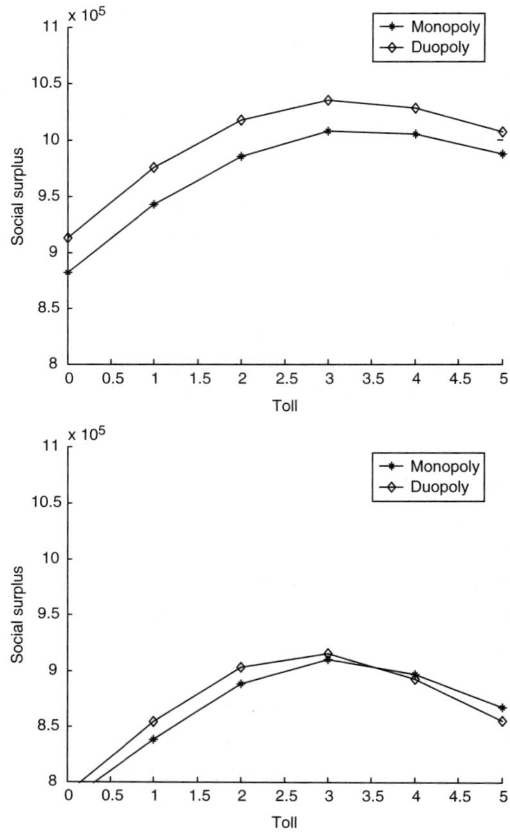

Figure 2.2 Changes in social surplus with toll (normal elasticity of bus/rail substitution above, increased elasticity of substitution below)

in the monopoly market. The optimal bus fare moves in the same direction as the toll charge in both monopoly and duopoly markets. However, the optimal rail fare falls with increasing toll in the duopoly market. This result demonstrates that the decongestion effect of the toll gives a competitive advantage to the bus operator, who can then increase bus fares relative to rail fares. In the monopoly market, rail fares do not respond in the same way. An increase in the elasticity of substitution reduces the difference between the monopoly and duopoly markets.

Figure 2.6 shows that consumer surplus falls with increasing tolls, because, with the exception of rail in a duopoly market, operators respond by increasing fares. When travellers pay more for transport services,

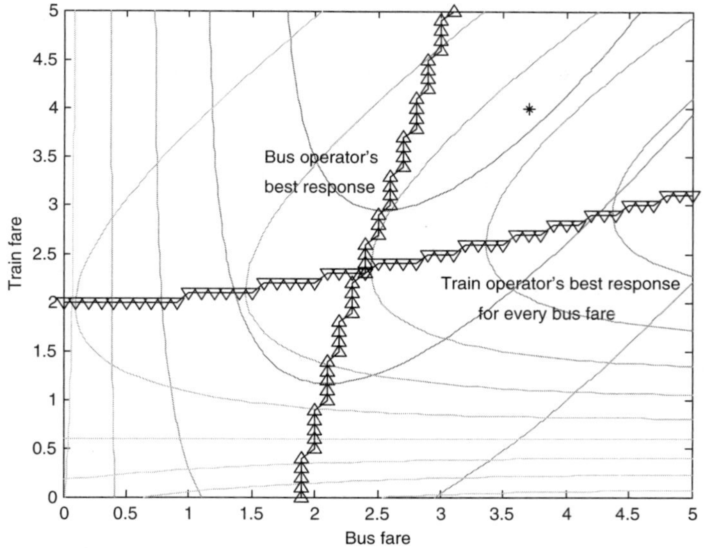

Figure 2.3 Optimal fares and contour lines of bus and train operators' profit at toll = £2

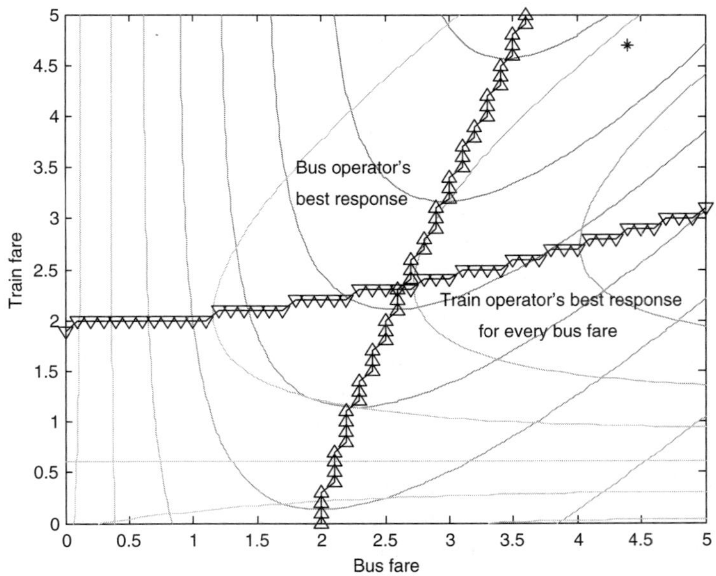

Figure 2.4 Optimal fares and contour lines of bus and train operators' profit at toll = £5

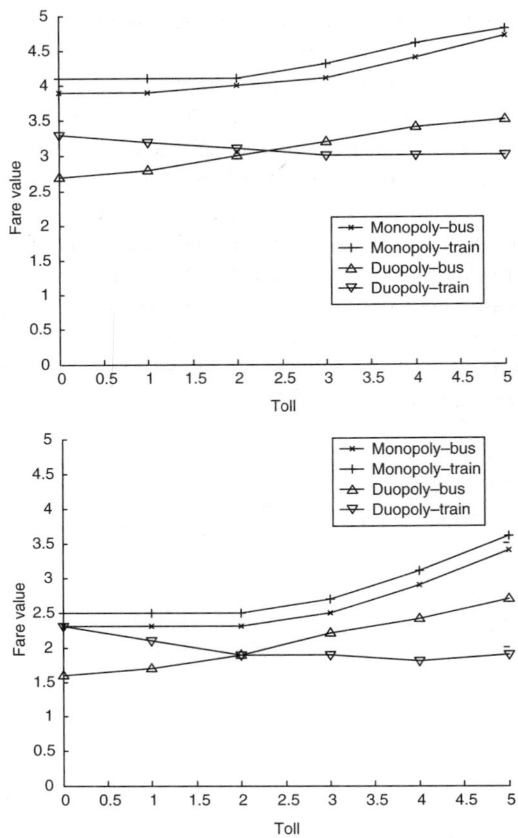

Figure 2.5 *Changes in transit fare with toll (normal elasticity of bus/rail substitution above, increased elasticity of substitution below)*

consumer surplus necessarily falls. Consumer surplus is lower for all tolls in a monopoly market. Consumer surplus is generally lower for a higher elasticity of substitution and also the effect of competition on consumer surplus is reduced. Figure 2.7 shows that operator profits mostly increase with tolls. The exception is rail in the duopoly market, where profits fall slightly with an increasing toll when the toll is low, presumably due to increased competition from the bus operator. Profits are generally lower with a higher elasticity of substitution. Figure 2.8 shows how government revenue changes with the toll. Government revenue is higher for a monopoly market, particularly for higher tolls. A revenue-maximising government

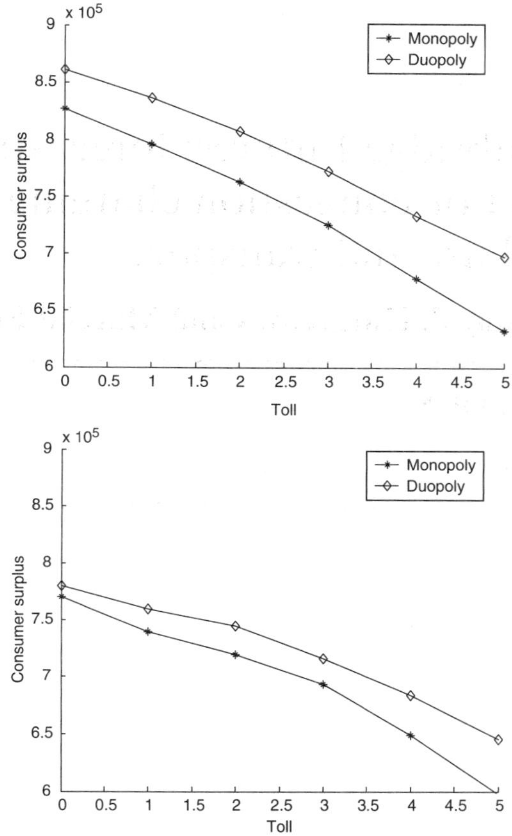

Figure 2.6 Changes in consumer surplus with toll (normal elasticity of bus/rail substitution above, increased elasticity of substitution below)

would set a toll that was higher than the toll yielding the largest social surplus. The effect of an increased elasticity of substitution is to reduce the government's ability to earn revenue by a congestion charge.

Figure 2.9 shows the changes in transit mode shares with respect to the toll (the residual represents the car mode share). Transit mostly gains as the toll increases. The exception is rail in a monopoly market, where there is a slight loss of patronage, particularly when the elasticity of substitution is high. When the toll is high, people shift to the transit mode. This happens to a higher degree in a duopoly market because transit fares tend to be lower.

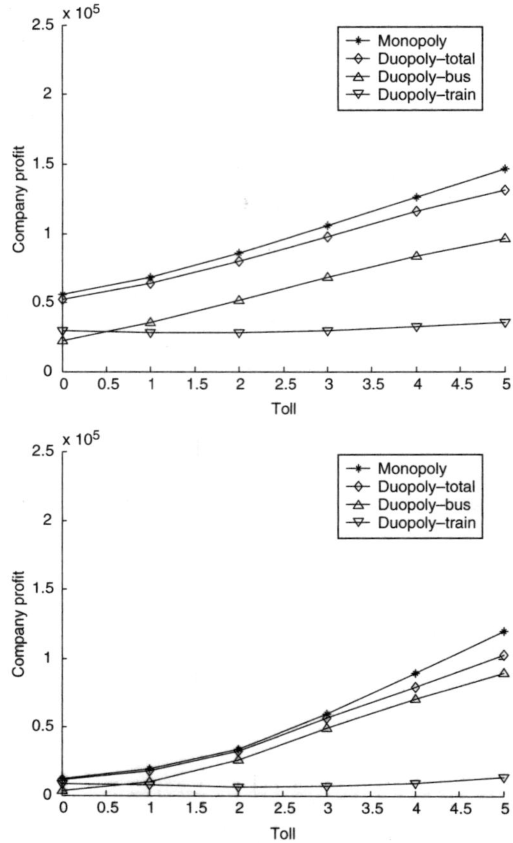

Figure 2.7 Changes in companies' profits with toll (normal elasticity of bus/rail substitution above, increased elasticity of substitution below)

4 CONCLUSIONS

The problem of determining the optimum congestion charge should take the response of the transit market into account. This chapter has shown that the problem can be formulated as a bilevel programme. In the model, total demand, mode shares and transit fares are computed endogenously as the congestion charge that maximises social surplus is sought. The results suggest that the effect of the congestion charge on transit fares depends to some degree on the nature of the transit market. It can be concluded that

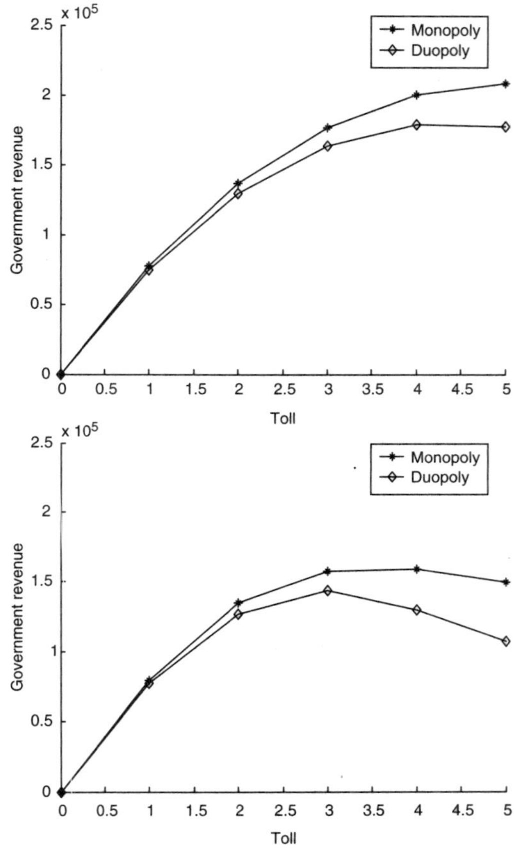

Figure 2.8 *Changes in government revenue with toll (normal elasticity of bus/rail substitution above, increased elasticity of substitution below)*

competition in the transit market is in general beneficial as it leads to a higher social surplus, except for high tolls combined with a high elasticity of substition. This benefit comes from competition within the transit market leading to lower fares.

The policy implication of the above analysis is that the level of competition in the transit market matters whether or not there is a congestion charge. However, as the congestion charge is increased, consumer surplus falls. Initially at least, this fall is more than compensated by increasing profits for transit operators. A bus operator stands to gain more than a train operator, because falling congestion in the road network increases

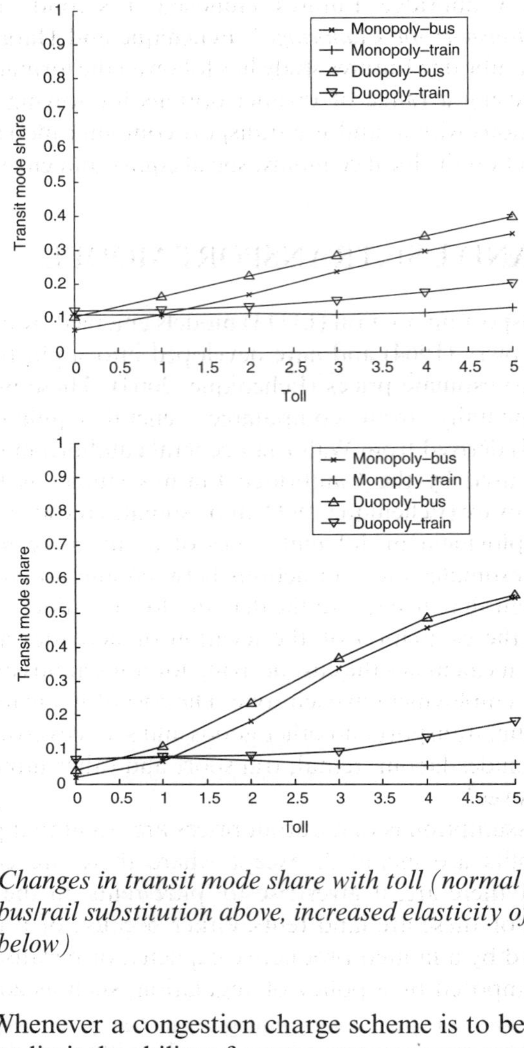

Figure 2.9 Changes in transit mode share with toll (normal elasticity of bus/rail substitution above, increased elasticity of substitution below)

bus speeds. Whenever a congestion charge scheme is to be introduced it is important to limit the ability of operators to convert consumer surplus into profits by increasing fares. This can be done by injecting competition into the transit market or by regulation. Potential for a serious conflict of interest arises when one public authority sets transit fares and the congestion charge. If the authority were to pursue a policy of revenue rather than social surplus maximisation, this analysis suggests that both bus and rail fares would be too high. In particular, rail fares would be too high as the authority would in effect be a monopoly transit operator.

The results also demonstrate that an absence of competition in the transit market can to some extent be ameliorated by increasing the elasticity of substitution between the transit modes. An improvement in the provision of passenger information, enabling passengers to respond more sensitively to fare and service differentials, shifts the balance of power from operators to passengers, enabling them to better resist monopolistic exploitation.

REFERENCES

Arnott, R. and K.A. Small (1994), 'The economists of traffic congestion', *American Scientist*, vol. 82, pp. 446–55.
Beesley, M.E. (1990), 'Collusion, predation and merger in the UK bus industry', *Journal of Transport Economics and Policy*, vol. 24, pp. 295–310.
Bertrand, J. (1883), 'Review of Cournot's "Rechercher sur la théorie mathématique de la richesse"', *Journal des Savants*, **67**, pp. 499–508.
Bishop, R. (1960), 'Duopoly: collusion or warfare?', *American Economic Review*, vol. 50, no. 5, pp. 933–61.
Department for Transport (2001), *Transport Economic Notes*, London: DFT.
Evans, A. (1991), 'Bus competition: economic theories and empirical evidence', *Transportation Planning and Technology*, vol. 15, pp. 295–313.
Evans, A. (1992), 'Road congestion pricing: when is it a good policy?', *Journal of Transport Economics and Policy*, vol. 26, no. 3, pp. 213–42.
Glaister, S. and D. Graham (2004), *Pricing Our Roads: Vision and Reality*, London: Institute of Economic Affairs.
Gómez-Ibáñez, J.A. and K.A. Small (1994), *National Cooperative Highway Research Program Synthesis 210: Road Pricing Congestion Management: A Survey of International Practices*, Transportation Research Board, National Research Council, Washington, DC.
Gwilliam, K.M., C.A. Nash and P. Mackie (1985), 'Deregulation of the bus industry in UK', *Transport Review* no. 5, pp. 105–32.
Lewis, H.G. (1948), 'Some observations on duopoly theory', *American Economic Review Papers*, vol. 38, no. 2, pp. 1–9.
Liu, L.N. and J.F. McDonald (1999), 'Economic efficiency of second-best congestion pricing schemes in urban highway systems', *Transportation Research Part B*, vol. 33, pp. 157–88.
Maskin, E. (1986), 'The existence of equilibrium with price-setting firms: some observations on duopoly theory: firm decision-making processes and oligopoly theory', *American Economic Review Papers*, vol. 76, no. 2, pp. 382–6.
May, A.D. and D.S. Milne (2000), 'Effects of alternative road pricing systems on network performance', *Transportation Research Part A: Policy and Practice*, vol. 34, no. 6, pp. 407–36.
McAffee, R.P., J.J. Simons and M.A. Williams (1992), 'Horizontal mergers in spatially differentiated noncooperative markets', *Journal of Industrial Economics*, vol. 40, no. 4, pp. 349–58.
Nash, J. (1951), 'Non-cooperative games', *Annals of Mathematics*, vol. 54, no. 2, pp. 286–95.

Perry, M.K. and R.H. Porter (1985), 'Oligopoly and the incentive for horizontal merger', *American Economic Review*, vol. 75, no. 1, pp. 219–27.
Salant, S.W., S. Switzer and R.J. Reynolds (1983), 'Losses from horizontal merger: the effects of an exogenous change in industry structure on Cournot–Nash equilibrium', *Quarterly Journal of Economics*, vol. 98, no. 2, pp. 185–99.
Scottish Executive (2004), Statistical Bulletin Series Trn/2004/4: Scottish Household Survey Travel Diary results for 2002.

3. Road pricing in Britain and its relevance to the United States: findings from two scenarios of national road charging in Great Britain and some reflections on governance

Terence Bendixson

1 INTRODUCTION

This chapter considers some implications of introducing variable, distance-based road charging in Great Britain. It looks at two pricing regimes involving different ways of spending the revenues. And, after noting that road charging could be used to suppress travel by car or provide additional transport capacity, it considers some of the implications of pay-as-you-go driving for governance.

The starting-point for this thinking is two reports produced by the Independent Transport Commission (ITC), a think-tank linked to the University of Southampton. Both reports were based on econometric modelling by Professor Stephen Glaister and Dr Daniel Graham of Imperial College London. The first (2003)[1] concluded that:

- The pricing of road travel in Britain is a muddle. What many people pay bears no relation to the real costs of their journeys.
- Given the practical limitations to road building, the country faces a choice between worsening congestion and road-use charging. Delay over the introduction of charging will make things worse.

Glaister, in his technical report, observed that under current rates of fuel duty, city areas and major inter-city routes tended to be undercharged while country areas were significantly overcharged.

The focus of the ITC's second (2006) report [2] was:

- The likely state of traffic speeds and flows in 2010, assuming that national road pricing is in place and taking into account current and planned road building.
- The effects of a regime of national road pricing which, in order to collect only as much money as would be gathered by continuing with fuel duties and vehicle registration, would lower costs for drivers using uncongested roads.
- The effects of another regime in which road charges would be used to raise £16 billion in addition to fuel duty.
- How road pricing could prompt some drivers to change the timing of their journeys to avoid paying the most expensive charges while others might commute with neighbours or colleagues in order to share the cost of driving at expensive times and places.
- How the weekly cost of road charges would vary for households in different regions.
- How road pricing charges for Greater London would be uniquely high.

2 WHY ROAD USER CHARGING?

National road user charging is not an end in itself. Its function would be to make the roads in Britain work better, contribute to the creation of wealth and make travel a more enjoyable experience. The answer to the question 'Why change from fuel duty to road pricing?' is thus 'to make Britain a better place in which to live, work and play'. If road charging could deliver such a future, if road journeys could be made more reliable, if the environmental damage caused by cars, vans and lorries could be reduced and if travel by other means from walking to going by bus could be promoted, it can be argued that Britain would be a more productive and better place in which to live. But how do we get from where we are to this possible land of promise? And what about governance issues? These are very big 'ifs'.

3 CONTEXT

Travel and Transport Trends: 1980–2004

The appetite of Britons for travel persists (Figure 3.1). Between 1980 and 2004, traffic on Britain's roads increased by 81 per cent while the distance

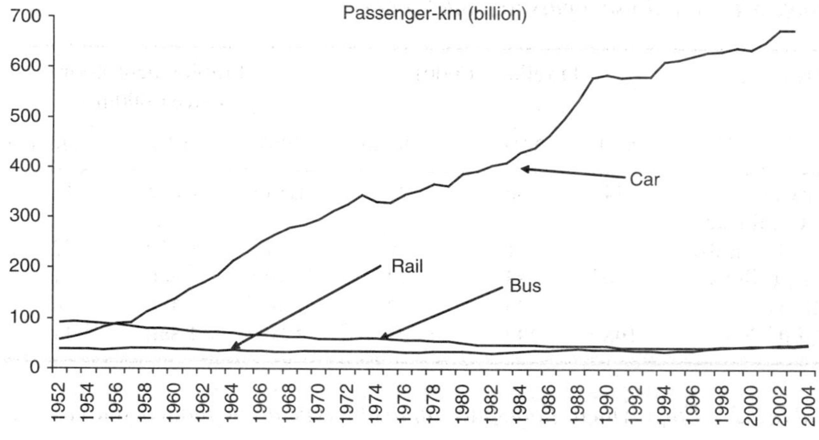

Figure 3.1 Travel in Great Britain, 1952–2004

travelled by buses and coaches went up by 49 per cent. Journeys on the national railways also increased – by 43 per cent.

The number of vehicles on the roads grew from over 19 million to over 32 million, and by 2003 three out of four households had at least one car. Over the same period the relationship between road traffic and economic growth changed. Before 1992 traffic grew faster than GDP. Since then, up to 2004, GDP grew by 42 per cent – twice as fast as road traffic.

Traffic speeds have, meanwhile, been falling on main roads. Over the same period, from 1980 to 2004, average speeds on trunk roads fell at all times. The greatest fall – from 55 to 52 miles per hour – was during the evening peak.

Motoring meanwhile continues to offer increased value for money. New models of car consistently offer higher performance and more equipment than those they replace but, thanks to increases in engine efficiency, longer maintenance intervals and little change in the real cost of fuel, the cost of running cars in 2004 was 'at or below' the 1980 level. Bus, coach and rail fares, by contrast, rose by 37 per cent.

Travel by bus declined in many cities but not in Greater London. In the capital, a combination of low car ownership, slow traffic speeds, controlled parking, high densities, fare concessions and dense bus and railway networks help to explain a decade of rising public transport travel. Travel by public transport also grew in Brighton, Oxford, Nottingham, Telford and other cities where the local authorities committed resources to assisting bus operators.

Investment

Investment in roads peaked in 1992 at £6.2 billion and fell to £3.5 billion in 2000, rising thereafter to £4.2 billion in 2003/04. Railway investment, which has quadrupled since 1995/96, exceeded road investment for the first time in decades in 2003/04. In that year the total was £4.7 billion. Figure 3.2 shows that motoring is becoming relatively cheaper.

No doubt some Britons believe that the country's only transport problem is inadequate roads and that, if only the government would allocate sufficient investment, it could slay the dragon of congestion. Ministers, however, seem increasingly sceptical about their power to play St George in a battle with traffic. In particular they express doubt about the practicality of meeting all demands for road space in and around the big cities in which so many Britons live. And, as anxiety about climate change increases, they are being forced increasingly to acknowledge the need for steady reductions in emissions of CO_2.

This is not the place for an extensive review of the evolution of thinking about road charging in Britain. Suffice it to say that ministers first considered it in the late 1960s and that, after many false starts, surges in the 1990s in economic and travel growth (coupled with low levels of investment in roads) forced it back on the political agenda. Following the election of a Labour government in 1997, Tony Blair's government included powers to introduce road charging in the Greater London Act 1999 and, for other local authorities in England and Wales, in the Transport Act 2000. (The Scottish Assembly followed with legislation in 2001.) Other important milestones were the successful introduction of a Central London congestion charge by Mayor Ken Livingstone in 2003 and a Transport White Paper in 2004 in which Prime Minister Blair gave road pricing his support.

The foregoing is intended to show how thinking within the Labour government has, in recent years, undergone significant change. By the time that Alistair Darling was replaced as Secretary of State for Transport by Douglas Alexander in May 2006, ministers had given clear signals that road pricing lay ahead. Electors and motorists willing, the aim was to move towards national road user charging by way of local or regional schemes. Ministers made clear that road charges, like fuel tax, might well be collected by private companies. Possible contractors were said to include insurance, satnav and cellphone companies.

Could some local authorities be persuaded to take on the political risk of introducing road charging in their districts? The government pinned its hopes on a Transport Innovation Fund (TIF) designed to offer financial carrots to local authorities willing to take the risk. The fund is forecast to be worth £290 million in 2008/09, rising to £2.55 billion in 2014/15.

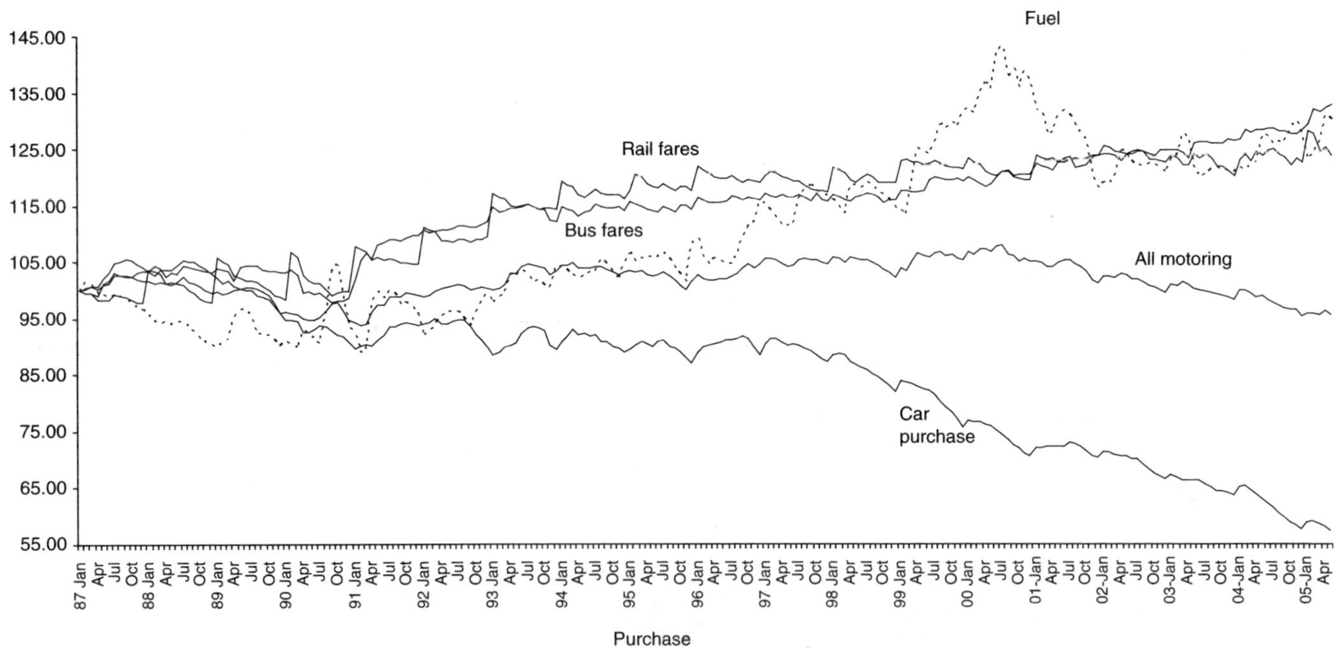

Figure 3.2 Motoring is becoming cheaper in relationship to real incomes (price indices relative to retail prices)

4 ROAD PRICING AND TRAVEL IN BRITAIN IN 2010: TWO SCENARIOS

The Model

Since they published their 2003 technical report 'Transport Pricing and Investment in England', Glaister and Graham have updated their model. They have expanded it to take in Scotland and Wales and incorporated 10,070 census wards to which are ascribed traffic characteristics. (The model does not simulate transport networks.) The base data now consist of incomes, traffic flows, fuel cost, environmental impacts and so on, forecast to 2010. Road costs, which cover road maintainance and depreciation, crashes, air pollution, noise and climate change, are estimated to average 2.54 pence per car-km. The sources of such data were the Department for Transport and other independent bodies.

How Much to Charge?

A central issue for road charging is how much revenue to raise. The starting-point is the amount needed to bring about a redistribution of car journeys that would turn the most congested roads into more free-flowing ones. In congested cities and suburbs and on busy inter-city routes, this amount has to be more than the amount drivers perceive to be their costs today. But this leads to another key issue: what should happen to the extra money collected on congested roads? Glaister and Graham list the following options:

- Improve road maintenance and increase capacity.
- Improve public transport alternatives.
- Pay for the initial and operating costs of a road charging system.
- Reduce fuel or vehicle excise duty.
- Devote the proceeds to other local or national public expenditure.

In consultation with the ITC, Glaister decided to explore the last two options (though in reality it is likely that paying for the charging system would be the first charge on revenue).

Two Scenarios

In both scenarios the charges would vary from vehicle to vehicle depending on contributions to congestion, road wear, environmental damage and road casualties. Charges would be highest on busy roads at busy times – in other words in cities, their suburbs and outer suburbs. They would also be high on some motorways and trunk roads at their busiest times. Charges of this kind would create a surplus of about £16 billion a year – minus the cost of running the system.

The first of the scenarios shows the effects of returning the surplus to drivers where and when traffic was light (Figure 3.3). Drivers in rural districts and at night would see their costs fall but there would also be economic and environmental gains in congested cities and suburbs. In the second scenario (Figure 3.4) the £16 billion would go to national or local government to be spent in accordance with their priorities. In this case the major beneficiaries would be taxpayers at large and the environment. But both scenarios would deliver overall economic benefits.

Spending road charging revenues in these two contrasting ways would have profoundly different side-effects. Making driving cheaper on uncongested roads, for instance, would boost driving in the country and benefit low-income country drivers who rarely have alternatives to cars. It seems likely that rural economies as a whole would benefit.

But there would be an environmental downside to this seemingly attractive scenario. With more driving taking place on country roads

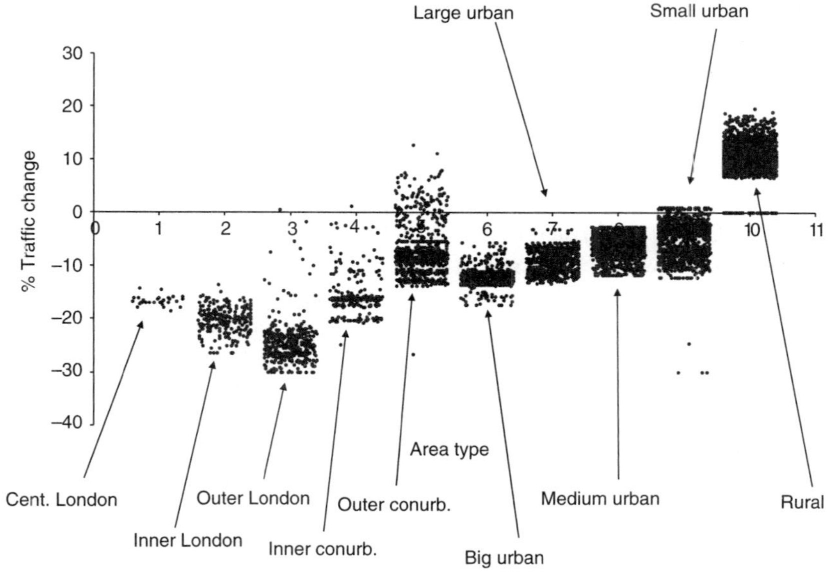

Figure 3.3 Scenario 1: how traffic in Britain would change, by electoral wards, if revenue collected on congested roads were returned to drivers on uncongested ones

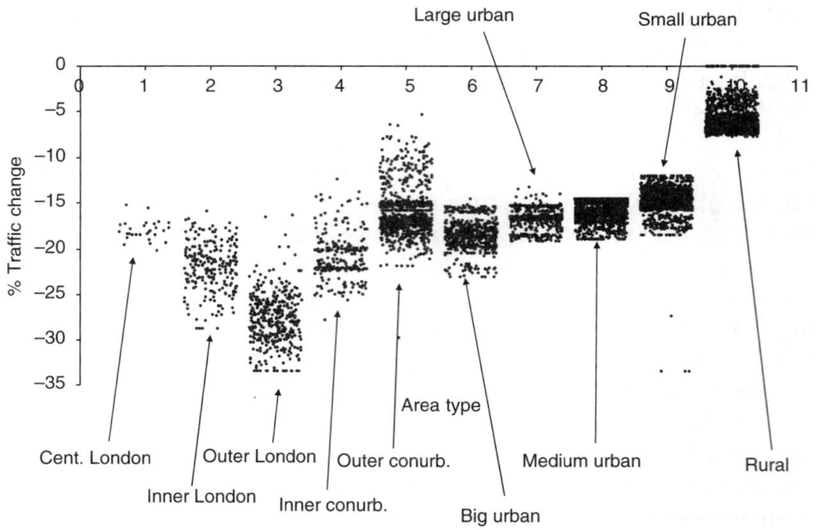

Figure 3.4 *Scenario 2: how traffic would change in Britain, by electoral wards, if road charges were used to raise an extra £16 billion and this revenue was allocated to general public expenditure*

there would, pending the arrival of 'green' vehicles, be more noise and air pollution and a greater contribution to greenhouse gases. More traffic could also add to road casualties (allowance for this is built into the charges) and, in the longer term, there could be decentralising land-use effects.

Given the way in which living costs tend to decline with distance from city centres, lowering the cost of rural driving would tend to add to the attractiveness of country living and so increase the number of city people trying to enjoy it. This land-use effect would, in turn, conflict with government policy for compact cities and minimising the use of the automobile.

In the second scenario, handing over the surplus revenue to central or local government would cause driving costs everywhere to rise and traffic levels everywhere to fall. Transport efficiency would rise in all urban and suburban districts and the contributions of road traffic to casualties, noise, air pollution and climate change would be reduced. Bus services would become faster and more reliable. Government would have new funds available which they could choose to spend on transport. The overall economic effect would be positive. The downside is that driving would, for the average driver, be more costly.

About four in five people in Great Britain would live with less traffic and a better environment while about one in five, mostly in country districts, would see more traffic and a slightly worse environment.

5 SOME EFFECTS OF THE CHARGING SCENARIOS

Driver Behaviour

Changing driver behaviour is at the heart of road charging. In the short term, drivers could change in four main ways. They could travel at different times, go at their usual times but share rides with other drivers; go by bus, train or other means or they could when it was practical; use conference calls; or work from home. (In the longer term, road pricing could be expected to persuade some employers, householders and even retailers to move to different locations.) The function of variable charges would be to persuade drivers to change in whatever ways were most convenient for them. It is not Stalinist planning but market economics. Therein lies its claim to efficiency.

Glaister and Graham looked at two types of behavioural change: switching time of travel and sharing rides. They are confident that both would happen but, in the absence of measured experience, uncertain by how much. One important finding nevertheless follows. The more drivers do change their behaviour and move away from peak times, or share rides, the lower would be the level of charge needed to bring about any given reduction in congestion. Thus drivers themselves, through their own driving decisions, would have power to influence how much it costs others, and themselves, to drive. Sharing rides could have an additional benefit. By reducing car use it would help cut CO_2 emissions.

London – A Special Case

As in so many things, Greater London and its surroundings are a special case. Within Greater London itself, nearly half of all households are without cars and would not be directly affected by road charging. Furthermore, even people in inner London households who do own cars often go by bus or Underground and would therefore benefit from any improvements to public transport. In the South West of England, by comparison, only 28 per cent of households are without cars.

Londoners are, furthermore, the most fuel-efficient drivers in Britain. Households with cars spend 30 per cent less on fuel than those in East Anglia, Wales, the West Midlands, Scotland and the North East. They may experience a lot of fuel-wasting congestion, but they do not drive so far as residents in other regions.

Suburbs

London's existing congestion charge is, of course, levied only in the centre of the city. Yet congestion is often just as serious, and certainly much more extensive, in the suburbs and outer suburbs of both London and Britain's other large cities. Furthermore, not only are levels of car-ownership and use high in such places, but residents in them make many journeys to places in other suburbs or in semi-rural districts to which buses or trains rarely run.

Under any charging regime, outer London boroughs such as Croydon and Hillingdon, and comparable places in outer Manchester and Birmingham, would be among those paying the highest road charges and making the biggest behavioural adjustments anywhere in Britain.

Moving further out from the main cities into the urban parts of such counties as Cheshire, Warwickshire, Surrey, Hertfordshire and Buckinghamshire, traffic could fall by about 8 per cent if no additional revenue was collected and by about 20 per cent if it was. Yet, as the ITC has found in its work on the future of suburbs, substantial population growth, and therefore traffic growth, are in prospect for precisely such places.[3]

Given the economic importance of today's suburbs and urban fringes with their airports, research laboratories, corporate headquarters and regional, drive-in shopping centres, and given too the levels of congestion found on their ring motorways (such as the M62 around Greater Manchester and the M25 around Greater London), there are good grounds for arguing that they are prime candidates for road pricing. Yet virtually nothing is known about how road charges might affect their economic performance and travel within them. What is known is that travel by public transport is rare. The average suburban British car owner makes fewer than 10 per cent of journeys by bus or train. Only a minority of suburban commuters go to a city centre by railway or to a suburban centre by bus. This raises the question whether, notwithstanding concerns about emissions and climate change, suburban road charging would need to be linked to plans to expand road capacity as well as, say, create park-and-ride services and cycle tracks radiating from schools. Such gaps in knowledge need filling.

Scotland and Wales

In Scotland and Wales cities such as Glasgow and Cardiff would see substantial reductions in congestion under both scenarios. But both countries are more notable for their many medium and small country towns and their very extensive tracts of open country. Under the first scenario, in which surplus revenue was returned to rural drivers, country districts would see some increases in traffic but, more significantly, a fall in motoring costs. A national road pricing scheme of this kind could therefore have significant economic benefits for the Scottish Highlands and the Welsh uplands.

Economic Efficiency

The two scenarios make clear that, whether or not road pricing is used to raise additional funds, it *is* economically efficient. If surplus revenue is returned to motorists, benefits of £6.8 billion a year are created, while if the surplus is allocated to public spending the benefits would be £9.7 billion.

What Should Happen to the Revenue?

Although these theoretical scenarios are only illustrative, they make abundantly clear that the destination of revenue from road charging is a matter of great importance. But who should decide this? And who should set the level of charges? Should it be the government? Or should the roads, like other public utilities such as gas and electricity, have charges set by a regulator? Such governance issues are clearly of the highest importance.

6 FROM LOCAL AND REGIONAL TO NATIONAL ROAD PRICING

In July 2004 the Secretary of State for Transport received the report of a working party called 'Feasibility Study of Road Pricing in the UK'.[4]

Since then, in a series of announcements, two different secretaries of state, while remaining uncertain about the political practicality of road pricing, pushed the idea of an all-party consensus on the issue. And some kind of consensus emerged. The Liberal Democrats committed themselves to road charging and, in the Spring of 2006, the Conservative policy machine was hard at work considering how to implement it. Meanwhile the government, after cancelling a proposed regime of lorry charges, seen by many as a first step towards national charging, made clear that, whatever the long-term prospects, charging would begin with pilot schemes that would be local or city-regional in scale.

In conjunction with these developments the Department for Transport created two new divisions, one technical and one policy orientated, designed to deliver road charging. The Treasury set up a TIF (see Section 3, above) aimed at local authorities willing to promote charging. And, finally, both the transport secretary and his officials made clear that, just as oil companies collect fuel tax, so it would be preferable for road charges to be collected by insurance, cellphone or other private contractors. The government had no intention of running a huge, risky, IT 'back-office' intended to handle charges from over 30 million vehicles.

7 GOVERNANCE

The London congestion charge was introduced largely because Mayor Livingstone, in addition to having powers to introduce it, ensured that he was elected with a mandate to do so. He put it in his manifesto. No other large British city has a mayor and none has an official with Livingstone's regional transport powers. All Britain's provincial conurbations are divided into district councils controlled by different political parties.

London's charge is also unique in another way. Because it covers only a tiny area in the midst of a metropolis served by a dense network of bus, underground and railway services, it is practical to suppress some car use on working days and propose no additions to road capacity. Some people, particularly Green party members and environmental transport lobbyists, are therefore assuming that London is a model for road charging elsewhere. Is this practical? Would suburban residents, who make 60 to 70 per cent of their trips by car, vote for such a regime?

The London charge, which is levied on top of, and not instead of, fuel tax, furthermore suggests that provincial schemes are also likely to involve drivers paying more than they do now. (How might fuel tax be reduced in a single region in such a small, dense country as Britain?) This is why the linking of pilot local or regional road charging schemes to reductions in property taxes is increasingly debated. Not only would such cuts be practical but, if combined with increases in road and public transport capacity, they could win voter support.

Some light about the willingness or otherwise of drivers to accept road charges is shed in a survey of opinion by the Royal Automobile Club in 2006.[5] Although the majority of motorists said they wanted tougher action against congestion, only 40 per cent were prepared to support road charging. Yet support went up to 60 per cent if the charges were linked to cuts in fuel or vehicle taxes or if the revenue was spent on the roads, and to 69 per cent if it was spent on public transport.

Late in 2005 the two words 'productivity' and 'competitiveness' surfaced in transport circles in Britain. Sir Rod Eddington (former head of British Airways) invited evidence on them in an enquiry for the Treasury and the Department for Transport. They crop up too in the government's January

2006 guidance on the TIF. This explains that the government is looking for demand management proposals or 'packages and schemes ranging from national to inter-urban and "exceptionally" local schemes that (would) contribute to national economic productivity objectives'. Some see in these words a green light for investment in inter-urban roads. Others see them pointing to regimes of charges offering widespread reductions in traffic congestion. Presumably it will lead to both.

Does all this mean that road charging is a foregone conclusion in Britain? No. Not only is it an acutely complex measure that requires the involvement of at least six government departments (Transport, Local Government, Environment, Home, Industry and Constitutional Affairs), but it is likely to necessitate local government reform and, if it involves more than city centre cordons, would be very unlike what has been pioneered in London.

So how might Britain move from interoperable regional schemes to (costs, technology and electors permitting) a national one? Here, perhaps, the London model is relevant. Assuming that incomes, car ownership and road travel continue to grow, assuming too that investment in road capacity continues to lag traffic growth, the outcome can only be, not daily gridlock (that is the stuff of newspaper headlines), but delays, unreliability and a worsening economic environment for British business.

By the time of the next British general election in, say, 2009, road congestion could become an electoral issue. Labour and the Conservatives alike could include in their manifestos commitments to national road charging and, whichever was elected, would have a mandate to introduce it. Would both parties, on climate change grounds, promise to use road charging to reduce car use? Or might the Conservatives promise to cure Britain of road congestion and build more roads? And, if they did, how would they square this with their environmental objectives?

8 ROAD UTILITIES UNDER A REGULATOR?

One result of the Conservative policies of former Prime Minister Margaret Thatcher is that all public utilities in Britain are private companies. Gas and electricity, in particular, are provided and distributed by companies that include a national grid and regional suppliers. Could this be a model for road utilities in the United Kingdom? Might the existing motorways and trunk roads be put in the hands of one operator, while other operators managed the roads in metropolitan and rural regions?

Simon Linnett, a member of the Independent Transport Commission, considered some of these issues in an ITC paper published in May 2004.[6]

He suggested that, under conditions where road charging was in operation, drivers would pay for a right to a journey, be encouraged to plan trips and be compensated if their reasonable expectations were not met.

Linnett concluded:

- Highway authorities would become a service industry of 'network managers' and not providers of road space.
- Key issues were the use made of any revenues and the re-design of highway authorities to take on new responsibilities.

Under the regulated utility model, existing metropolitan district, county and national highway authorities would be transformed. One possibility would be to assign their road assets to non-profit distributing trusts. Alternatively the assets might go, as in the gas and electricity industries, to private companies. Proud counties, long the guardians of the 'Queen's highway', would be loath to see the privatisation of such a key function, but there is no fundamental reason why Parliament should not do it.

What about the charging revenues? Would they be decided by a government-appointed regulator? Could they be set to bear down on congestion and provide funds for maintaining and expanding the infrastructure? To what extent would fuel duty continue to be collected? And could sufficient transparency be provided to satisfy motorists?

Just as with the water companies, which have to provide adequate supplies of H_2O and repair leaks in their networks, road companies would be faced by similar concerns. And so, every three to five years, they would go to the regulator with proposals to deliver a set of traffic speeds, widen certain roads, upgrade traffic management systems, build bypasses or tunnels, introduce new park-and-ride services, create cycling networks around high schools and so on. The regulator, who would be supervising perhaps 10 different operators, would be in a position to compare their performance and establish best practice as a norm.

Any physical expansion of the infrastructure, be it for bigger junctions, park-and-ride sites or new roads, would necessitate permission from the relevant planning authority. This would provide local authorities with the opportunity to inject their transport policy thinking into the work of the operators – although there would no doubt be informal consultation too. Additional public control over the road operators could be provided by means of appointed (or elected?) 'roads boards', representing bodies such as motoring clubs, hauliers, port and railway operators, pedestrians and cyclists. These could be given a statutory right to be consulted by the roads companies.

9 ROAD FRANCHISES?

The history of mobile telephony in Britain points to one way in which the roads could be privatised and financed. In the cell-phone case the government, as proprietor of the necessary wavelengths, sold franchises by auction. The bidders borrowed money by securing it on income from future sales and proceeded to invest it in the necessary infrastructure. Might not road franchises be financed on the same basis?

The M6 toll motorway and the Dartford river crossing on the M25 motorway provide precedents in the highways industry. In both cases operators bid to provide and manage infrastructure serviced by income from its future use.

An important aspect of this concept is that it envisages road user charging less as an instrument for reducing travel (as is envisaged by environmentalists) than as one for financing the expansion of networks. Reductions in road congestion would, at the same time, cut emissions per kilometre travelled.

The creation of road utilities would leave plenty of work for the Department for Transport to do. It would, for instance, need to establish and supervise the office of the road regulator and have a role to play in enforcement and in ensuring the interoperability of the toll collection technology used by different road operators.

This preliminary discussion of some road governance issues leaves many questions unanswered. No doubt Britain will proceed to road charging, if such a decision is made, by pragmatic steps starting from known conditions. Regulated utilities and franchising could provide some such known foundations.

10 CONCLUSIONS

Efficiency and Productivity

Drivers in Britain already pay to use the roads. They pay through vehicle excise duty and fuel tax and they pay heavily – though not more so than they used to. Currently (September 2006) for every 90 pence paid for a litre of lead-free petrol, the tax is about 60 pence while for diesel at 95 pence per litre, the tax is slightly more. Those who consume the most fuel therefore pay the most tax, yet fuel duty, unlike the cost of hotel rooms, air fares and many other goods and services, does not go up at peak times and come down at slack ones. There is no obvious financial inducement for drivers to travel outside peak periods, share rides, walk, use park-and-ride schemes or otherwise minimise their contributions to jams and delays.

This is inefficient. Hauliers, drivers going to business meetings and others whose time is valuable, get bogged down in congestion. Highway authorities receive false signals about demand.

Finally, although fuel tax and vehicle excise duty have to be paid for all motor vehicles used on the roads, such taxes, like taxes in general, are in no way tied to related expenditure. This is not to say that fuel tax is without its advantages. From the Exchequer's point of view it is easy to set and cheap to collect. Tax farmers such as oil companies do the work and raise huge sums at low cost. In this way fuel tax *is* efficient.

Changing from road taxation to pay-as-you-drive road charging (or combining the two – as is more likely) and the consequences of so doing, are currently under debate in Britain. However it needs to be stressed that the purpose of variable road charging would not be to force people to give up their cars. It would be to increase the efficiency of road use, capitalise on the potential of Britain's major cities (and on the so-called 'agglomeration effect' of their extensive labour and services markets) and *add to the productivity of the British economy.* Politicians could well focus on the single objective of reducing congestion, as Livingstone did in London, but that would be for reasons of keeping the message simple.

Road charging could also be designed, as is assumed in the Glaister scenarios, so that drivers of cars and lorries paid for the environmental damage caused by their vehicles. This would, indirectly, put pressure on the motor industry to build cleaner vehicles but would not, by itself, be sufficient to achieve that end.

Finally, road user charging, by making traffic flow more smoothly, would reduce CO_2 emissions and thus transport's contribution to climate change. But again it would not alone be effective in bringing about a shift to ultra low-emission or non-carbon fuelled vehicles. Other parallel policies would be required. Road charging alone would not resolve all transport's economic and environmental problems.

Governance

The possibility that road user charges might be introduced raises two key governance issues: who would set the charges and who would collect them?

In this chapter it is assumed that drivers would accept pay-as-you-go charges only if the present set-up, where fuel duties vanish into general funds at the Treasury, is changed. It is assumed too that drivers would want more transparency and a fair relationship between what they pay and what is spent on road maintenance and investment.

The regulated public utility model looks promising. Following the structure of the gas and electricity industries, government would put the roads in the hands of a trunk operator and a group of area operators covering metropolitan and rural regions. Such franchise holders could be non-profit distributing trusts or companies that would compete to manage specific sets of roads. A regulator, who would have the advantage of being able to compare the performance of several operators, would set traffic speed targets and levels of charges.

Should an operator wish to build a new road, expand an existing one or create a park-and-ride scheme, an application would have to made to the relevant local authority for planning permission. This would ensure local democratic input. The M6 toll motorway and the tolled Dartford crossing on the M25 motorway may be seen as precursors for such futures in British road governance.

Lessons for America

Only a rash policy analyst would argue that there is much for Americans to learn from Britain's modest experience of road charging. The sum total of this experience is the small congestion charging zone in central London, a tiny one in the historic city of Durham that has not been considered in this chapter, and a substantial body of theoretical work going back to Reuben Smeed in the 1960s.

The primary lesson from central London is that charging is effective in reducing car use and promoting travel by bus and that political opposition is modest where residents are few and where most visitors, be they workers or visitors, arrive on foot or by public transport. The other big lesson is the need for strong political leadership backed up, if possible, by an electoral mandate.

The great *terra incognita* for road pricing is the suburbs. Britain has, as yet, no experience of charging where car use is intense, journeys are made in all directions, and where only city centre destinations tend to be served by bus and rail services.

This leads on to the other key issue of what use would be made of the revenue from road charging. In central London it is satisfying a 'green' agenda by being directed at improvements in bus and Underground services. However, it is unlikely that such a use alone would satisfy suburban electors where, though improved bus services and safe cycle routes to schools could attract some travel, part of any revenue from road pricing is likely to be needed to increase road capacity.

It follows that road pricing should not automatically be assumed to be an effective tool for reducing vehicle emissions. Such an objective is likely to require parallel policies designed to bear down on vehicle engine size, fuel consumption per mile, and the burning of fossil fuels.

Just as in the US, where high occupancy toll lane construction is driven by a need to find private finance for roads, so in suburban Britain, congestion charging, far from being a mechanism for reducing car use, as Green campaigners hope, could, as in cities in Norway, become a form of road tolling aimed, at least partly, at financing road investment.

NOTES

1. ITC (2003), 'Transport Pricing: Better for Travellers', reviewed modelling by Professor Stephen Glaister and Dr Daniel Graham of Imperial College London, of the spatial effects of different road charging scenarios. Their technical report, 'Transport Pricing and Investment in England' (2003), and the ITC's report may be downloaded from www.trg.soton.ac.uk/itc.
2. ITC (April 2006), 'Paying to Drive', reviewed 'Road Pricing in Great Britain: Winners and Losers' also by Glaister and Graham (2006). Both may be downloaded from www.trg.soton.ac.uk/itc.
3. ITC (2004), 'Suburban Future', review of technical report, 'The Future of Suburbs and Exurbs', technical Professor Marcial Echenique and Mr Rob Homewood, ITC 2003. Downloads from www.trg.soton.ac.uk/itc.
4. UK Department for Transport (July 2004), 'Feasibility Study of Road Pricing in the UK', London.
5. RAC (2006), Report on Motoring: 'The Future of Motoring: A clear road map or a collision course?', June, reviewed a survey in January and February 2006 of 1,000 'regular drivers' by MORPACE International Limited.
6. ITC (2004), Occasional Paper No. 1, 'Beyond Congestion Charging', by Simon Linnett. Downloads from www.trg.soton.ac.uk/itc.

4. National road pricing in Great Britain: is it fair and practical?

Stephen Glaister and Daniel J. Graham*

1 INTRODUCTION

In June 2005 the then UK Secretary of State for Transport, Alistair Darling, addressed the Social Market Foundation, suggesting that the basic principle of a national road pricing scheme would be to charge road users according to the use they actually make of the network at different times and in different places. The Labour Party 2005 Election Manifesto also indicated that the possibility of a national, comprehensive system of road pricing should be investigated as an important component of future transport policy: 'because of the long term nature of transport planning we will seek political consensus in tackling congestion, including examining the potential of moving away from the current system of motoring taxation towards a national system of road pricing'.

More recently, the Secretary of State for Transport, Douglas Alexander, noted:

> [C]ongestion is getting worse in our major towns and cities and on some parts of the strategic road network. If we do nothing, it could damage our long-term economic growth. But doing nothing is not an option. That is why I am clear that we need . . . to take road pricing off the drawing board and make it work for road users.

While politicians are fond of saying that 'doing nothing is not an option', and while government support looks encouraging, for the last couple of decades governments have in fact done very little. Indeed, the response to political pressures has been to withdraw plans to provide new road capacity and to reverse previous policy that duties on road fuels should be steadily increased in order to mitigate traffic growth: fuel duty has not even been increased in line with inflation since 2003.

Traffic congestion has inevitably got worse, and on current policies it will continue to do so. It would be perfectly possible to tolerate this and to continue to 'do nothing'. For many people for much of the time, traffic congestion is a minor inconvenience. But for others and for industrial sectors that rely on fast, predictable movement of goods, congestion is becoming a major nuisance and is causing damage to the economy.

Although road pricing is controversial, it should be welcomed. It is hard to imagine how the prospects for future national traffic growth can be managed without it, especially in places such as the outer suburbs of Greater London. Although the principle of road pricing is official policy for all three of the major parties, we are a long way away from having a detailed, practical proposition that could be put to the public.

There are a number of major issues that will have to be resolved before either Parliament or the public could be asked to endorse a national road pricing policy. Any road pricing system needs to define clearly what happens to any revenues collected. Other preliminary considerations include who would be granted reductions or exemptions and the costs of implementing and operating any scheme. Reductions and exemptions would have direct consequences on efficacy and acceptability.

Decisions would also have to be made about the practical meaning of 'a national scheme', and the governance arrangements between local and national authorities. In addition, policy makers need to assess how other areas of policy – such as the finances and capacity of public transport, land-use policies and the case for providing more road capacity – would be affected by road pricing.

Sitting above all of these is the matter of what are the objectives of a policy for road pricing, if any, beyond mitigating congestion? There are a number of candidates:

- to make more efficient use of the existing network in the name of 'national productivity' or 'the public good' (these are not quite the same thing);
- to fund new roads or new public transport;
- to provide a new source of local or national income;
- to manipulate residential and commercial densities;
- to replace land-use planning controls; and
- to help mitigate 'social exclusion'.

Road pricing would bear on all of these objectives but in different ways depending upon how the outstanding issues were resolved.

This chapter aims to inform the debate by illustrating the potential winners and losers and equity implications of two national road charging scenarios. We have modelled a fully revenue-additional scheme as well as a revenue-neutral scheme to examine how changes in traffic, charges for road use and speed would differ by region, levels of deprivation, urbanisation and other aspects of driver behaviour.

By evaluating the distributional consequences of the two 'polar policies', this chapter discusses whether road pricing can be made to work effectively in the UK context.

2 PUBLIC ATTITUDES AND ISSUES TO BE RESOLVED

Acceptance by the public is a prerequisite for the introduction of any scheme. In part this will be determined by whether people perceive traffic congestion to be a significant problem. Surveys of public attitudes confirm that congestion is bad enough to upset some people while simple arithmetic suggests that it is going to get worse (Royal Automobile Club, 2006; Lyons et al, 2004; Green and Stone, 2004). Government has responded by supporting the principle long advocated by those with faith in the power of markets: national road pricing. The mayor of London, Ken Livingstone, has already taken the risk and demonstrated that it can be made to work and that it can make a difference to congestion. This section sets out some of the questions to be addressed if we are to progress to a national scheme.

Once a detailed policy proposal has been established it will be necessary to explain it to the general public in order to gain sufficient support. A recent survey of attitudes of the general public conducted for the Department for Transport (DfT) confirmed that some people (though not all) are finding traffic congestion to be an increasing problem:

> A substantial minority of respondents experienced congestion frequently. Almost a quarter (23%) said they experienced congestion all or most of the time on their road journeys.
>
> People are more likely to consider congestion in towns and cities to be a serious problem to them personally (18% said this was a very serious problem; 33% a serious problem) than motorway congestion (13% and 20% respectively).
>
> A far higher proportion (87%) of respondents considered road congestion to be a serious problem in the country as a whole, than who reported frequently experiencing congestion themselves.
>
> The majority (76%) of respondents felt it very or fairly important for the Government to tackle congestion relative to its other priorities (DfT, 2006).

The DfT's survey found that:

> [L]evels of support for the principle of road pricing do, however, vary depending on exactly how the question is phrased and the context in which it is asked. These varying levels of support reflect the complexity of the issue and a degree of uncertainty about exactly what such a system would entail and how it would impact on individuals (DfT, 2006).

An essential prerequisite of gaining support will be presentation of a complete package – otherwise the public will see the disadvantages to themselves without a balanced view of the overall benefits and costs.

How Would the Revenues Be Used?

The issue of how revenues would be used is perhaps the most crucial issue for securing public support and one of the least explicitly discussed to date. The DfT survey found that:

> [S]upport for individual elements of a road pricing system ('paying more to drive on busy roads than quiet roads' and 'paying more at busy times than quiet times') was relatively low at around 25%. Conversely, support was higher if 'there would be no overall increase in the amount of taxation paid by motorists as a group' (44%) or 'as long as any extra money raised was spent only on roads and transport' (61%) (DfT, 2006).

Some observers talk as if it is obvious that road pricing could only be introduced if it were designed in such a way that the total cash taken from road users as a whole (or, perhaps, private motorists, or commercial traffic) were not increased. Others appear to assume that road pricing could be introduced only if there were new expenditures to improve roads or alternatives: perhaps by improving public transport (sometimes referred to as 'complementary measures').

Assumptions are made about how any revenues would be used. These include:

- to improve road maintenance and road capacity;
- to improve public transport; to defray the investment and operating costs of the pricing system;
- for other local or national public expenditure purposes; and
- to reduce fuel duty or Vehicle Excise Duty (the tax disc).

Since the money can only be spent once it would not be possible – as is sometimes implied – to avoid having to make difficult choices to allocate the revenues between these different uses.

The alternatives

The following fragments from Alistair Darling's speech to the Social Market Foundation illustrate how an uncritical listener might gain the impression that the revenues could be spent more than once:

> Our objective is not to put people off the roads. It is to enable us to get more out of the network. . . . this could be achieved by more than half of the traffic paying less than they would otherwise be paying in fuel taxes . . . In any overall package of measures to tackle congestion, improved public transport is absolutely essential . . . (Alistair Darling, Speech to the Social Market Foundation, 2005).

In our modelling of road pricing explored in this chapter, we concentrate on two alternatives: either the revenue is all held for general local or national expenditure purposes (we call this 'revenue additional') or it is all returned to the *national* community of charged road users by rebating fuel duties ('revenue neutral').

Tax revenue neutrality is calculated from a national Exchequer viewpoint. The charges would not be neutral from the point of view of most individuals or groups of individuals. For instance, it would change the balance between cars and commercial vehicles. With the revenue-additional policy some, or all, of the money could be made available for national or local transport purposes such as 'complementary measures'. But we assume that the benefits are general, rather than imputing them to any particular individuals, and their value is appraised at £1 per £1 of revenue. Under the revenue-neutral policy there is, by definition, no new money available. In practice a specific proposal may fall anywhere on a spectrum between these two polar opposites: in other words involving a degree of revenue additionality.

Current legislation requires that all revenues raised by a local authority through congestion charging must be spent on transport purposes within the area for a period of 10 years. Net revenue from the London Congestion Charging Scheme (LCCS) is a contribution to Transport for London's general budget: it is not specifically rebated to the road users that pay it: it is not, in our sense, revenue neutral.

Exemptions and Discounts

Another significant issue to be resolved is the pattern of discounts and exemptions. There will always be a long list of people arguing for concessions and any practical policy will have them. The LCCS has many concessions including:

- residents of the charged area;
- the less able;

- police and emergency services;
- some 'essential service workers';
- motor cycles;
- public transport vehicles;
- commercial vehicles;
- taxis; and
- alternative fuel vehicles.

A particularly important discount in the current LCCS is the 90 per cent rebate which is granted to those resident in the charged area. This was offered in order to gain support for the initial scheme introduced in February 2003. It was possible because the resident population of that area is so small. But the western extension of the area from February 2007 greatly increases the number of residents enjoying this discount. No scheme on a national scale can afford to give discounts of this magnitude to residents. As most private car and freight vehicle trips are short and close to home, the objective of reducing congestion would be given away.

The principle by which commercial vehicles are to be charged is an issue that will fundamentally affect the balance of advantage as between cars and commercial vehicles. Traditional traffic engineering practice is to work in terms of 'passenger car unit equivalents'. Thus an articulated lorry would be imputed a value of about three passenger car units to represent the fact that this vehicle consumes more road capacity than a private car. In the case of the LCCS, it was originally proposed that commercial vehicles would be charged at three times the rate of a private car, but eventually a concession was offered so that all vehicles are now charged at the same rate.

While it is tempting to offer discounts and exemptions to secure support, each one dilutes the effectiveness of the policy. And once granted they are extremely difficult to rescind.

Scheme Cost

In any discussion of direct taxation or charging it is important to include proper consideration of the costs of collecting the revenues. With indirect taxes whose primary aim is to raise revenues, the usual argument is that a tax has the disadvantage that it inserts a 'wedge' between the cost of production and the value to the end consumer (expressed as the price they are willing to pay). This wedge creates a 'deadweight loss' from the tax. In the case of road pricing where charges are set to align user prices with the true congestion and perhaps with external environmental costs imposed by their actions, there is, happily, a 'deadweight gain'. But there is still a cost of installing and operating the systems which are necessary to collect the charges and to enforce them.

There would be, for example, capital costs associated with hardware and software and revenue costs associated with billing and enforcement. These cannot be neglected: as in the case of the LCCS they could be of the same order of magnitude as the revenues and the economic benefits (TfL, 2006). The estimates prepared for the Road Pricing Feasibility Study illustrated this point (DfT, 2004). If the costs exceed the benefits then the overall case for the policy becomes doubtful.

Unfortunately, it is not possible to begin to estimate the likely costs until the technical requirements have been specified – exactly where and when charges would apply and to whom. Apparently small differences in specification can have large cost implications, as was illustrated by the abortive attempt to design a distanced-based lorry-charging scheme.

Further, some costs tend to be neglected in policy discussion, though they are no less real than the financial costs visible to the public sector. These include the costs to firms and private individuals of compliance. In addition to their equipment costs these can include worry and inconvenience due to the need to comply with an extra piece of bureaucracy.

Both economic benefits and costs will vary considerably by the nature of the location – for instance, the degree of urbanisation – and so will the balance between them. It follows that road pricing might not be worthwhile in some circumstances even though it is worthwhile in others. Different technical solutions may be appropriate in different places (Glaister and Graham, 2004).

It is likely that some costs can be mitigated by piggy-backing road charging onto other services and technologies.

The Case for Building More Road Capacity

For a variety of reasons, Conservative and Labour governments alike have found it difficult to construct new road capacity, despite the fact that any conventional appraisal demonstrates that some road schemes offer unbeatable value for public money. It is the failure to expand road capacity as the demand for it has been allowed to grow which has created the justification for the high levels of road prices presaged in the government's Road Pricing Feasibility Study and our own work.

But suppose that a proper system of national road pricing were put in place, would there still be an argument for building new road capacity? Clearly, the demands would be less, because the prices are designed to reduce the demand to match the available capacity at the peak times.

Currently, prices charged for use of the road system (via fuel duty and so on), the level of capacity provided (the number and size of roads) and the sources of funding are all decided independently of one another under a centralised, administered system. The same is now true of the railways. But for a rationally managed infrastructure enterprise these three things should be determined *simultaneously*. If 'proper' road pricing were in place then the revenues generated by any piece of road infrastructure in relation to the costs of providing extra capacity would provide a clear signal about the justification for making the expansion.

Current research aims to estimate the costs of expanding capacity under certain circumstances and then to throw light on the extent to which new capacity might be justified if road pricing were in place (Glaister and Archer, 2006).

The Meaning of a 'National' Scheme

Along with addressing these issues, any road pricing system would require us to be clear about the meaning of a 'national' scheme. Under the provisions of the Transport Act 2000, any local authority has the powers to introduce a road pricing scheme in its area. Although several have considered doing this, apart from the small scheme in Durham, none looks likely to do so without additional support from national government. That is why, convinced of the need to use road pricing as one of the tools to manage traffic growth, the government has started consideration of a policy at national level.

However, it is not yet clear what the meaning of 'national' might be, as issues about technical standards, taxation and finance, as well as governance would need to be addressed prior to implementing a road pricing system.

Common technical standards

Some features of a national scheme would be unexceptionable and obvious. Legislation to enforce a common technological standard is one. Although not essential, it would seem sensible to legislate so that the vehicle equipped to use roads in one part of the country would be compatible with equipment in use in another. This does raise the question as to whether standards should apply across the UK or across the whole of Europe. If the latter, then the delay to securing agreement might in itself be an obstacle to a national scheme. Similarly, it might make sense to create a single 'back office' operation to which any local scheme could subscribe. This would avoid unnecessary duplication in the creation of software and systems and exploit any economies of scale (although we are not aware of any evidence on the existence or the nature of economies of scale).

National fuel duty

Fuel duty is presently set according to a national standard. If this remains the case and if an element of revenue neutrality accompanies the introduction of road pricing through the reduction of the fuel duty, then this points to the requirements for the scheme to be 'national'. However, fuel prices at the pump do vary geographically, and in other parts of the world road fuels carry differential rates of local taxation. Further, one could envisage schemes for rebating fuel duty on fuel purchased locally even though duty had been levied at the standard national rate.

National and local government

The more substantive question is what would be the relative roles of national and local government in deciding precisely which roads should be charged, determining what times of day the charges would apply, setting the rates of charge, administering the collection and enforcement of the charges, being accountable for the revenues and deciding how those revenues should be spent.

One view is that 'national' implies that all of these things would become the responsibility of Whitehall – just as, in essence, they are now. The magnitude of such a task and the inevitable lack of local knowledge in Whitehall would require a strictly rule-based system.

For instance, this might take the form of a formula defining the rates of charge at a particular time of day in terms of the linear distance between the vehicle's current position and the centre of the nearest urban area. Since this could be a continuously variable *rate* of charge it would have the considerable advantage of avoiding the problems caused by having to define precise zones or boundaries, thus reducing the requirements for precision in recording location, and removing the problem caused to people situated on the 'wrong side' of a boundary.

It seems more likely that considerable powers and discretion would be left with local government in view of the large amounts of money involved; the bitterness of road users concerning the mismatch between what they currently pay through duties and the government expenditure on roads; the general lack of trust that the electorate now has towards government promises on the new systems of taxation and charging; and in the general move towards devolution of powers away from Whitehall. That is certainly the spirit of the current policy of encouraging local authorities to work out 'pilot' schemes funded under the Transport Innovation Fund.

Implications for local government finance

But if, in a national scheme, local authorities are left in control of substantial amounts of road charging revenues, consideration of the local government finance regime will be inevitable. What adjustments should be made to the grants made by Whitehall to local authorities if they start to collect this new money? This question would be particularly acute under a revenue-neutral package because, as we shall show, large city regions would stand to lose several billions of pounds a year. On the other hand, if the city regions were allowed the use of a substantial part of these revenues they could become a crucial contribution towards the portfolio of new locally generated incomes that devolutionists see as vital to the future democratic viability of city regions (Glaister, Travers and Wakefield, 2004).

The need to reform governance structures

But this raises another issue: are the current local governance arrangements appropriate? The geographical coverage of a sensible road pricing scheme must correspond to a suitable set of traffic patterns. Apart from Greater London, few local authorities are suitably configured. For instance, the West Midlands Passenger Transport Authority does have a suitably defined territory but the powers over highways reside with the seven constituent local authorities. It was only possible to introduce congestion charging in London because the directly elected mayor can be held to account for the charges, the budgeting is transparent and the net revenues are manifestly applied for transport purposes within the authority.

We do not believe that a national road pricing scheme could be successfully introduced until this issue is satisfactorily dealt with. There are several possibilities. Trunk roads and motorways could be handled by Whitehall under arrangements similar to the present Highways Authority. These are strategic roads – many of them with serious congestion problems – and the more important ones may not fall naturally to any local authority. But the vast majority of roads are administered by local highway authorities. The government is discussing the possibility of replicating the London system in other city regions and that could solve the problem in those regions.

A less radical change would be to capitalise on the existence of the passenger transport authorities (Travers and Glaister, 2006). When first created in 1968 they were envisaged as powerful bodies responsible for passenger transport in the metropolitan areas. But over the years their power has diminished. They could be reinvigorated as all-embracing transport authorities with such highways powers as they would need in order to implement road pricing. New authorities might be created to serve cities such as Bristol, and the Southampton–Portsmouth area.

Other solutions could be seen by some to have the advantage of removing these decisions from national and local politics and establishing the required trust of those who would be paying the charges. One solution would be to create public trusts – as with the US federal roads, some British ports and the institutions that funded many roads in the first place – which would be legally accountable for the revenues and for spending them in accordance with defined objectives.

Once properly priced, roads would be little different from other infrastructure utilities such as gas, water and electricity. It would be possible to think in terms of creating a similar regulatory regime.

The urgency for London

So far as London is concerned there are some special considerations. A significant portion of the nation's road congestion occurs in the London area. It is hard to imagine how a long-term transport policy for outer London could be managed without some form of road pricing: it would manage the growth in traffic and also provide sufficient net revenue to service tens of billions of pounds of prudential borrowing, thus funding a substantial part of the new transport infrastructure that is needed for London. 'Tag and beacon' technology is now available to achieve this. If London waits for the introduction of a national scheme it risks having to wait for a long time. Worse, it risks losing some of the revenues to other parts of the country if the national package were to include an element of revenue neutrality.

The Interrelationship with Other Policy

The successful introduction of national road pricing would have profound implications for other areas of government policy although there is no sign that these are currently being actively considered. Although the effect on public transport patronage would not be great at the national level, it could be an issue in certain local markets. Buses would need to carry more people, but would gain capacity because of higher road speeds. The passenger demand for railways in the commuter markets in London and other big cities would be increased. But these are situations where crowding on rail is already severe. Conversely, if a revenue-neutral package made road use cheaper in rural areas, then the competitive position of railways and other public transport would be further weakened in markets where they are already weak.

Planning and land-use policy would also be affected. Some planning policies are used to prevent excessive road congestion – for instance, restrictions on out-of-town shopping centres. If congestion had been dealt with directly by road pricing then this motive for planning restrictions would fall away. In the long term, road pricing would inevitably affect densities because the costs of transport are one of the main determinants of density. Those parts of government seeking to influence densities should have an interest in the policy on road pricing.

Political leadership

Road pricing may be an excellent idea in principle. But there is a long way to go before a coherent and complete policy is formulated sufficiently precisely for the public to understand what is proposed and form their opinion. The LCCS came about because of several special circumstances including the creation of a brand new local government, the existence of a carefully worked-out scheme, and an independent, radical candidate for mayor who was willing to take the necessary political risk.

Considerable leadership would be required from a prime minister if a scheme were to be attempted at a national scale. Apart from anything else, in order to produce a firm proposition, it would be necessary to secure the agreement of a long list of Whitehall departments, each with its own direct interest, which includes: the Department for Transport, Her Majesty's Treasury, the Department for Communities and Local Government, the Department for Food and Rural Affairs, the Department of Trade and Industry and the Department for Constitutional Affairs.

The Need for Clarity

It will be extremely important to maintain clarity about the objectives of a national scheme and to keep things simple and manageable. Road pricing is certainly about mitigating congestion and making more efficient use of the existing network.

But what does 'efficient' mean? Does efficiency include saving leisure and commuter time or is it more to do with national productivity and those things that contribute to GDP as measured? Is road pricing considered a new source of local or national income and if so, is that to be used to fund new roads, new public transport or for some other public purpose? To what extent is road pricing regarded as a measure to protect the environment? Is it part of the debate about poverty and social exclusion? Is it to be used as a tool to manipulate land uses and densities?

No single policy can be expected to deal with such a long list of objectives and, indeed, some may contradict others. The abortive scheme for distance-based charging for lorries illustrates how an idea can become too difficult, too expensive and too risky if sight is lost of a simple primary objective.

Clearly, until these issues are settled the various interests would not be in a position to judge their level of support for any road pricing policy.

3 WINNERS AND LOSERS OF ROAD PRICING

One important question surrounding the introduction of a charging scheme is how would it impact on the population: who would stand to win or lose? The distributional consequences of road user taxation have attracted interest in the literature over the years (Evans, 1992; Blow and Crawford, 1997; Santos and Rojey, 2004; Poterba, 1991; Richardson and Bae, 1998). It is, however, difficult to generalise about whether road user taxation is regressive or progressive *per se*, because so much depends on the limits and design of any one particular system and on the specific context within which it is implemented.

But what we can show is how road users living in different areas of the country, experiencing different levels of deprivation, are likely to fare under the two polar options for national pricing schemes. We also look at how households which are not private car users will be affected.

Modelling the Effects of Road Pricing

In modelling the consequences of various road user charging policies, we have analysed a simple policy of making a charge per vehicle-kilometre for the use of all roads at a rate that reflects the level of congestion and environmental damage (Glaister and Graham, 2004). This charge rate will vary by the current traffic level, the size of the vehicle, capacity of the road and the nature of the locality.

There are certain things to note when looking at our conclusions. Our approach recognises three fundamental linkages. First, varying prices will change traffic volumes which, in turn, will lead to variations in important dimensions of quality, such as speed, as congestion changes. Second, varying prices and travel times will affect the time of day chosen by drivers to travel, the propensity for drivers to share cars and whether drivers in fact chose another mode of travel. Third, varying prices, taxes and subsidies will change the burden on the public purse and may change the funding available for new infrastructure from both public sources and privately funded investment. Differing pricing regimes will create changes in patterns of demand, and consequently changes in the case for investment in infrastructure.

In developing the model we needed to make a series of simplifying assumptions. The model does not assume any explicit transport network and makes no attempt to represent origin-to-destination trip patterns. Consequently, we are not able to distinguish between changes in the number of trips and changes in average trip length. Our modelling worked throughout in terms of costs and charges per vehicle-kilometre and average traffic flows (passenger car units per hour). The model was not capable of representing certain types of charging schemes, such as workplace parking charges, cordon or area schemes.

The costs that are to be imputed to environmental damages such as air pollution and climate change are uncertain but they are important determinants of the pricing policies considered in this study. There are some important factors that have not been – or cannot be – quantified. Some of these omitted factors may be detrimental to the environment or create social costs: for example, severance of communities by roads. Others create benefits: for example, better accessibility to family, leisure pursuits or employment opportunities.

Under our model, once road pricing is introduced and traffic levels settle down to a new equilibrium, congestion and pollution fall and the road user charge generates revenue. The model shows that road users generally would be worse off because either they would be paying more or they would be deterred from travelling. But, it also shows that there would be more than enough revenue to compensate them so that everyone can end up better off if compensation were considered desirable. This is a reflection of the fact that the facility (the road network) is being more efficiently used, so the overall economic value of the system is increased.

This suggests one reason for the crucial importance of what happens to the revenues from road user charging. In practice, government may not choose to use the revenues to compensate the particular individuals paying the charge. If this is the case, road users in general will be disadvantaged. That is why the legislation to ensure that the LCCS proceeds must be applied to transport purposes in London was crucial to securing public support.

In reality (and in our model) charged roads will be used by drivers making trips to which they will attach a range of values of time saving. When charges are introduced, drivers who attach higher values to the trips they are making will be more inclined to stay on the road, pay the charge and enjoy the benefit of higher speed. Therefore the scarce and valuable resource, road space, is reallocated to those drivers who place a high value on the trips they are making.

This is a further source of economic efficiency. Even without compensation, some road users may gain overall. There might also be economic welfare benefits that go beyond the benefits arising purely in the transport sector. For example, the overall costs of living in a congested area (say, the South East) will be more closely matched by the costs paid by the individuals who live there. Individuals can take economically more efficient decisions about where they live, where they locate and the conduct of their businesses.

We emphasise that road charging is about charging for *trips*, rather than charging individuals. The point is to adjust trip making so as to create an incentive for the highest value trips – as judged by trip-makers themselves – to have the use of the road, while the lower value trips divert to other, neighbouring times. In other words it is about discouraging individual journeys rather than discouraging certain individuals from using their vehicles *per se*. There is certainly a correlation between incomes and trip values but it is by no means perfect. There are occasions when individuals on low incomes will attach a high value to saving time and so will enjoy a benefit of a faster peak trip.

Winners and Losers

For the purposes of our model and this chapter we ignore the costs of concessions and collecting revenues to allow us to concentrate on the biggest issues: what is to be done with the revenues from road user charging and how do decisions about this affect who gains and who loses? Would the revenues be used by the Exchequer for general purposes, made available in the locality in which they are collected for use for transport purposes or returned to the generality of road users across the nation in some form?

To point up the differences we have considered two alternatives. In one, the 'revenue-additional' case, the revenues are either used by the Exchequer for the general benefit, or they are used by an administration local to the area in which the revenues are collected for local benefit. In the other, the 'revenue-neutral' case, fuel duties are reduced in such a way that the sum of fuel duties and road pricing revenues is held constant: thus the national road-using community as a whole (including freight vehicles) would pay the same in total with or without road pricing. Table 4.1 summarises the evaluations of our two 'polar' policies.

Both approaches produce overall net benefits, the revenue-additional policy rather more. They both produce a saving in environmental costs, the revenue-additional policy substantially more because, in addition to achieving a more efficient (that is, lower environmental cost) pattern of usage of the road network, it reduces total national volume of traffic.

Table 4.1 Economic performance of revenue-additional and revenue-neutral policies (£bn per annum)

	Change in motorist benefit	Savings in environmental costs	Change in tax and charge revenue	Net benefit
Revenue additional	–8.2	2.1	15.8	9.7
Revenue neutral	6.3	0.5	0	6.8

There is a crucial difference between the two policies: with the revenue-additional policy, motor vehicle users as a group are definitely considerably worse off. The environmental cost savings and the tax revenues both represent benefits to others (and to road users in the other aspects of their lives) and they are more than sufficient to outweigh the disbenefits to road users overall.

This illustrates the basic economic efficiency proposition in favour of road pricing – that, in principle, the benefits represented by the environmental savings and the revenues are more than sufficient to compensate those who pay the charge. The proposition stands irrespective of whether compensation is actually made. Within the population of road users there will be some, typically those with high values of time savings, for whom the benefits of higher traffic speeds exceed the money charges, so they will be better off even though not compensated. However, as a group, road users are made worse off.

By contrast, with revenue neutrality, road users as a group are made better off. In effect the compensation is made through the reduction in fuel duty. Some – those with high valuation of time savings – will be made considerably better off. Some individual road users will be made worse off but, overall, the gains will outweigh the losses.

Representing Results at Ward Level

The 10,070 census wards of Britain form the basic unit of analysis used in the following representations. Figure 4.1 displays an estimate of the average traffic volume changes experienced in each census ward under a revenue-additional policy. Figure 4.2 displays the corresponding revenue-neutral policy.

Two points are immediately apparent from these two figures. Under either policy there is a marked difference between the impact on urban areas and that on the much larger rural areas and there is a contrast between the experiences of the southeast region and the rest of the country.

The busy urban areas experience similar traffic reductions under either policy – congestion is treated aggressively in both cases. But at the other end of the scale the policies have very different implications. With the revenue-additional policy the rural areas experience a small reduction in traffic: while there is no charge for congestion in rural areas, we do allow for a relatively small charge reflecting the environmental damages.

In contrast, under the revenue-neutral policy, rural areas experience a 22–26 per cent increase in traffic. This is because the revenues earned in the urban areas are used to reduce the cost of fuel – approximately equivalent to a complete removal of the duty on road fuel (but not of value-added tax: VAT).

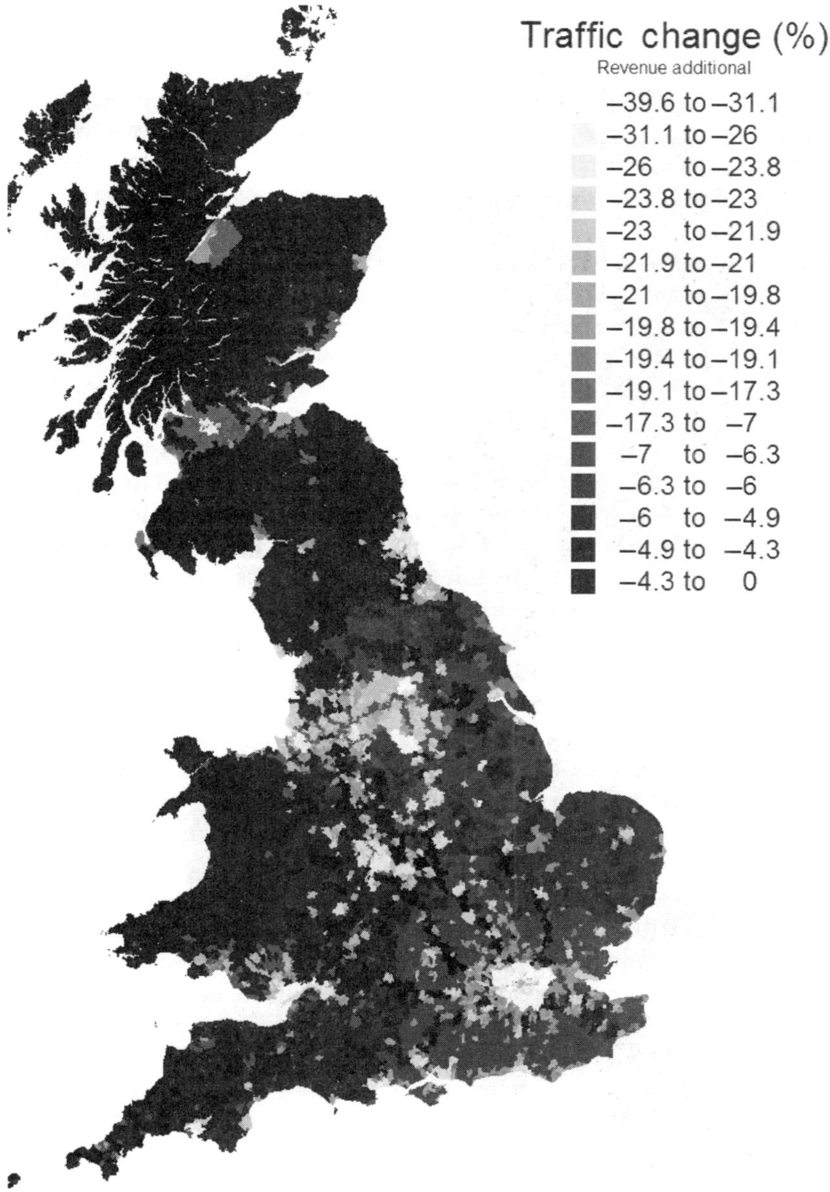

Source: Authors.

Figure 4.1 Average percentage traffic changes by census ward, GB, 2010, additional revenue: £16 billion per annum

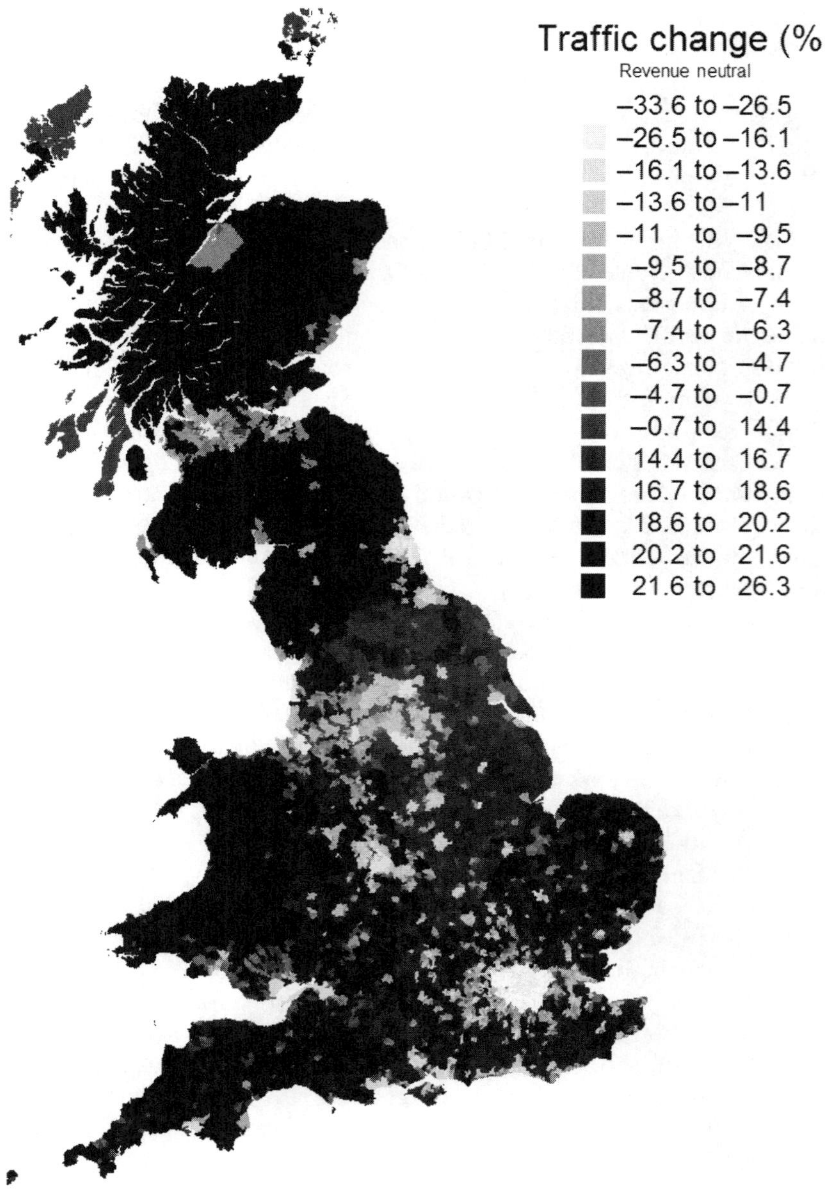

Source: Authors.

Figure 4.2 Average percentage traffic changes by census ward, GB, 2010, revenue neutral

This estimate of the traffic increase is a simple arithmetic consequence of the form of the demand relationship assumed and the empirical estimates we have used of how motor vehicle users respond to changes in fuel prices. However, this is equivalent to very considerable changes in fuel price and, as a consequence, these estimates should not be taken too literally given that this is not an approach government is likely to adopt.

This points up the major feature of the revenue-neutral policy: it would transfer considerable sums of money from urban to rural areas. Unless compensation were made through a change in the local government finance regime the residents of the urban areas would, as a group, be made worse off – particularly most of those paying the road charges. Since a majority of the population lives in or near urban areas, the consequence would be that a large number of people would be made worse off and a small number would be made better off, some of them considerably so. In England, 79 per cent of the resident population would live in areas where money and time costs of using roads would have increased on average, whereas 21 per cent would enjoy lower costs. In Great Britain the corresponding figures are 77 and 23 per cent, respectively.

There are important differences between the impact of the two policies in the suburbs. For instance, in the areas just outside Greater London, under the revenue-additional policy, traffic would fall by around 20 per cent. Under the revenue-neutral policy it would fall by around only 8 per cent. There are similar implications in the suburbs of the West Midlands and in the Liverpool–Manchester–Leeds–Sheffield area. This distinction is particularly significant if it is expected that there will be long-term growth of population in areas such as this.

A large increase in traffic in the rural areas should not necessarily be regarded as a bad thing on the crucial proviso that this traffic is genuinely paying the full cost of the congestion and other damages inflicted on others. Then, the benefits to the extra traffic will outweigh the costs to others.

If the revenues of road user charging are disbursed to the road-using community in the form of rebates on fuel duties under a revenue-neutral scenario, important distributional questions are raised. Some of the damage costs – noise, air pollution and climate change – fall on people other than road users and they would not be compensated. As we have already noted, so far as damage to the environment is concerned, the revenue-additional policy is a more effective remedy.

Under current policy, government is committed to a 'compact city' policy in the belief that, compared with traditional development, it would be more fuel efficient, it would promote healthy walking and cycling lifestyles and enable the poor to be better served by services including public transport. Whether such policies are fully supported by evidence is an open question (Cheshire, 2006). But if a revenue-neutral policy made rural driving cheaper than now, it might attract additional people to enjoy it and so might be in conflict with a compact city policy.

In the revenue-additional case, speeds all improve. But in much of the country the change is negligible because the traffic is free flowing anyway, so change in traffic volumes causes little change in speed. The situation is actually very similar to the revenue-neutral policy because the large traffic increases occur on uncongested roads. There are a few places where speeds do fall as the traffic increases (for example, in the area around Cambridge and other parts of East Anglia) but the average speed reduction is more than 1 per cent in only about 2 per cent of the wards.

However, they do reflect the average changes in money costs of motoring to private individuals. Again, these are averages across the week: within that there will be times when there are much lower charges and peak periods when they are much higher.

Figure 4.3 classifies the census wards by the 10 area types (listed in Table 4.2) and then relates each type to the percentage traffic change. In these diagrams each point represents one of the 10,072 census wards positioned to show the area type in which it is situated and the average traffic change it would experience.

There is not much difference in traffic change by area type under the revenue-neutral scheme (as shown in Figure 4.3) and the revenue-additional version (not reproduced here), except that the revenue-neutral one is shifted vertically. Referring to the revenue-neutral case, the traffic reduction in outer London – the outer boroughs such as Hillingdon and

Table 4.2 Area types

Area types	Description	Population
1	Central London	
2	Inner London	
3	Outer London	
4	Inner conurbation	
5	Outer conurbation	
6	Urban big	> 250,000
7	Urban large	> 100,000
8	Urban medium	> 25,000
9	Urban small	> 10,000
10	Rural	

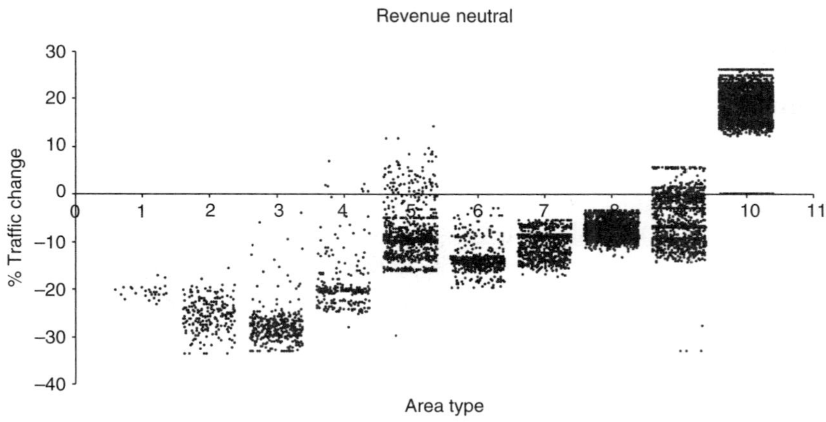

Source: Authors.

Figure 4.3 Census wards and traffic change by area type, Great Britain, 2010

Croydon – is the greatest in Britain and typically greater than in inner London. There is significant traffic reduction in other inner conurbations. Most outer conurbations also have traffic reductions but a few have increases. The four types of urban area (as distinct from conurbation) generally have small traffic reductions on average. There is a large population resident in areas of this kind. Finally, the rural areas experience a traffic increase.

Figure 4.4 shows a similar plot for Britain, but for the price change rather than traffic flow. Comparing this with the traffic changes in Figure 4.3, it is interesting to note that the price increases in central, inner and outer London are much higher than elsewhere and high relative to the traffic reduction achieved. This is because traffic speeds are already low in the London area so time costs are high and money price is a smaller proportion of total cost. Therefore the price must be raised more in absolute terms in order to secure a given traffic reduction.

In the case of a revenue-additional policy, road users as a group would be worse off. The extra revenues would amount to about £16 billion per annum – though concessions might reduce this. If these revenues were returned to the local communities from which they came then road pricing could lead to important overall gains for the communities, although the net effect on road or transport users will depend upon how the money is spent. So long as the costs of collection do not consume too much of the revenues, there would be a new and significant stream of annual income that local

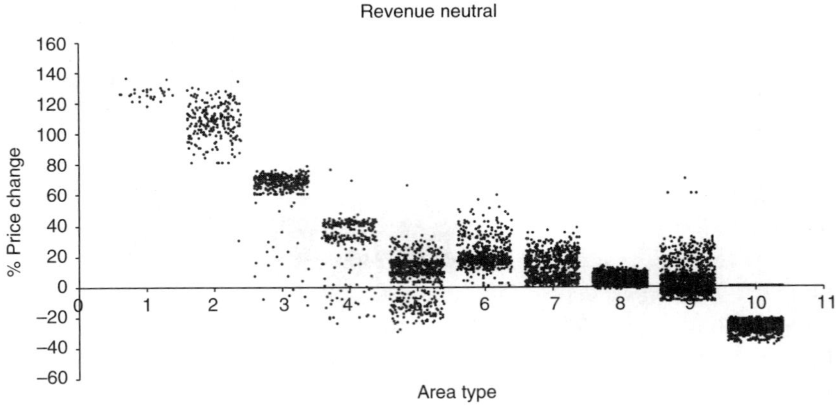

Source: Authors.

Figure 4.4 Census wards and price change by area type, GB, 2010

authorities could use either for revenue support or to service prudential borrowing. That could be used for capital finance for some of the items they cannot fund presently.

The revenue-neutral policy would generate somewhat less overall net benefit. But it would make road users as a whole better off because the revenues are returned to them and the road network is used more efficiently. A major feature of the revenue-neutral policy is that it would transfer considerable sums of money from urban to rural areas, particularly from London. Unless compensation were made, such as a change in the local government finance regime, the residents of the urban areas would, as a group, be made worse off. Since a majority of the population lives in or near the urban areas, a consequence could be that a large number of people would be made worse off and a small number would be made better off.

These average calculations need to be treated with caution because they conceal important variations. For instance, under a revenue-neutral scenario, car users in urban areas at uncongested times would be paying less, even though, averaged across the week, car users in urban areas were paying more. The revenue-neutral proposal has important presentational attractions. However, there would be no net revenue to defray the costs of the scheme or to spend on the 'complementary measures' that may be important in winning general support.

4 WOULD ROAD PRICING BE FAIR?

Not surprisingly, surveys of public attitudes confirm a clear concern that road pricing might not be 'fair'. This is often expressed as a concern that the poor should not be adversely affected.

In the DfT's survey, '32% felt that road pricing would be fair, 50% thought it would be unfair'. The main reason given for perceived unfairness was the inability of people to change their behaviour. The lack of adequate alternatives, cost and the potential for disproportionate impact on people on lower incomes were also mentioned relatively frequently (DfT, 2006).

A full analysis of this issue would be extremely complex and it would require data that probably do not exist. However, some general indications are possible. We have ward data which allows us to compare our model to the government's official measures of deprivation in order to understand how fair road pricing would be.

In doing so we drew conclusions about the relationships between changes in traffic and the level of deprivation of wards as a whole. This does not mean that this will be the common experience of all the individuals in the ward. There will, of course, be considerable variation in the circumstances of individuals within any ward. In particular, some of them will be private car users and some will not.

This section also addresses the point that car users will be differently affected from, for example, public transport users. We are relating the experience of road users in wards with various levels of deprivation, irrespective of whether or not those users are themselves deprived. In drawing conclusions we are implicitly assuming that car users living in deprived wards tend to suffer more deprivation than car users living in less deprived wards.

The 2004 deprivation index is a composite of seven 'domains':

- income deprivation;
- employment deprivation;
- health and disability deprivation;
- education, skills and training deprivation;
- barriers to housing and services deprivation;
- crime rates; and
- living environment deprivation (which includes air quality and road traffic accidents). (Office of the Deputy Prime Minister, 2004.)

Figure 4.5 displays the geographical distribution of the deprivation index for England. (As in all the maps in this document each shade of grey represents approximately the same number of census wards.) Note how

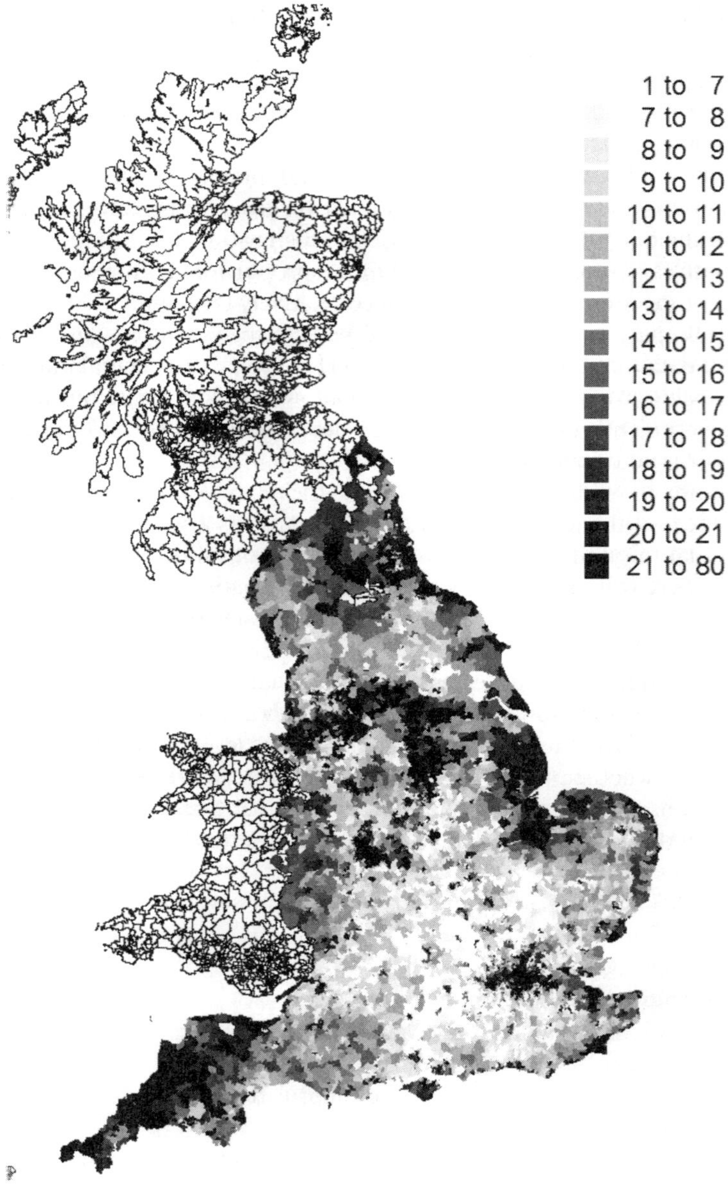

Note: Low numbers indicate low deprivation. High numbers indicate high deprivation.

Source: Authors.

Figure 4.5 Deprivation index for England

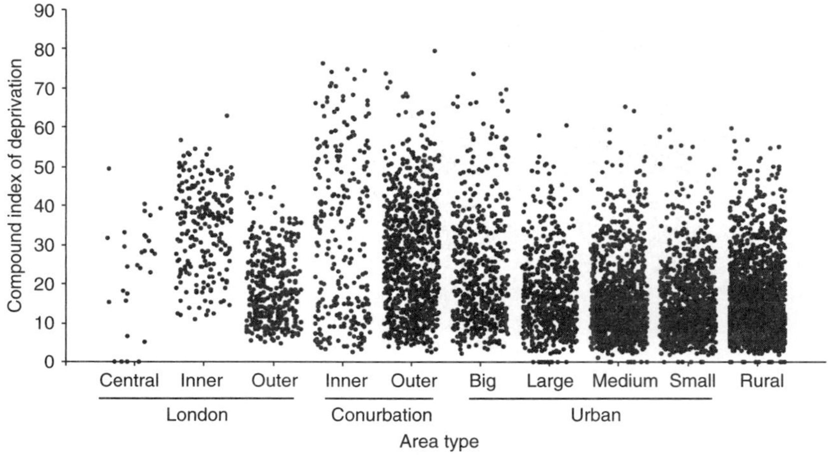

Source: Authors.

Figure 4.6 Compound deprivation index of census wards and area type, England

high deprivation occurs in both inner city wards and remote ones, whereas low deprivation is prevalent in non-urban 'middle England'.

Figure 4.6 illustrates the relationship between area type and the degree of deprivation of census wards in England. We had expected to find a strong relationship, with considerably more deprivation in the large conurbations and less in the rural areas. Since road pricing would definitely involve higher charges in large urban areas, we expected that there would be a strong relationship between road pricing and deprivation. Figure 4.6 suggests that, while there is indeed a relationship between the type of area and degree of deprivation, it is not very strong: high deprivation is to be found in most types of area.

To examine the link between road pricing and deprivation, Figures 4.7 and 4.8 show the changes in traffic levels by degree of deprivation for both the revenue-additional and revenue-neutral policies.

In both cases the wards fall into two groups. Those in the first group, typically rural areas, have less deprivation and have a relatively small traffic reduction or a traffic increase in the revenue-neutral case. Those in the second group have a bigger traffic reduction, spread across the scale of deprivation and do not appear to have any particular relationship with deprivation.

The implication is that while the rural areas tend to have less deprived wards and suffer less reduction in traffic under road pricing, once the rural

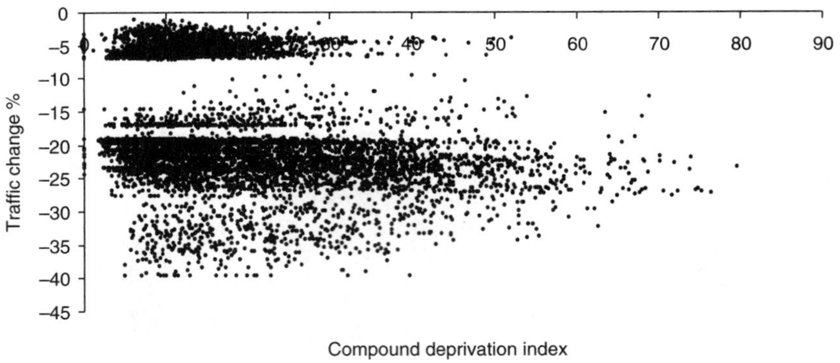

Source: Authors.

Figure 4.7 Percent traffic change and compound deprivation index, England: revenue additional

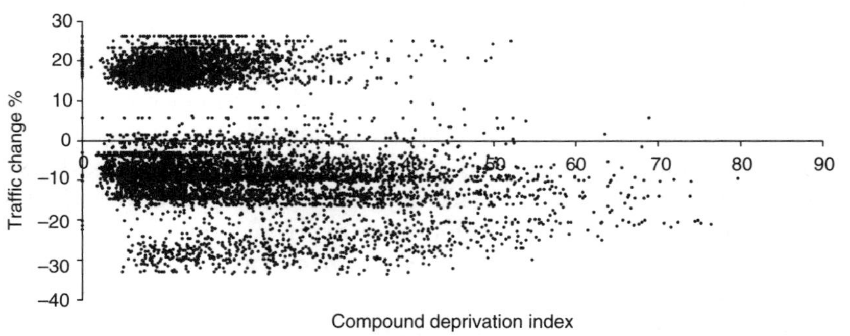

Source: Authors.

Figure 4.8 Percent traffic change and compound deprivation index, England: revenue neutral

areas are excluded, there is no obvious, systematic relationship between deprivation and the degree of traffic reduction.

All the Domains Together

The several domains of deprivation are not, of course, independent of one another: for example, income deprivation can depend on levels of employment deprivation. Because of this it can be misleading to consider the relation between traffic change and any one of the domains in isolation: the one domain may be acting as a proxy for one or more other domains with which it is correlated.

By analysing the relationship between traffic change and each of the domains of deprivation (while holding all the other domains constant) we can show that employment, housing and education deprivation all show a significant *positive* relationship with traffic change. Thus wards showing high deprivation on any of these three measures will, other things being equal, tend to have smaller traffic reductions – because of smaller price increases. In the case of the revenue-neutral policy they are more likely to experience price reductions and traffic increases. To the extent that reduced travel costs are helpful in mitigating these types of deprivation, road pricing will be more helpful on these factors than as measured by the other domains.

Differences in Urbanisation

Our previous work (Glaister and Graham, 2004) on the implications of transport pricing has suggested that under today's overall rates of fuel tax, city areas and major inter-urban routes tend to be relatively undercharged while the country areas are significantly overcharged. In other words, broadly speaking we might expect a positive association between level of urbanisation and price based on the costs of additional road use.

Figure 4.9 charts levels of urbanisation in English wards against the price change that would result under the revenue-additional pricing scenario. Note that since this scenario assumes additional taxation, prices are increased everywhere. Although the degree of urbanisation represents an index, the lower end of the scale clearly relates to deep rural areas, while the upper end relates to particularly heavily urbanised places, high resident population and high employment close by. Many, but not all of these will be London wards.

Figure 4.9 confirms that there is a strong relationship between the level of urbanisation of a ward and the price change that would result, given our road pricing scenarios. Indeed, for the most heavily urbanised wards (which sit in a group on the right-hand side of the figure), the average price increase is somewhere between 13 and 23 pence per car-kilometre. Rural areas, on the other hand, are clustered in a group on the left-hand side of the horizontal axis and the price change experienced here is very small.

However, this link between price and the level of urbanisation is unlikely to have a simple bearing on equity, as the most urbanised wards – which would see the greatest price change – do not necessarily tend to be the most deprived.

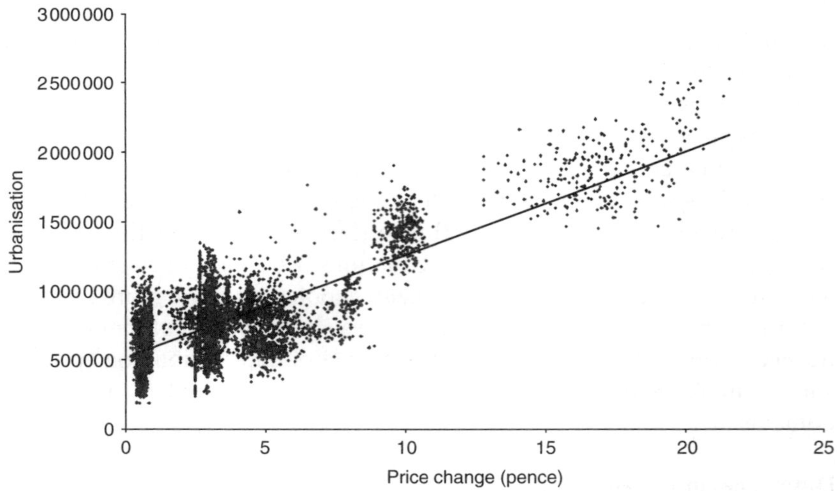

Source: Authors.

Figure 4.9 Ward urbanisation and price change (pence per vehicle-km), revenue-additional scenario

The measure of urbanisation we use is likely to provide a reasonably good proxy for the volume of traffic flows in any ward, and so could also be interpreted as showing that *in general* there is no tendency for deprived areas to have more traffic than non-deprived areas.

Thus, while there are certainly heavily urbanised wards with heavy traffic which have high income deprivation, equally there are heavily urbanised wards that have low income deprivation. For instance, it is easy to think of deprived wards in Westminster and the eastern edges of the City of London which share boundaries with some of the least income-deprived wards in the country.

The Effect on Typical Trips

Obviously the trips made by the residents of any ward will not necessarily be contained within the boundaries of that ward. So if we want to analyse the effects of pricing in relation to deprivation, we need to represent price, speed and generalised cost changes over a wider area than the ward of residence.

In fact the wards of Britain are relatively small, the average width being only four kilometres. Table 4.3 shows average trip lengths by trip purpose

Table 4.3 Average trip lengths by trip purpose, GB, 2004

Purpose	km	Purpose	km
Commuting	13.7	Visiting friends at home	14.8
Business	34.4	Visiting friends elsewhere	9.7
Education	4.8	Entertainment	12.6
Escort education	3.5	Sport: participate	10.1
Shopping	6.8	Holiday: base	82.2
Other escort	8.4	Day trip	23.5
Personal business	7.1	Other	1.8
All	11.1		

Source: DfT (2006).

for Britain. Of course these average figures will vary considerably across the country, for instance, commuting journeys in London and the South East are typically very much longer than in other regions of Britain. But the table does demonstrate that average trip lengths for most purposes will tend to take travellers outside of their ward of residence.

As a large proportion of trips made from any ward will take travellers outside the boundaries of that ward, we are interested not so much in price and speed changes in each of the wards but in how prices and speeds will change on average *around* the wards.

Whether looking at London wards (which would tend to experience relatively high rises in price in our scenarios) or at wards outside of London, our results show a fairly diverse spread of levels of income deprivation. Thus, even if we consider a wider range of trips outside a ward, there is no clear relationship between average price changes in the revenue-additional scenario and the level of income deprivation, and so there are no clear implications for equity in our model of road pricing (Glaister and Graham, 2006).

But could there be an impact on equity even after accounting for the fact that different wards have different levels of urbanisation?

Looking at the relationship between income deprivation and changes in prices and speeds, there is some conflicting evidence (Glaister and Graham, 2004). For all trips within 15 km of a ward, other things being equal, drivers from more deprived areas will tend to pay more and have greater increases in speed.

Yet considering trips in wider areas, for example within 30 km and 50 km of the ward, the data show that it is drivers from the least deprived areas that tend to pay more and have the highest speed increases. Despite the statistical significance of some of the results, the evidence from our model is inconsistent, suggesting overall the existence of only weak effects on equity (Glaister and Graham, 2004).

Thus while drivers in more urbanised areas will pay more and so see greater changes in traffic and speed, there is no evidence of a systematic relationship between price change and deprivation.

This is because congestion, which has the largest influence on price change under the revenue-additional and revenue-neutral scenarios, tends to be highest in the most urbanised locations: the more urban a place, the more traffic congestion, the higher the road pricing charges and the subsequent increase in traffic speed. But with no direct link between urbanisation and deprivation, there is no unambiguous evidence to suggest any implications for policy.

Accounting for car-owning households

One important factor that we have not considered in the above analysis is that the number of car-owning households, and therefore the number of people affected by the price and speed changes, may vary across the wards.

For example, while London wards typically have the highest increases in generalised cost, they tend to have less residential and more commercial land use. In addition, car ownership is low in London.

This in turn may affect the relationship between income deprivation and the change in costs faced by motorists as a result of road pricing. Indeed, accounting for the number of car-owning households does change the distribution of costs, and suggests that more deprived areas are relatively worse off. But as we have found consistently in this section, the relationship is weak.

Effects on Household Budgets

As we have already noted, the effect of road pricing on any individual will depend upon the extent to which they happen to be car users: individuals that do not own cars and do not use them as passengers will be much less affected. Indeed, they are more likely to be public transport users and therefore more likely to benefit from potential improvements in speed and reliability of bus services if congestion is reduced.

This section uses data from the 1996–97 Family Expenditure Survey (FES) to address these issues. It can only be regarded as the roughest of sketches because of the age of the data. The FES gives detailed information on all items of household expenditure, the households being classified in a variety of ways. They have the important advantage that they record the proportion of households that record zero spending on each item, so we can differentiate those that bought motor fuel and so would be affected by road charges from those that would not.

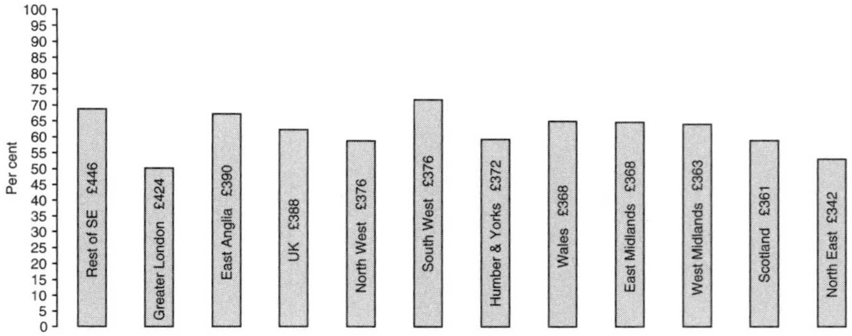

Figure 4.10 Percentage of households buying fuel, FES, 1996–97

Figure 4.10 displays the proportion of households that bought motor fuel in the FES, categorised by standard region. The bars are ordered in declining total weekly household expenditure (in 1996–97), and the total is shown in pounds sterling on each bar. The expenditure figures have been converted from May 1996 prices to May 2005 prices using the Retail Prices Index. This will understate total 2005 expenditures because of growth in real incomes. Over this period the real price of fuel rose by about 16 per cent, a little less than the growth in real incomes.

Assuming that households that do not buy fuel would not be affected by road pricing, about half of all households in Greater London would not be directly affected, in spite of relatively high incomes. At the other extreme, only 28 per cent of households in the South West would not be affected. This is probably because of the much superior availability of public transport in London, so car use is lower even though incomes are higher. Further, those households that do buy fuel in London spend less on it than those in any other region, as shown in Table 4.4.

The striking thing about all of these figures is the extent to which charges are higher in London than anywhere else. Although about half London households would be unaffected (and would benefit from improved public transport service quality), the other half would on average be spending an additional 4.8 per cent of their total household budget on the charges: £20 per week out of a total of £424 per week (and would benefit from less traffic congestion). This approximately doubles their current outgoings on fuel.

Aside from London, there is considerable variation across the regions. Road users in the North East would spend a particularly high proportion of their household budgets on the charges. This is the consequence of car use being relatively rare in this region and those who have cars spending more on fuel than the national average.

Table 4.4 Estimated household spending on motor fuels by households that buy it in 2005, and additional spending on road charges (£ per week), by region: revenue additional

	Rest of SE	Greater London	East Anglia	UK	North West	South West	Humber & Yorks	Wales	East Mid-lands	West Mid-lands	Scotland	North East
Spending on fuel	25	21	25	24	21	23	22	26	23	25	24	23
Additional spending on charges	2	20	2	4	2	2	2	3	3	3	2	6

There is no obvious, simple relationship between ranking of regional incomes (corresponding to the order of the bars in Figure 4.10) and the proportion of incomes that would be spent on revenue-additional road charges. Despite both areas spending a relatively high proportion of their weekly income on road charges, London and the North East are almost at opposite ends of the spectrum in terms of income.

Employment, housing and education deprivation all show a significant *positive* relationship with traffic change. Thus wards showing high deprivation on these measures will, other things being equal, tend to have smaller traffic reductions – because of smaller price increases.

Indeed, in the case of the revenue-neutral policy they are more likely to enjoy price *reductions*. To the extent that reduced travel costs by car are helpful in mitigating these types of deprivation, road pricing will be less damaging on these measures than on the other measures of deprivation.

The true effects on households would be determined by the charges in the areas through which they drove, rather than where they live. The indication is that there is no systematic relationship between ward income deprivation and the speed and price changes that might arise from road charging. The same appears to be true for Wales and Scotland. Adjusting for variations in rates of car ownership does not change this result.

5 CONCLUSIONS

The results presented here illustrate a well-known proposition: that altering charges to make them reflect social costs more accurately can generate new economic value. In our context they do this by making road users face up to the congestion and environmental costs they impose on others and by giving road users incentives that guide them towards more intelligent use of scarce highway capacity.

Most conventional taxes are imposed because of a need to raise revenues, and they have the distinct disadvantage that they distort the relationship between cost and value to the end user. By contrast, well-designed road charges with low enough collection costs improve the match between social cost and value to users so that the overall cost is *less* than the benefit represented by the value of the revenue raised.

So, in principle, unlike most taxes, road pricing can both raise revenue and 'do good' in the round. But, as with a tax, who gains and who loses depends crucially on who pays the tax and how the benefits of the revenues are disbursed.

Who might gain and who might lose plainly depends crucially on a number of characteristics of the policy. We have analysed a simple policy of making a charge per vehicle-kilometre for the use of all roads at a rate that reflects the level of congestion and environmental damage. This charge rate will vary by the current traffic level, the size of the vehicle, capacity of the road and the nature of the locality.

This is an idealised scheme and practical schemes would probably be simpler. The nature of the simplification – such as a cordon scheme or an area scheme like the LCCS – could significantly alter the incidence of charges on particular groups of individuals. Any concessions granted in a practical scheme will also have direct implications. We abstract from this issue by assuming that there are no concessions.

Also we have ignored the issue of how much these charges might cost to collect. This must not be neglected in practice – every £1 spent on hardware or administration is £1 of benefit lost, to be set against the traffic and environmental gains.

In an earlier study we showed how different technologies dictate different relationships between geographical coverage and cost. We argued that rather than attempting to do everything, it might be better to accept a part of the available gross benefits with a less than complete geographical coverage. This is an issue that we do not consider in this chapter, but it is a vital component of practical policy design: and the trade-offs change rapidly as technology advances.

We have analysed the impact of road pricing by the degree of urbanisation. The projected traffic reduction in outer London is the greatest in Great Britain and typically greater than in inner London. There would be significant traffic reduction in other conurbations. Under a revenue-neutral policy, smaller urban areas (as distinct from conurbations) generally have small traffic reductions on average. There is a large population in areas of this kind. The rural areas experience a traffic increase. A revenue-neutral charging scheme could be greatly beneficial for poor rural car users given that car-ownership rates are high in rural areas: in rural areas the poor do run cars.

A revenue-additional scheme, by contrast, would be no better for poor rural drivers than existing taxes and charges. And given the impossibility of providing more than sketchy rural bus services, this would mean that a revenue-additional scheme would hit the rural poor perhaps harder than their urban counterparts who may be able to walk the shorter distances or catch a bus and who might benefit from the revenues being channelled back into their areas.

Not everybody is a car user and those who are not would stand to benefit from the clearer roads and improved bus services. Car use in London is much lower than the national average because of the superior public transport. But road charges would, on average, be substantially higher in London. The combined effect is that under either kind of revenue policy, private car users in the London area would spend a higher proportion of their household budget on motoring: the extra might be a twofold increase, from the 5 per cent of household budgets presently spent on fuel to 10 per cent.

This neglects important benefits in terms of the value of higher road speeds and it does not take account of the benefits from the charge revenues if spent in London. For those who are not car users there would be no increase in charges, and the benefit of clearer roads. Under the revenue-additional policy, for most of the other regions the additional spending would be between 1 and 2 per cent of household budgets for those who use cars, but it might be 3 per cent in the North East.

We have argued that road pricing cannot be considered in isolation from a range of other, controversial policy areas. Public transport policy would be directly affected. So would local government finance and, crucially, the geographical span of the present local governance arrangements would be brought into question. We have discussed several solutions to this, some incremental and some radical and several of them consonant with proposals under consideration in other parts of government to do with devolution and policy towards 'city regions'.

For London – which accounts for a significant portion of the nation's congestion problem – adequate governance arrangements are in place and an attractive package could be defined and implemented quite quickly, should a mayor seek a mandate. But the public will have to be engaged in a number of fundamental debates before a similar position could be secured for national road pricing.

It follows from this complexity that the proposal for national road pricing is unlikely to come to much unless there is clear leadership on the topic. To gain public understanding and acceptance the key is to work up and present the complete package of charges and benefits and to demonstrate its superiority to the 'doing nothing' alternative.

NOTE

* This chapter draws on research commissioned by the Independent Transport Commission and funded by the Rees Jeffreys Road Fund, the Joseph Rowntree Foundation and the Esmee Fairbairn Foundation. We are grateful for guidance and comments from the members of the Independent Transport Commission and from its Secretary, Terence Bendixson.

REFERENCES

Blow, L. and I. Crawford (1997), *The Distributional Effects of Taxes on Private Motoring*, London: Institute of Fiscal Studies.

Burris, M.W., K.K. Konduru and C.W. Swenson (2004), 'Long-run changes in driver behavior due to variable tolls', *Transportation Research Record*, 1864, 78–85.
Cheshire, P.C. (2006), 'Resurgent cities, urban myths and policy hubris: what we need to know', *Urban Studies*, **43**(8), 1231–46.
Department for Transport (2004), *Feasibility Study of Road Pricing in the UK*, London: DfT.
Department for Transport (DfT) (2006), *Public Attitudes to Congestion and Road Pricing*, London: DfT.
Evans, A.W. (1992), 'Road congestion pricing: when is it a good policy?', *Journal of Transport Economics and Policy*, **26**, 213–43.
Glaister, S. and C. Archer (2006), *Investing in Roads*, London: Independent Transport Commission.
Glaister, S., T. Travers and J. Wakefield (2004), *London – on the Move or in a Jam?*, London: Development Securities PLC.
Glaister, S. and D. Graham (2004), *Pricing Our Roads: Vision and Reality*, London: Institute of Economic Affairs.
Green, E. and V. Stone (2004), *Public Attitudes to Road Pricing in the UK: A Qualitative Study*, prepared for the DfT, BRMB Social Research.
Labour Party Manifesto (2005), *Britain Forward Not Back*, London: Labour Party.
Lyons, G., G. Dudley, E. Slater and C. Parkhurst (2004), *Evidence-Base Review – Attitudes to Road Pricing*, Centre for Transport and Society.
Office of the Deputy Prime Minister (2004), *The English Indices of Deprivation*, London: Office of the Deputy Prime Minister.
Poterba, J.M. (1991), *Is the Gasoline Tax Regressive*?, Cambridge, MA: NBER Working Paper 3578.
Richardson, H.W. and C.-H.C. Bae (1998), 'Road pricing and income distribution', in K. Button and E. Verhoef (eds.), *Road Pricing, Traffic Congestion and the Environment*, Cheltenham, UK and Northampton, MA, USA: Edward Elgar.
Royal Automobile Club (2006), *The Future of Motoring: A Clear Road Map or Collision Course*, Norwich: RAC.
Santos, G. and L. Rojey (2004), 'Distributional impacts of road pricing: the truth behind the myth', *Transportation*, **31**, 21–42.
Transport for London (TfL) (2006), *London Congestion Charge – Impacts Monitoring, 4th Annual Report*, London: TfL.
Travers, T. and S. Glaister (2006), *Improving Local Transport: How Small Reforms Could Make a Big Difference*, London: Local Government Association.

APPENDIX 4A THE WORKINGS OF THE MODEL

The analysis summarised in this chapter is mainly based on spatial (geographical) units, not on individuals or households. We have not had access to measures representing variation in individual incomes or in wages and salaries to provide a match with the output from our road pricing model. Instead, we consider traffic, price, speed and cost changes from charging scenarios in relation to the spatial distribution of measures of deprivation for small areas of Britain. In effect, we use the deprivation measures as a proxy for spatial variation in *relative* poverty and affluence.

An important consequence of using spatial units rather than individuals or households is that we cannot offer firm conclusions about whether any pricing scheme is truly regressive or progressive. But what we can show is how road users living in different areas of the country, experiencing different levels of deprivation, are likely to fare under national pricing schemes. We also make a preliminary exploration of the implications of the fact that households which are not private car users will be differently affected.

This chapter is based on research reported in full technical detail in Glaister and Graham (2003 and 2006a). We used a simple and conventional approach to model the consequences of the various road user charging policies. The 'generalised' cost to a user of a specific mode, in a particular place at a particular time of day is a measure of the money value of the total of all the costs faced per person-kilometre (for example, the money cost or fare, plus the cost of in-vehicle time, plus the cost of waiting time, plus charges, plus other relevant costs).

Our approach assumes that the demand for any one mode is dependent on the generalised cost of using that mode and on the generalised cost of using all other modes. This represents the propensity to switch between modes (including not travelling) in response to changes in relative money charges or congestion.

In order to establish our representation of the initial, 'base' situation, data are required on travel demands, values of time, vehicle operating costs, traffic speeds, external costs and travel demand elasticities. Our study relates to the nine English government office regions, Wales and Scotland. The data are further divided by type of road, a variety of urban and rural area types and 19 times of the week. This yields 5,562 categories and for each of these there is a 'busy' and a 'not busy' direction, giving a total of 11,124 'cases'.

Detailed road traffic flow data were provided by the UK Department for Transport (DfT). These data are used to create a 'base' set of figures to represent the situation in 2010. The data represent flows of private cars, buses, light, heavy and articulated goods vehicles. The data for private cars are further disaggregated into six journey purposes.

Public transport demand data are derived from published sources, principally Transport Statistics for Great Britain (DTRL, 2001a) and Regional Transport Statistics (DTLR, 2001b). We derive estimates for bus and rail passenger-kilometres by region and for average bus and rail fares paid. While bus fares vary by region, rail fares did not because we could not secure satisfactory rail receipts data by region. A national average was used for rail.

Financial quantities are expressed in 2005 prices. Values of time have been taken from the DfT's Transport Economics Note and adjusted to allow for inflation and expected real income growth to 2010. The costs of congestion are represented by the money value of time lost by the drivers, passengers and vehicles themselves. This per hour value of time varies by vehicle type and trip purpose.

However, the per hour value of time is oversimplified in several respects: it does not vary by geographical location – as it ought to do to reflect variation in local wage rates and productivities; it does not distinguish between time saved in the course of 'leisure' and time saved on the journey to work – in accordance with an unsatisfactory 'equity time value' convention used by the DfT; it does not represent changes in predictability or reliability of journeys – it deals only with the average journey time.

Each of these is an important limitation and the model would be substantially improved by removing them. The representation of changes in reliability is becoming a particularly urgent matter as networks become more congested and less predictable.

The DfT's Transport Economics Note provides vehicle operating cost formulae. Fuel efficiency gains were applied. Fuel is assumed priced at £0.80 per litre for cars and commercial vehicles and at £0.34 per litre for public service vehicles, after rebate of fuel duty.

We assume that bus average loads stay constant so that the total capacity adjusts in step with volume of patronage; we also assume that bus costs vary in direct proportion with patronage.

For rail we were unable to determine a defensible assumption on how rail costs might vary with rail traffic. We therefore assume that train services and hence train costs would be unchanged throughout, changes in patronage being accommodated by changes in average train loadings. In cases where rail demand falls this may be realistic. In cases where it rises then it is unrealistic because the railway is already at or near full capacity in many cases; for instance, in the London peak commuter market.

Relationships between speed and flow for each road and area type are crucial to the computation of the costs of congestion because they represent the way that speeds reduce as traffic increases. We used ones suggested by the UK Department for Transport.

We derive the elasticities from a variety of sources to represent the propensities of the various types of traveller to change their volume of travel or to switch mode of travel in response to changes in travel costs and journey times. Graham and Glaister (2002a, 2002b, 2004) provide a survey of evidence on price elasticities of car traffic and freight traffic.

The most important of these is a long-run elasticity of car traffic with respect to fuel price of – 0.35: that is, if the price of fuel rises by 10 per cent car traffic will fall by 3.5 per cent. Bus elasticities are derived from Dargay and Hanly (1999) and rail elasticities from the Association of Train Operating Companies (ATOC, 2001). London specific elasticities are provided by Grayling and Glaister (2002).

Time Switching

For each region, area type and road type the model represents travel at 19 different times of the week as shown in Table 4A.1.

We have not been able to find good evidence to guide us on the magnitudes of time switching likely to occur in practice. Small (1982) and Burris et al. (2004) report some relevant empirical evidence but it is not a great deal of help in our context. Therefore our approach has been to postulate several alternative magnitudes of switching and to investigate the sensitivity of our results (see Glaister and Graham, 2006a, for details).

Commercial vehicles are assumed not to switch times of travel. This is a simplification because, in reality, commercial vehicles do have substantial flexibility. Some current night-time deliveries could revert to daytime to

Table 4A.1 Times of the week represented in the model

Period	Day	Time	Period	Day	Time
1	Mon–Fri	00:00–06:00	12	Saturday	00:00–09:00
2	Mon–Fri	06:00–07:00	13	Saturday	09:00–14:00
3	Mon–Fri	07:00–08:00	14	Saturday	14:00–20:00
4	Mon–Fri	08:00–09:00	15	Saturday	20:00–24:00
5	Mon–Fri	09:00–10:00	16	Sunday	00:00–10:00
6	Mon–Fri	10:00–16:00	17	Sunday	10:00–15:00
7	Mon–Fri	16:00–17:00	18	Sunday	15:00–20:00
8	Mon–Fri	17:00–18:00	19	Sunday	20:00–24:00
9	Mon–Fri	18:00–19:00			
10	Mon–Fri	19:00–22:00			
11	Mon–Fri	22:00–24:00			

take advantage of lower labour costs and greater convenience for customers. Equally, some peak deliveries could divert to off-peak times to take advantage of lower road charges.

Our analysis suggests that time-of-day switching could be a significant – though not overwhelming – factor in designing road pricing schemes. Substantial benefits can be obtained through persuading a few users to change their time of travel, thereby securing a more efficient use of the limited highway capacity.

Response of Car Occupancy

Increasing charges would give an incentive to increase average occupancies. This could be an important phenomenon because increased average occupancies mean that the same number of people would be carried while consuming less road space and therefore causing less congestion. The Department for Transport's Feasibility Study (2004) confirmed that this consideration should not be neglected.

As with time-of-day switching we do not have suitable empirical evidence to guide us as to the propensity of people to switch between being drivers and being passengers – though casual observation suggests that it may be quite low. We approached the problem by hypothesising several different propensities and evaluating the difference it makes to our results (see Glaister and Graham, 2006a, for details).

Traffic Data

For traffic data we take passenger car unit (PCU) per hour values for each type of road averaged over defined periods of the day, and allocate this value to the wards in correspondence to their associated region and area types. We then multiply the ward PCU/hr values by the length of each road type in the ward to calculate PCU kilometre per hour values. This is our measure of ward traffic flows.

For speed and price data the procedure differs because we have to account for the fact that speeds and prices are associated with different traffic flows.

Income Deprivation

For England and Wales the income domain indices have been constructed at the super output area level, which defines approximately 32,500 spatial units in England and 1,900 in Wales. The Scottish index has been constructed at the data zone level which defines 6,500 zones. This is a much finer level of spatial disaggregation than we can achieve in the spatial representation of results from our pricing model. For this reason, we have constructed weighted average income domain scores for larger geographical units defined by the Census area statistic ward disaggregation of Britain. There are 7,970 wards in England, 1,219 in Scotland and 881 in Wales. To construct the aggregated scores we use the ward share of population in each of the smaller areas as the weight.

5. Cambridge Futures: forecasting the effect of congestion charging on land use and transport

Anthony J. Hargreaves and Marcial Echenique

1 BACKGROUND

Cambridge Futures is a non-profit-making group of local business leaders, politicians, local government officers, professionals and academics who have been looking at the options for growth in and around Cambridge in the United Kingdom.

Cambridge Futures was founded in 1996 to inform the debate on whether and how Cambridge should be allowed to grow. At that time, Cambridge still had planning policies in place (Holford and Wright, 1950), which introduced a 'green belt' urban growth boundary that constrained the size of this historic and attractive city, to a population of around 100,000. The Mott Report (1969) resulted in a slight relaxation of planning policy by allowing the development of a science park on the northern edge of the city. This was enormously successful, largely due to the growth of hi-tech companies spinning off from the research of the world-renowned Cambridge University. This began a rapid growth in employment (Segal, Quince and Wicksteed, 1985), which has led to steep increases in house prices, and increasing amounts of commuting as more and more workers need to find housing beyond the green belt.

The first Cambridge Futures study tested several options for the future physical form of the Cambridge area (Echenique, 1999). The most notable outcomes of the study were the public recognition that the city needed to be allowed to grow, and that there was less public opposition to expanding into the green belt than had been expected. Cambridge Futures influenced the Regional Planning Guidance (Go-East, 2000), and many aspects of the study have since been included in the Cambridgeshire and Peterborough Structure Plan (Cambridgeshire County Council, 2002).

However, regardless of the land-use scenario, it was clear that transport was the most pressing problem facing the expansion of Cambridge.

Consequently, Cambridge Futures launched a second study in year 2000, *What Transport for Cambridge?* (Echenique and Hargreaves, 2003). This second Cambridge Futures study has followed the format of the earlier work by proposing a range of distinct options for solving the problem, testing the options with a land-use transport computer model, and assessing their impact on the local economy, social equity and environment.

2 THE LAND USE TRANSPORT MODEL

Land-use/transport interaction (LUTI) models emerged from the pioneering work of Lowry (1964) and have developed into sophisticated models that are able to estimate prices (Echenique, 2004). These models do not differ fundamentally from computable general equilibrium models (Bröcker, 2004) derived from Walrasian general equilibrium theory.

The model used by the Cambridge Futures studies is based on the MEPLAN software (Echenique, 1992) and estimates the location of households and employment in different zones of a city or region (Williams, 1979). It also estimates the transactions between and among households and employment that give rise to the demand for transport.

As part of the estimation of the location of activities and the trade between them, it calculates the cost of living for households and the cost of production for employment in each zone. The cost of living for households includes housing, transport and other goods and services costs. The cost of production includes labour, rental, transport and other input costs, measured by employee.

The basic assumption is that output prices are equal to input costs and no excess profits are included, except where there are constraints in production. If there are, it gives rise to 'pure rents' in the Marshallian sense. Typical of these are land rents, either because of a physical constraint imposed by a limited productive capacity, or because an artificial constraint is imposed by a policy of regulation, such as zoning of land uses.

The model uses an input–output framework as illustrated in Figure 5.1. The model estimates a demand coefficient matrix A in space, giving rise to all transport demands. The sectors are employers disaggregated by industrial sector, retail, education and households disaggregated by size, socio-economic group and car ownership. The factors are products and services, such as labour, retail goods, dwellings and floorspace.

- Section A_1 of the matrix a^{mn} represents the demand for factor m to produce factor n. This area is normally included in standard

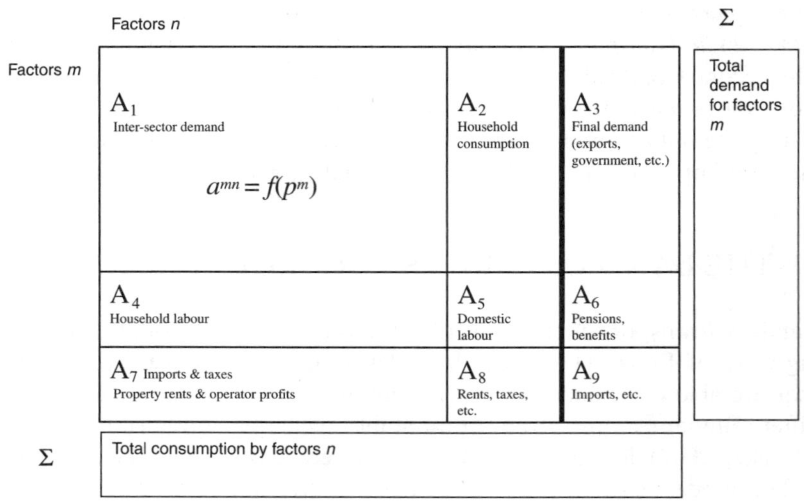

Figure 5.1 Demand coefficient matrix (input–output)

input–output models (Leontief, 1951). It represents the sales of factors produced by sector *m* and consumed by sector *n*, which is a function of the price *p* of factor *m*.

- Section A_2 of the matrix a^{mn} represents the demand for factor *m* by household *n*, in other words the consumption by a household of products or services *m*.
- Section A_3 of the matrix a^{mn} represents factors *m* (products and services), to be exported outside the area under consideration. Normally, both sections B and C are considered to be the final demand in standard input–output models that also includes investments and government consumption. It is described as the exogenous sector, that is to say, it is determined outside the model. In the model used in this chapter, only section C is estimated exogenously.
- Section A_4 represents the sale of labour or other income received by households type *m* from a factor *n*. For example, labour produced by households of socio-economic group *m* that is consumed by employer sector *n*, or a dividend paid by industry sector *m* and received by a household group *m*.
- Section A_5 represents the sale of labour from a household in socio-economic group *m* to a household in socio-economic group *n* (for example, domestic labour).

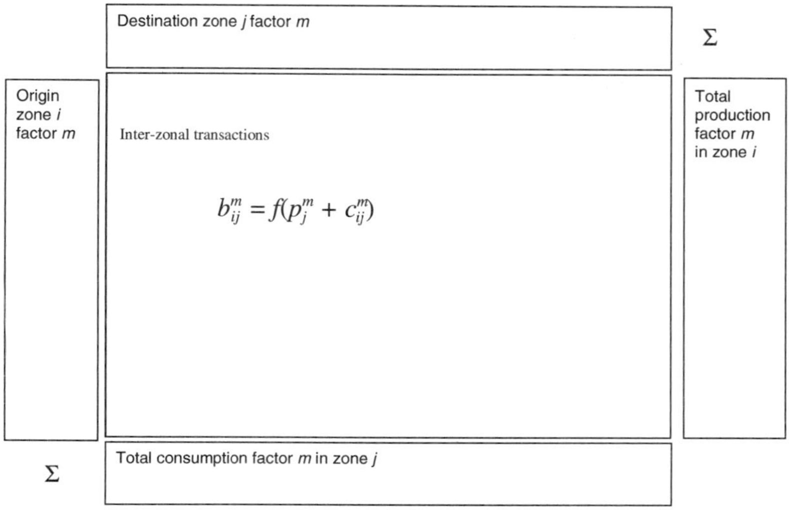

Figure 5.2 Trade coefficient matrix (spatial allocation)

- Section A_6 represents the sale of labour or other income received from exogenous factors, such as pensions and other benefits from government and so on. This section is exogenously determined and is part of the final demand.
- Section A_7 represents the imports from outside the area and payments to the exogenous factor such as taxes to the government. In this section, rental of property or land is included.
- Section A_8 represents the payments by a household such as taxes, rental and so on.
- Section A_9 represents payments by the exogenous factor to itself, such as imports for the government or for investments (exogenous).

The demand for factor *m* is estimated by the above transaction matrix, a^{mn}. The trade coefficients *b* for factor *m* are estimated for each zone *j* and allocated to zones of production *i* by a trade matrix *T* (Figure 5.2), as a function of the price and transport costs of the factor.

The model contains a series of steps that are illustrated in Figure 5.3 and can be summarised as follows:

1. The initial demand for factor *n* in zone *j* (normally termed as 'final demand' in input–output analysis) is exogenously determined and represents export to the outside world and other income received, Y_j^{n*}. To this initial demand an intermediate demand, Y_j^{n**}, calculated below

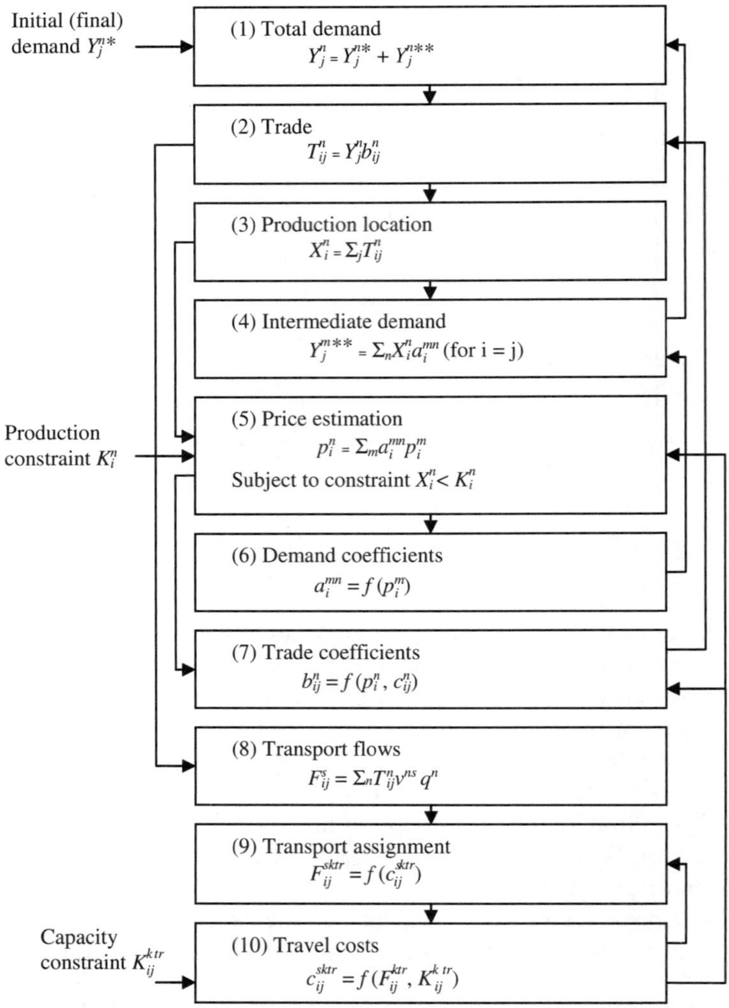

Figure 5.3 Model operation

(step 4), is added to give the total demand for factor n in zone j, Y_j^n, that is,

$$Y_j^n = Y_j^{n*} + Y_j^{n**}.$$

2. Estimation of the trade between zones of production i and consumption j for factor n, T_{ij}^n. This estimation is based on trade coefficients b_{ij}^n, calculated in step 7, that is,

$$T_{ij}^{n} = Y_{j}^{n} b_{ij}^{n}.$$

3. The resulting trade estimation gives the location of the factor n in the zone of production i (for example, households and employment), X_i^n by adding the trade over all zones of consumption j, that is,

$$X_i^n = \sum_j T_{ij}^n.$$

4. This leads to the estimation of the intermediate demands, Y_j^{m**} (that feed back into step 1), by applying a demand coefficient, a_i^{mn}, to the production X_i^n, that is,

$$Y_j^{m**} = \Sigma_n X_i^n a_i^{mn} \ (\text{for } i = j).$$

5. Estimation of output prices of factor n in zone of production i, p_i^n. The output price of each factor depends on the demand coefficient estimated below (step 6) and the consumption price of input m in zone i, p_i^m, that is,

$$p_i^n = \Sigma_m a_i^{mn} p_i^m.$$

The consumption price of m in zone i, p_i^m, is estimated by the aggregated cost of buying and transporting the input m from zones of production j. The aggregate cost is calculated as the log sum of the factor prices and its transport costs (Williams, 1977), that is,

$$p_i^m = -1/\beta^m \log \Sigma_j \exp[-\beta^m (p_j^m + c_{ij}^m)].$$

The price includes the effect of competition between zones. The degree of market imperfection is in the parameter, β^m, calibrated. For those factors that are constrained in their production or transport (for example, land), K_i^m, the prices are adjusted until the market clears. The initial price, p_i^{m*}, is adjusted to a new price, p_i^m, giving rise to 'pure' rents that bring the demand in line with the constrained supply, that is,

$$p_i^m = p_i^{m*} + 1/\beta^m \log(X_i^m / K_i^m) \quad \text{for } X_i^m > K_i^m.$$

6. Estimation of the functional relationships a between factors m and n in each location i, a_i^{mn}. This represents the input–output or demand coefficients, which are elastic with respect to the consumption price of factor m in each zone i, p_i^m. For households, the demand function uses the concept of a utility budget U^n that represents the expected standard of living for each category of household n (for example, preferences).

This is an indirect utility function (Varian, 1992) that estimates the consumption of factor m by adding to a minimum consumption, a^{mn}_{min}, a variable amount based on the consumption price of the factor m in zone i (see step 5), p^m_i, and the parameters that give rise to the elasticities of demand estimated with respect to the price of factor m for each household group n, α^{mn}, that is,

$$a^{mn}_i = a^{mn}_{\min} + U^m(\alpha^{mn}/p^m_i)/[\prod_m(\alpha^{mn}/p^m_i)^{\alpha^{mn}}].$$

7. Estimation of the trade coefficients, b^n_{ij}, that are elastic with respect to the production price of factor n in zone i (p^n_i), the transport cost (c^n_{ij}) for factor n between i and j and include other costs, representing all other considerations. The trade model is a logit model derived from random utility theory (Domencich and McFadden, 1975), that is,

$$b^n_{ij} = \exp[-\beta^n(p^n_i + c^n_{ij})]/\Sigma_i \exp[-\beta^n(p^n_i + c^n_{ij})].$$

8. Estimation of transport flows. The trade matrices, T^n_{ij}, from step 2, can be transformed into flows of passengers and freight type s, F^s_{ij}, between zones of origin i to destination j by applying a trip rate or value to volume ratio, v^{ns}, and a scalar, q^n, to convert the trips to the time period modelled in the transport model, that is,

$$F^s_{ij} = \Sigma_n T^n_{ij} v^{ns} q^n.$$

9. These flows are assigned to modes k of transport, time of day t and to routes r using multinomial logit models that take into account the generalised cost of travel by mode, time of day and route for each flow type s.
10. The resulting volumes of traffic can be compared against the capacity of transport links, and generalised costs of travel are adjusted to reflect congestion in each link.

The process above is repeated several times until equilibrium is reached, that is to say, no changes in prices occur.

The options were tested using the LUTI model of Cambridgeshire County Council that had been set up to test the Cambridgeshire and Peterborough Structure Plan policies. This model uses the MEPLAN model described in the previous section, and also includes a SATURN traffic model of the Cambridge urban area to improve the representation of traffic congestion. It models the morning peak period from 7 am to 10 am.

3 THE TRANSPORT OPTIONS

The Structure Plan 2016

The Structure Plan includes over 40,000 new dwellings and jobs in the Cambridge sub-region by 2016. As a result, in a major change in policy, around 17,000 of these new dwellings and a large proportion of the new employment sites will be located on the edge of the city, expanding the Cambridge urban area beyond the city boundary into the green belt of South Cambridgeshire.

The other notable aspect of the Structure Plan is development along the A14 transport corridor to the north-west of Cambridge where there is a commitment to widen the A14 to dual 3 lanes, and operate guided buses that would run on a track outside Cambridge and then as a normal bus through the city. There will be new settlement of 6,000 dwellings on this transport corridor by 2016 (Figure 5.4).

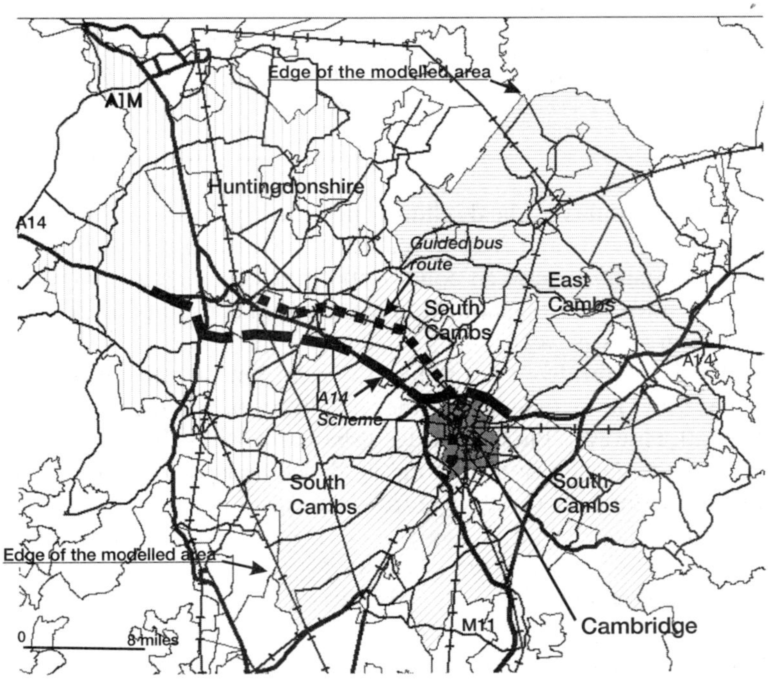

Figure 5.4 The modelled area and Structure Plan A14 corridor transport proposals

Table 5.1 Land-use inputs to model zones

Area	Dwellings ('000)			Employment floor space ('000m²)		
	2001	2016	% change	2001	2016	% change
City of Cambridge	44	56	27	1,145	1,275	11
South Cambs	55	71	29	1,048	1,279	22
East Cambs	30	34	13	499	500	–1
Hunts	67	76	13	1,409	1,486	6
Total	196	237	21	4,101	4,540	11

Table 5.2 Employment floorspace inputs by type in the modelled area

Area	Primary & industrial ('000)			Office ('000)			Retail ('000)		
	2001	2016	% change	2001	2016	% change	2001	2016	% change
Cambridge	399	391	–2	346	450	30	399	433	8
Rest	2,118	2,115	0	430	687	60	406	454	12
Total	2,517	2,506	0	776	1,137	46	805	887	10

Table 5.1 summarises the dwellings and industry floorspace input to the model between 2001 and 2016 to represent the Structure Plan policies. To simplify comparison between policy options, the demographic inputs, and the trips in and out of the modelled area, are kept constant for each option for the given forecast year.

By 2016 Cambridge is expected to have almost half of the office floorspace, and more than half of the retail floorspace in the modelled area (Table 5.2). However, the Cambridge urban area will still have less than a quarter of the dwellings in the total modelled area, so there will be even more commuting into Cambridge from the surrounding areas, worsening congestion.

Design of the Options

Cambridge Futures therefore examined a range of options that would provide additional transport measures to those already committed to in the Structure Plan.

The aim is to test distinct transport policies to demonstrate the limits of what each approach could achieve in isolation to inform the debate about transport policy direction for Cambridgeshire. Cambridge Futures tested:

1. The deposit draft Structure Plan land-use and transport policies (referred to as the 'reference case' because it acts as a benchmark against which to compare the options).

Cambridge Futures then tested the following options that would be additional transport measures to those already committed to in the Structure Plan:

2. Public transport (building a guided-bus tunnel under the city centre and extending the guided bus network).
3. Road building (in the form of an orbital highway).
4. Demand management (represented by congestion charging).
5. Combined option that examines how all of the options would work when tested together (see Figure 5.5).

Public transport option

The main feature of this option is a tunnel under the historic core of Cambridge to facilitate fast and reliable guided bus services to pass under the city centre and avoid the delays and environmental impacts that result from operating bus routes through the narrow streets of this historic city. The option includes additional park-and-ride sites on the perimeter of the city, and would also expand the proposed guided bus system to include a loop around the north-east edge of Cambridge.

Highways option

This includes a link road around the south and east of Cambridge connecting the A14 trunk road with the M11 motorway to form an orbital route around Cambridge and some widening and junction improvements on radial routes outside the city.

Congestion charge option

The congestion charge zone (Figure 5.5) would have a toll to cross the cordon of £3.50 per day and a charge of £0.50 per day for car trips wholly within the cordon (in 2003 prices). People travelling into Cambridge would be able to access all of the park-and-ride sites and some outer employment sites without being charged. The level of charging is similar to the £3 per day in 2000 prices tested by the CHUMMS study of the A14 transport corridor (Go-East, 2001).

For comparison, the charge for the central London cordon scheme was £5 per day in 2003, and has an annual operating cost of £90 million (TfL, 2006), and an implementation cost around £180 million. It is assumed

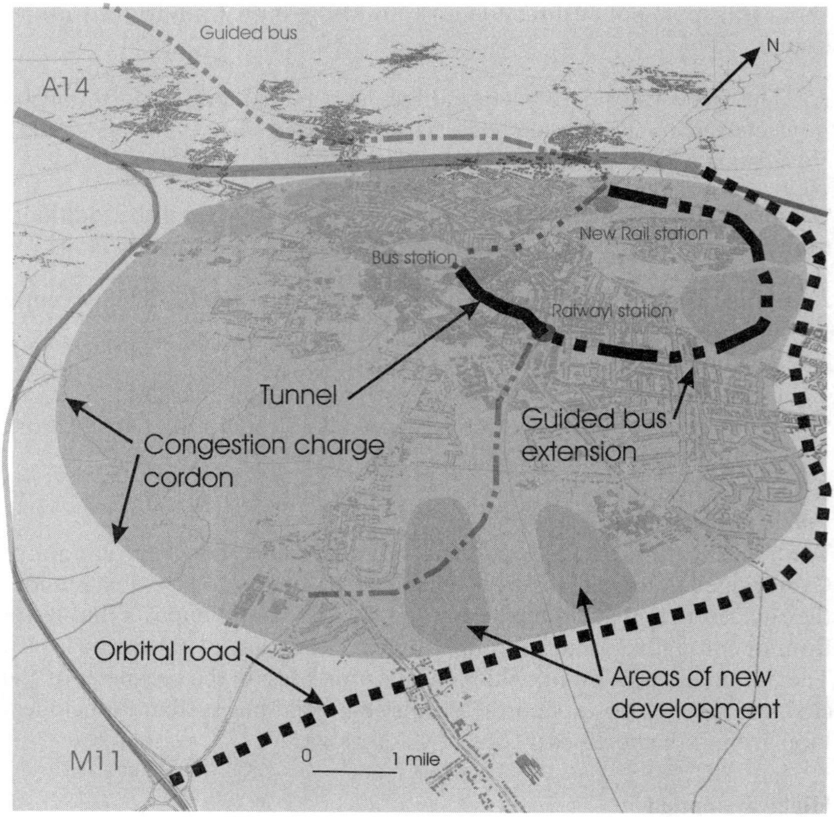

Source: Cambridge County Council (2002).

Figure 5.5 The 'combined' option

that a congestion charging scheme for Cambridge would have an implementation cost of only £15 million and an operating cost of £15 million per annum in 2003 prices, because this option tested for Cambridge does not include the extensive traffic management measures implemented for the London scheme and would require fewer cordon crossing points.

Combined option

This option tests how all of the individual options would work together in combination by combining the public transport, orbital highway and congestion charging measures into a single option. (Combinations of fewer individual options have not been tested due to the limitations on the resources available for this study.)

There are synergies between the individual options that make them work together well in combination. The orbital road would help to reduce through-traffic, offering drivers the opportunity to travel to edge of city employment centres without congesting the urban roads. The expanded and improved public transport system with faster services through Cambridge would provide an attractive alternative to the car, and the orbital would also give drivers good access to the park-and-ride site that best serves their destination.

4 RESULTS

Land-use Comparison between 2016 'Reference Case' and 2001 'Base Case'

The Cambridge urban area has far more jobs than households, resulting in more workers commuting into the city than living and working in Cambridge (Tables 5.3 and 5.4).

The projected growth inputs for employment are mainly in the higher earning sectors of tertiary (hi-tech industry and higher education), and private services. The model allocates most of the office-based employment to the office floorspace within and around Cambridge (Table 5.3).

This results in the professional and managerial households (SEG1 and 2) increasing at a faster rate than the skilled and unskilled households (SEG3 and 4), which adds to the upward pressure on house prices, and hence the cost of living in Cambridge and South Cambridgeshire (Cambs) (Table 5.4).

Table 5.3 Reference case allocation of jobs (000), and growth (%) from 2001 to 2016

Area	Basic			services			Total
	Primary	Secondary	Tertiary	Retail	Private	Public	
City of Cambridge	4	9	20	17	8	50	109
	3%	–4%	30%	31%	39%	34%	28%
South Cambs	7	16	8	11	10	20	72
	4%	–6%	149%	38%	128%	12%	26%
East Cambs	5	4	1	4	2	7	23
	4%	–6%	20%	30%	19%	–4%	3%
Hunts	8	18	4	13	5	29	76
	5%	3%	21%	18%	27%	9%	10%
Total	25	47	33	45	24	106	280
	4%	–2%	45%	28%	60%	19%	20%

Table 5.4 Reference case allocation of households (000) and growth (%) from 2001 to 2016

Area	Households					Total
	SEG1	SEG2	SEG3	SEG4	Inactive & unemployed	
City of Cambridge	15	13	8	8	21	65
	28%	30%	23%	28%	48%	34%
South Cambs	18	14	12	9	19	72
	28%	28%	17%	24%	47%	30%
East Cambs	5	4	5	3	10	27
	–5%	–4%	–5%	–4%	25%	4%
Hunts	14	12	13	11	19	69
	1%	4%	3%	5%	29%	9%
Total	52	42	38	31	69	233
	16%	17%	10%	14%	38%	20%

However, house prices would fall slightly after adjusting for inflation in the outer districts of East Cambridgeshire and Huntingdonshire (Hunts) where the supply of new dwellings exceeds the growth in jobs and households. (Note that the number of inactive and unemployed households, which is input into the model, increases at a faster rate than other households due to the expected increases in students and retired households).

The model outputs show that the Structure Plan policies would result in a steep increase in the household cost of living and production costs for employers, which would make the Cambridge sub-region less competitive and put the growth strategy at risk, and the population growth would result in worsening traffic congestion and a 30 per cent increase in emissions (Table 5.5).

Land-use Comparison between the Options and the Reference Case

The public transport option would have no effect on the location of jobs but there would be a slight dispersal of employed households (2 per cent) from Cambridge and slight increase in non-working households (1 per cent) within Cambridge due to the improved access into the city by public transport from surrounding areas.

The highway option would also have very little effect on the location of jobs. However, the creation of an orbital route around the edge of Cambridge would result in 2 per cent of all household types dispersing into the surrounding district of South Cambs.

Table 5.5 Comparison between the 2016 reference case and 2001

	Social cost of living	Economic production costs	Environmental CO_2 from traffic
Cambridge urban area	+ 73%	+ 60%	+ 30%
Total modelled area	+ 39%	+ 24%	not available

Table 5.6 Difference in households between the congestion charge option and 2016 reference case

Area	Households (% change)					
	SEG1	SEG2	SEG3	SEG4	Inactive & unemployed	Total
City of Cambridge	2	3	2	1	–5	0
South Cambs	–5	–6	–4	–4	1	–4
East Cambs	1	0	1	2	3	2
Hunts	2	2	2	3	3	2
Total	0	0	0	0	0	0

The congestion charge option would have a much larger effect on the location of households and employment. This would change the demographic mix of households, with working households moving into the city from the surrounding district of South Cambs to avoid paying the cordon charge, bidding up the rental values and displacing the inactive and unemployed households (Table 5.6).

The higher rental values in Cambridge and transport costs would increase the cost of living and production costs for employers. This would result in around 5 per cent fewer retail, private and public service jobs in Cambridge (Table 5.7). (The model assumes that the location of the service sector employment is more responsive to changes in costs than the basic sector employment.) The model redistributes these jobs into the outer districts. The total number of jobs in the modelled area is constrained to be the same for each option but in reality some of these jobs may leave Cambridgeshire.

Table 5.8 shows that the combined option would result in around 5 per cent reduction in households within Cambridge and dispersal of these households to the outer districts. The improved transport links into Cambridge would reduce the demand to live within the city and reduce the rental values of dwellings in the city.

Table 5.7 Difference in employment between the congestion charge option and reference case

Area	Location of employment (000 and % change)		
	Primary, secondary & tertiary	Retail private & public services	Total
City of Cambridge	0	–5	–3
South Cambs	0	3	2
East Cambs	0	5	2
Hunts	0	5	3
Total	0	0	0

Table 5.8 Difference in households between the combined option and the reference case

Area	Households (% difference)					
	SEG1	SEG2	SEG3	SEG4	Inactive & unemployed	Total
City of Cambridge	–5	–2	–4	–5	–4	–4
South Cambs	0	–2	0	2	2	0
East Cambs	7	8	5	4	2	5
Hunts	2	2	2	3	1	2
Total	0	0	0	0	0	0

Table 5.9 shows that the combined option would have a very similar effect on the allocation of employment as the congestion charge alone, with a reduction of 5 per cent in service sector employment within Cambridge.

Transport Comparison

The transport schemes in the 2016 reference case would be insufficient to cater for the growth in population and employment and there would be increasing traffic delays and lower average speeds than in 2001 (Tables 5.10 and 5.11).

The highways option of supplementing the Structure Plan measures by creating an orbital route around Cambridge would improve accessibility but would have very little effect on reducing traffic delays within the city.

The public transport option would not reduce traffic delays in Cambridge and expanding the guided bus network would only be cost effective where

Table 5.9 Difference in employment between the combined option and 2016 reference case

Area	Location of employment (% change)		
	Primary, secondary & tertiary	Retail private & public services	Total
City of Cambridge	0	–5	–3
South Cambs	0	4	3
East Cambs	0	2	1
Hunts	0	3	2
Total	0	0	0

Table 5.10 Car travel details for the urban Cambridge area

Options	Travel time	Travel distance	Travel delay	Average speed
2001 Base year	100	100	100	100
2016 Reference case	137	125	167	91
2016 Public transport option	135	124	164	92
2016 Highways option	145	148	168	102
2016 Congestion charge	113	122	118	108
2016 Combined option	116	135	115	117

Table 5.11 Car travel details for the full modelled area

Options	Travel time	Travel distance	Average speed	Number of trips	Fuel consumption
2001 Base year	100	100	100	100	100
2016 Reference case	123	117	95	117	131
2016 Public transport option	123	117	95	116	131
2016 Highways option	125	119	95	118	152
2016 Congestion charge	113	113	100	113	121
2016 Combined option	115	115	100	114	131

Note: An average speed greater than 100 is a benefit relative to the year 2001 base. For the other factors, a score greater than 100 is a disbenefit relative to the year 2001 base.

Table 5.12 Mode split of trips entering Cambridge, morning peak period (%)

Mode	Option (2016)				
	Reference case	Public transport	Highways	Congestion charge	Combined
Car	53	51	54	39	39
Bus modes & rail	37	39	36	47	49
Cycling & walking	10	10	10	14	12
Total (%)	100	100	100	100	100

there are large concentrations of development. Elsewhere the settlements around Cambridge are too dispersed to support a fast and frequent service. However, good quality public transport, park-and-ride, cycling, and walking facilities are essential within Cambridge because the compact historic road network does not have the capacity to accommodate traffic growth.

The congestion charge and combined options would reduce delays and result in a substantial mode shift from car to public transport for trips entering Cambridge; mainly to the guided bus system and park-and-ride (Table 5.12).

5 ASSESSMENT

Transport Assessment

Table 5.13 shows the annual social rate of return on the investment cost for transport users, operators and government.

The overall benefits of the congestion charge and the combined options would accrue only to the operators of the toll, with a disbenefit to car users. The overall rate of return assumes that the revenue would be reinvested for the public benefit. The tolls would substantially reduce the amount of traffic entering Cambridge, and it therefore may be possible to implement a fast and reliable public transport scheme with less tunnelling, which would greatly reduce the capital cost of the combined option.

Wider Socio-economic Considerations

Investing in additional highway or public transport improvements would reduce the cost of living and employers' production costs (Table 5.14). However, the highways option would increase energy consumed by transport.

Table 5.13 Transport assessment of the options (costs and benefits in £m)

Item		Public transport option	Orbital highway	Congestion charge	Combined option
Investment cost		325	120	15	460
User benefits	Tolls			−41.4	−38.6
	Cost savings	1.4	1.1	3.3	5.7
	Time savings	2.6	6.4	20.6	28.2
	Total	4.0	7.5	−17.5	−4.7
Operator benefits	Revenues	0	0	41.4	38.6
	Costs	0	0	−15.0	−15
	Total	0	0	26.4	23.6
Government benefits	Total	−0.1	1.3	−2.8	−1.3
Total benefits		3.9	8.8	6.1	17.6
Social rate of return		1%	7%	40%	4%

Note: This assessment does not include the relative comfort of different modes, or externalities such as noise, pollution and accidents. Also, the rate of return for the congestion charge option is very sensitive to the assumptions made about the implementation and operating costs.

Table 5.14 Comparison on indicators for Cambridge between the options and the reference case

Options	Social	Economic	Environmental
	Cost of living	Production costs	CO_2 from traffic
Public transport	−8%	−6%	0%
Highways	−17%	−11%	+16%
Congestion charge	+9%	+11%	−8%
Combined option	−20%	−11%	0%

Implementing congestion charging without additional infrastructure investment would have environmental benefits due to the reduction in traffic, but with the detrimental social and economic impacts of increasing the cost of living and cost of production by around 10 per cent above the reference case. These increases in costs would have an inequitable effect on low-income households and are also likely to result in the loss of some jobs and population from the Cambridge sub-region. Quantifying this would require further testing with a wider-scale and less constrained model.

Both the highways and the combined options would substantially reduce the cost of living and production costs below the reference case to maintain the economic competitiveness of the Cambridge economy and support the growth of jobs and population within the sub-region.

A full assessment of these wider socio-economic benefits would need further modelling and analysis. However, they are likely to be substantially greater than the social surplus of the transport assessment because the Cambridge cluster of the university and hi-tech companies are reportedly worth £57 billion of the annual UK GDP (Library House, 2006).

The combined option would achieve these socio-economic improvements without increased car use, thereby supporting economic growth in a way that is consistent with the UK government sustainability targets of improving energy efficiency (Stern, 2006).

6 CONCLUSIONS

The Cambridge Futures study shows that a congestion charge cordon would have a detrimental effect on the Cambridge local economy and on social inclusion unless it is part of a package of transport measures to improve accessibility into the city from the surrounding areas.

Improving transport links into Cambridge would increase the number of dwellings and employment sites within easily commutable distance of the city, and thereby reduce rental values. The combination of lower property prices and less congestion would reduce the cost of living and production costs in the sub-region.

Investment in highways, such as an orbital route, would be the most cost-effective method of improving accessibility from the surrounding districts, given that many of the employment sites are around the periphery of Cambridge. However, this would result in longer distances travelled by car and more energy use and emissions.

A combination of measures that includes the congestion charge along with transport investment to improve accessibility from the surrounding districts would provide similar socio-economic benefits to highways improvements but without the increasing distance travelled by car, and with less traffic congestion within Cambridge. A public transport system that provides an attractive alternative to the car would be an important part of this strategy.

This Cambridge Futures study demonstrates the importance of an efficient transport system to maintain the competitiveness of the Cambridge sub-region and that a combination of transport measures could help achieve the forecast levels of growth in a sustainable way. Further research would be needed to assess the most appropriate combination of transport measures.

REFERENCES

Bröcker, J. (2004), 'Computable general equilibrium analysis in transportation economics', in D.A. Hensher, K.J. Button, K.E. Haynes and P.R. Stopher (eds), *Handbook of Transport Geography and Spatial Systems*, Amsterdam: Elsevier, pp. 269–90.

Cambridgeshire County Council (2002), *Cambridgeshire & Peterborough Joint Structure Plan Review, Deposit Draft Plan*, Cambridge: Cambridgeshire County Council.

Domencich, T. and D. McFadden (1975), *Urban Travel Demand: A Behavioral Analysis*, Amsterdam: North-Holland.

Echenique, M.H. (1992), *Technical Introduction to MEPLAN*, Cambridge: Marcial Echenique & Partners.

Echenique, M.H. (1999), *Cambridge Futures*, Cambridge: University of Cambridge Department of Architecture.

Echenique, M.H. (2004), 'Econometric models of land use and transportation', in D.A. Hensher, K.J. Button, K.E. Haynes and P.R. Stopher (eds), *Handbook of Transport Geography and Spatial Systems*, Amsterdam: Elsevier, pp. 185–202.

Echenique, M.H. and A.J. Hargreaves (2003), *Cambridge Futures 2: What Transport for Cambridge?*, Cambridge: University of Cambridge, Department of Architecture (www.cambridgefutures.org).

Go-East (2000), RPG6 'Regional planning guidance for East Anglia', Government Office for the East of England Region, November.

Go-East (2001), *Cambridge to Huntingdon Multi Modal Study (CHUMMS)*, Final Report, Cambridge: Government Office for the East of England Region, July.

Holford, W. and H.M. Wright, (1950), *Cambridge Planning Proposals*, Cambridge: Cambridge University Press.

Leontief, W.W. (1951), *The Structure of the American Economy*, Oxford and New York: Oxford University Press.

Library House (2006), *The Impact of the University of Cambridge on the UK Economy and Society*, Cambridge: Library House Consultancy (www.libraryhouse.net).

Lowry, I.S. (1964), *A Model of Metropolis*, Santa Monica, CA: Rand.

Mott Report (1969), Cambridge: University of Cambridge Press.

Segal, Quince and Wicksteed (1985), *The Cambridge Phenomenon*, Swavesey, UK: SQW Ltd.

Stern, N. (2006), *Stern Review: The Economics of Climate Change*, London: HM Treasury (www.hm-treasury.gov.uk).

Transport for London (TfL) (2006), *Central London Congestion Charging: Impacts Monitoring, 4th annual report overview*, Mayor of London, June (www.london.gov.uk).

Varian, H.R. (1992), *Microeconomic Analysis*, 3rd edn, New York: Norton.

Williams, H.C.W.O. (1977), 'On the formulation of travel demand models and user-benefit measures', *Environment and Planning A*, **9**, 285–344.

Williams, I.N. (1979), 'An approach to solving spatial-allocation models with constraints', *Environment and Planning A*, **11**, 3–22.

6. Road user charging in the UK: the policy prospects

Martin G. Richards

1 INTRODUCTION

Road pricing has been on the political agenda in the UK for over 40 years since Reuben Smeed chaired a committee charged with studying the technical feasibility of 'improving the pricing system relating to the use of roads, and on relevant economic considerations' (Ministry of Transport, 1964). But successive governments have decided that it was not a policy for their time.

It was put back on the agenda in 1997, by the new Blair government, in the form of local congestion charging, and was adopted by Ken Livingstone as a central element of his manifesto for the 2000 London mayoral election. But, having rejected him as the official Labour candidate, the government distanced itself from the principle.

However, with the success of Livingstone's London Congestion Charging Scheme (LCCS), and a realisation that the government's bold claim in its 2000 Ten Year Transport Plan that it would reduce congestion by 2010 was not going to be achieved, charging returned to the government agenda, with the commissioning of a road pricing feasibility study (DfT, 2004a). Although the brief was a national scheme, the government made it clear that it did not intend to introduce charging on its roads – the national motorway and trunk road network – for several years, although it wanted local authorities to introduce charging schemes.

This chapter is intended to provide an analysis and assessment of the prospects for the introduction of road user charging in the UK.

2 ROAD PRICING: SMEED TO LIVINGSTONE

The Smeed Committee was commissioned under a Conservative minister of transport, but the follow-up fell to a Labour minister, Barbara Castle, who concluded 'it is not certain that road pricing can offer a solution; it will, in any case, not be an immediate one' (Ministry of Transport, 1967).

There followed a number of studies, including the GLC Supplementary Licensing Scheme (GLC, 1974) and Area Control Study (GLC, 1979), the London Assessment Studies (Glaister, 1991) and a series of studies by the London Planning Advisory Committee (May and Gardner, 1990), all of which made the case for the introduction of road pricing, at least in central London. However, politicians of the day were not convinced, deciding against introducing charges.

These were followed by the London Congestion Charging Research Programme (DTp, 1995a; Richards et al., 1996). John MacGregor, one of the transport secretaries in office during the study, told the Chartered Institute of Transport 'the decision on the best form of traffic restraint will not be an easy one . . . the Government must not shrink from this . . . if the costs of congestion are not to grow'. However, in announcing publication of the study report, the then Transport Secretary, Sir George Young, was less committed, saying 'we have no current plans to introduce congestion charging . . . or to bring forward proposals for the primary legislation', concluding that technology and complexity prevented early implementation, a remarkably similar conclusion to Barbara Castle's some 30 years earlier (DTp, 1995b).

By 1997, Blair's New Labour government was in power, with John Prescott responsible for transport (and much else). Encouraged by advisers strongly committed to road pricing, Prescott pursued it as a key element of the new transport policy, with a commitment to enacting the necessary legislation (DETR, 1998).

Despite Prescott's commitment, Blair's advisers in Downing Street were determined not to risk the loss of support from 'Mondeo Man',[1] and thus to avoid transport policies that might be seen as anti-car. Prescott became increasingly frustrated, complaining about 'the faceless wonders of Downing Street' (*The Times*, 1999). In the end, Prescott's Ten Year Transport Plan was published, complete with road user charging (DETR, 2000). Not only had he won against Downing Street, but he had also gained a major concession from the Treasury. The net revenues from local charging schemes could be retained locally for investment in transport. The Treasury has long opposed earmarked, or hypothecated, taxes, but it had become evident that without the ability to retain the revenues, there would be no takers. It can be argued that, with its strong control of local government finance, the Treasury was giving little away, and might have seen it as an opportunity for an additional source of revenue, another of Chancellor Gordon Brown's 'stealth taxes'.

The Ten Year Plan assumed that by 2010, eight congestion charging schemes would be introduced in 'our largest towns and cities', together with 12 workplace parking schemes, generating a total of £1.2 billion for English local authorities outside London. The plan also had a target to reduce congestion below its current level, 'particularly in large urban areas', thereby emphasising the expected responsibilities of local authorities. However, the plan, and the complementary 2000 Transport Act, made no provision for the introduction of charging on the government's own trunk roads and motorways except those within local schemes.

Within less than a year of publishing the plan, Prescott had lost his transport responsibilities to Stephen Byers. Not only was Byers a sceptic, his ministerial colleague, John Spellar, was reported to have sought to frustrate implementation of the LCCS (*The Economist*, 2003). In the end, only Durham used the charging powers of the 2000 Transport Act, for a small scheme in the historic city centre; Livingstone used the 1999 Greater London Authority (GLA) Act powers.

Loathed by key figures in the Labour party Livingstone had defied, and from which he had been expelled, the success of the LCCS and other mayoral policies led Blair to welcome him back to the party in time for the 2004 mayoral election. By then, Alistair Darling was Transport Secretary. Although doubtful about Livingstone's plans, once the London scheme had been proved effective and generally acceptable, he was persuaded that 'road user charging has to be considered as part of sensible management of our roads . . . we would be failing future generations if we did not find out if this is feasible and examine what gains could come from it' (DfT, 2003a). With congestion getting worse, rather than better, as intended under the Ten Year Plan, Darling commissioned a road pricing feasibility study (DfT, 2003b).

3 THE ROAD PRICING FEASIBILITY STUDY

The study, overseen by a Steering Group of stakeholders and civil servants, was charged with advising 'the Secretary of State on practical options for the design and implementation of a new system for charging for road use in the UK' (ibid.).

In its report, published in July 2004, the Steering Group concluded that 'road pricing would help unblock roads to the overall benefit of the economy and the environment. The time savings and reliability benefits that we would get in return for the prices we pay are potentially large' (DfT, 2004a). Although publication of the study was greeted by headlines that focused on the highest charge rates considered, research for the study indicated that, with a nationwide charging scheme based on marginal social cost, less than 20 per cent of vehicle-km would incur a charge greater than 5 p/km, and only some 3 per cent would pay a charge over 20 p/km.

The Steering Group concluded that a national charging system would need to relate charges to location, the time of day and the distance travelled, but that appropriate satellite-based systems would not be available at an affordable price 'until at least 2014' (ibid.). Further, fitting the 33 million vehicles in the UK with the necessary onboard units was seen as particularly challenging.

The costs of implementing and operating a national charge system would be substantial. The study estimated total start-up costs of between £10 billion and £27 billion, increasing to between £23 billion and £62 billion when allowing for optimism and annual operating costs for the full system at between £2 billion and £3 billion, possibly as high as £5 billion with optimism bias (DfT, 2004b). To set these costs in context, total vehicle excise and fuel duty revenues were some £28 billion, and the planned 2006/07 Department for Transport capital budget was some £7 billion. While acknowledging that the estimates produced 'very large numbers', the Steering Group noted that they are 'lower than the potential value of the benefits' and that they should be set against the £60 billion a year spent on private motoring (DfT, 2004a). Yet it expressed reservations about a mixed system in which direct charges apply to the congested parts of the network, with fuel duty providing the base charge.

4 LORRY ROAD USER CHARGING

Under pressure from the haulage industry to deal with 'unfair' competition within the UK from operators from other EU countries running on cheaper fuel purchased on the continent, the Chancellor initiated the Lorry Road User Charging scheme. The plan was for a distance-based lorry road user charge, for all goods vehicles in excess of 3.5 tonnes on all UK roads, regardless of nationality (Treasury, 2002).

Government, and its potential suppliers, had invested heavily in the scheme, which was at an advanced stage of procurement when, in July 2005, it was cancelled. In announcing the cancellation, Darling explained that as 'plans for distance-based lorry charging [should be taken forward] as part of the wider work on national road pricing . . . to develop a single, comprehensive, cost-effective system . . . the current procurement for lorry road user charging will not continue' (Hansard, 2005). The logic of this explanation has been widely questioned, as the scheme could have been a useful pathfinder for a nationwide charging scheme for all vehicles. However, there had been a reluctance to provide information on its likely costs and value for money (McKinnon, 2004; Richards, 2005), and the likely explanation is that it had become too expensive. Cancellation of the scheme raised questions about the strength of the government's real commitment to charging, with the Commons Transport Committee concluding that, given the government's commitment to national road pricing, the decision was 'an embarrassing muddle' (House of Commons, 2005a).

5 THE POLICY RATIONALE

The rationale for charging schemes implemented under the 1999 GLA or 2000 Transport Acts is clear: to reduce congestion. Although the potential to raise funds was probably very attractive to Livingstone, during the development of the scheme he was very careful not to identify this as a justification.

The government's rationale for the possible nationwide and local schemes is also to reduce congestion. However, Livingstone's plans to vary the congestion charge by the emissions rating of vehicles, with a £25 charge for the worst polluters (GLA, 2006), has added emphasis to a debate about environmental charges. It was one considered by the Conservative 'Quality of Life' policy review led by John Gummer, with the work on transport led by former Transport Minister Steve Norris. Considering the impact of motor vehicles on the environment a key issue, Norris proposed increasing the annual Vehicle Excise Duty very substantially for new vehicles with high levels of polluting emissions, possibly to £1,700 for the worst, and also increasing fuel duty very substantially (LTT, 2006a). His basic premise was that taxes should be high on 'bad' things like pollution, and low on 'good' things like employment, with the increase in revenues from vehicle and fuel duties used to reduce employment taxes, as well, possibly, as Council Tax or VAT. While substantial increases in fuel costs would have a substantial effect on demand, reducing congestion, Norris accepted that there could be a congestion charge where congestion remained.

The House of Commons Environmental Audit Committee criticised the government's failure to introduce substantial differentials between higher and lower carbon-producing cars (House of Commons, 2006a). The Committee was clearly attracted by the government's Sustainable Development Commission's proposal for increasing Vehicle Excise Duty for vehicles producing the most carbon to £1,800, compared with £210 in 2006 (SDC, 2005). Although Transport Minister Stephen Ladyman did not rule out 'doing more in the future' he questioned whether increasing tax would be effective (BBC, 2006a).

The Committee's report was followed by publication of the Liberal Democrats' new tax policy, which included a proposal for a 'sharply graduated' Vehicle Excise Duty on new cars based on carbon emissions, rising to

£2,000 for those with the highest emissions (Liberal Democrats, 2006). They also committed to increasing fuel duty in line with inflation and to increasing it to offset in part any fall in oil prices.

Compelling though some may find the principle of using charges, whether through fuel duty, annual vehicle fees or structured direct road user charges, to reduce polluting emissions, it is an idea largely rejected by Chancellor Brown. John Major's Conservative government introduced a Fuel Duty Escalator in 1993, to encourage both a reduction in traffic and the introduction of more fuel-efficient engines. Initially set at an annual rate of 3 per cent, in real terms, Brown increased it to 6 per cent in his first Budget. But, in response to fuel price protests (Lyons and Chatterjee, 2002), he abandoned it in 2000; and between then and 2006 he increased fuel duty just twice, and then only in line with current inflation. His chief secretary, John Healey, told Parliament that changes in the real value of fuel duty between 1999 and 2006 equated to a 7 p per litre reduction in fuel duty, in real terms (Hansard, 2006).

Although Brown made it clear that he considered that the fight to prevent climate change is a 'moral duty' of the developed world, he justified freezing fuel duty at a time of rising oil prices on the grounds of the need for 'a balanced judgement between the needs of the economy and the protection of citizens', regarding high oil prices as enough of a burden for motorists (BBC, 2006b). And although he restructured annual Vehicle Excise Duty in 2006 on the basis of emissions, the increase for the worst polluters – to £210 – was so modest as to be totally ineffective.

The Blair government's record on vehicle taxation led *The Economist* to question its real commitment to road user charging, accusing it of 'short term cowardice and long term courage' by announcing its plans for moving towards a national road pricing system in the same week as it deferred a planned increase in fuel duty and accused it of being timid 'when facing down the noisy motoring lobby' (*The Economist*, 2004).

A third rationale is to replace the funding of highways from general taxation (in the UK, motoring tax revenues go into general taxation receipts) with direct user charges. But this does not seem to be on any political agenda.

Whatever the policy rationale, there needs to be a complementary rationale for the use of the revenues, a requirement addressed later.

6 THE TRANSPORT INNOVATION FUND

Announcing publication of the Road Pricing Feasibility Study, Darling also announced the creation of the Transport Innovation Fund (TIF). Intended to provide local and regional agencies with 'incentives to develop and deploy coherent, innovative, local and regional transport strategies', TIF was planned to 'support the costs of such packages – which will include road pricing, modal shift, and better bus services' (DfT, 2004c). Starting at £290 million in 2008/09, it is scheduled to increase to £2,550 million by 2014/15 (DfT, 2006a).

The creation of TIF recognised that if local authorities were to be persuaded to implement local charging schemes, they would need some financial incentives. As a first step, local authorities were invited to submit bids for TIF 'pump priming funds' to study the feasibility of schemes combining demand management (including road pricing) with better public transport, to address congestion and improve local travelling conditions. Bids from seven areas were successful:

- Bristol, Bath, North Somerset and South Gloucestershire;
- Cambridgeshire;
- Durham City;
- Greater Manchester;
- Shrewsbury;
- Tyne and Wear; and
- the West Midlands.

In announcing these, Darling said, 'We are looking forward to working with these authorities to develop practical solutions to congestion problems, and support the development of a national road pricing scheme' (DfT, 2005), and Transport Minister Stephen Ladyman called for local authorities and government to be 'bold together' (*Transport Times*, 2006a). That seemed to anticipate a call for bids for a second round of pump priming studies, which stressed the need for inovation (DfT, 2006b).

When TIF was first announced, in 2004, it was widely anticipated that the funds would be focused on support for demand management. However, TIF was also to support:

- local mechanisms raising new funding for transport schemes, and
- regional, inter-regional and local schemes beneficial to national productivity,

the first of which can reasonably be linked to congestion, given limits on local authorities' fund raising. But, by early 2006, the annual allocation to demand management schemes had fallen to £200 million, most of the fund available for 2008/09, but less than 10 per cent of that planned for 2014/15, with the balance allocated to national productivity schemes (DfT, 2006a). With £200 million unlikely to support more than one major scheme at a time, there have been suggestions that the government was only looking for a few (possibly three) schemes (NCE, 2006).

Despite the case for investing in transport schemes to improve national productivity, the apparent reduction in emphasis on demand management either called into question the true government priority for local charging schemes or emphasised an incomplete understanding of what local authorities must be able to deliver to secure adequate local support for the implementation of a charging scheme.

7 HYPOTHECATION AND THE USE OF REVENUES

Hypothecation of local charging scheme net revenues for funding transport had been seen as central to obtaining local support. However, addressing a conference in December 2005, Ladyman expressed reservations about the principle, which he described as 'unreasonable' (*Transport Times*, 2005). Government thinking was that if much of the necessary funding for scheme implementation was provided through TIF, the government could reasonably expect to receive the revenues. However, as well as serving to strengthen Whitehall's grip on local expenditure, and thus policy, this revealed an apparent failure to understand local politics, a weakness identified by the Chartered Institute of Logistics and Transport (CILT) (CILT, 2006). Some, if not all, of the 'pump priming' authorities made it clear that without hypothecation the government would have no pathfinding local schemes; and by mid-2006 it seemed that government had recognised this. It also acknowledged that consideration was being given to using some of the revenues to support a reduction in local council taxes (*The Times*, 2006a).

However, the use of revenues from a national charging scheme, or from environmental charges, is not resolved. There is an argument that a national road pricing scheme designed to manage demand relative to capacity should be revenue neutral, with revenues offset by an equal reduction in existing motoring taxes. Although this would probably make it relatively attractive to some, it implies that the costs of running the system are borne by taxpayers generally, not road users. It also creates a difficulty for the distribution of revenues, if some of the revenues were retained locally (particularly in areas with higher charges).

This raises a crucial question of how national road user charges should be set and the net revenues allocated. The Feasibility Study Steering Group concluded:

> [T]he institutional structure used to operate the system and to regulate (or otherwise control) the charges will need to gain and retain the trust of road users. It will need to demonstrate transparent treatment of what use is made of the

> money that road users pay. . . . clear objectives and criteria would need to be established for setting charge levels. (DfT, 2004a)

Although others have stressed the need for an independent regulator, the principle is one likely to cause concern in government, particularly in the Treasury, which has increasingly sought to control all public finances. Yet, if national road pricing is to achieve public acceptance, it is likely that Westminster and Whitehall will have to relinquish some of the control amassed over recent governments.

The use of revenues from an environmental charge is more complex. In so far as they are a charge on ownership, it can be argued that they should be a true tax, and there may be a greater public acceptance of such treatment. However, acceptance of high levels of fuel duty is likely to require the transparency called for by the Feasibility Study Steering Group, with a clear determination of what parts of fuel duty are an 'environmental' tax, a (possible) road user charge and a source of general taxation.

8 THE DARLING DAYS

It is widely accepted that Alistair Darling was appointed transport secretary with the brief to keep transport out of the headlines. This reflected the tumultuous times leading to the resignation of his predecessor, Stephen Byers, and uncertainty within the government about what do with transport, with it becoming increasingly evident that the Ten Year Plan was failing.

Despite the commitment of the Ten Year Plan to congestion charging (DETR, 2000), by 2003 the Commons Transport Committee concluded that the government had distanced itself from the implementation of charging schemes (House of Commons, 2003). Although Darling had told it that the government's position was determined by technology rather than political cowardice, the Committee questioned the government's political nerve to support 'bold experiments in reducing congestion' and accused it of sitting on the fence and failing to lead.

The decision to commission the Road Pricing Feasibility Study in 2003 most probably reflected a recognition that Britain was never going to build its way out of worsening road congestion, and that a study managed by an 'independent' Steering Group could help defer the need for difficult decisions. Publication in 2004 was complemented by a White Paper, *The Future of Transport*, which stated that 'Government will lead the debate on road pricing, working with stakeholders to establish and explain how and when pricing might provide the reliability and standards road users want', and in its Foreword, Blair explained: 'the key is how, not how much, motorists pay for road use. We will do the work necessary to allow the hard decisions to be taken nearer the time' (DfT, 2004c).

As Darling might have hoped, the need for early decisions could be avoided, as the Steering Group concluded that a national road pricing scheme was unlikely to be 'technologically feasible' for ten years, which Goodwin described as 'close enough not to be fantasy, distant enough to be on some other government's agenda' (LTT, 2005). While Darling was happy to accept the Steering Group's advice, his counterpart in Holland was determined to introduce a national road pricing system by 2012, using the satellite technology the Steering Group had concluded would not be feasible until 2014 (LTT, 2006b), and Norwich Union were using it for their pay-as-you-go car insurance. But that was for national road pricing; technology was not seen to be a barrier to local charging schemes, which would be encouraged as pathfinders, supported by TIF.

Although it seemed that the government had got off the fence the Commons Transport Committee had accused it of sitting on, it appeared to get back on it, with the follow-up expected in the autumn of 2004 postponed until 'well after polling day', expected in May 2005 (*Financial Times*, 2004). The party machine must have been relieved, given the response of the popular press when Darling did talk about road pricing. A report on an interview with London's *Evening Standard* in November 2004, covering a variety of transport issues, focused on charging with a double page inner headline: 'Pay every time you drive your car', and a leader headed 'Transport is still a mess, Mr Darling', declaring that 'the proposal will infuriate drivers' (*Evening Standard*, 2004). A few weeks later, reporting Darling's address to a conference, the *Daily Express* carried the front page headline '£1.34 a mile driving tax' with an inner headline 'it's just another stealth tax – this time on drivers' and a leader headed 'Pay-as-you-go motoring will take away our liberty' (*Daily Express*, 2004).

Perhaps mindful of these responses and of the forthcoming General Election, Darling was cautious in front of the Commons Transport Committee in February 2005, explaining, 'I would be very wary about saying it [road pricing] is inevitable . . . there is still an awful lot of things . . . to sort out before you can say, "Yes, we are definitely doing road pricing" ' (House of Commons, 2005b).

Yet, even after the election, progress appeared to be cautious, leading the CILT to title a report intended to contribute to the public debate Darling had called for 'Frustratingly Slow' (CILT, 2006). Darling's diffidence on charging was displayed in criticism of the Forth Estuary Transport Authority's proposals for variable Forth Bridge tolls, to help manage demand, when he was campaigning for the Labour candidate for the

Westminster seat of nearby Dumfermline in a 2006 by-election. Since transport is the responsibility of the Scottish Parliament, not Westminster, Darling's intervention was particularly striking.

His equivocation was also demonstrated by his approach to charging on trunk roads and motorways, telling the Commons Transport Committee: 'I am not convinced you could say to people when everything is "free" today . . . that you are going to charge them for going down a road tomorrow . . . which is why I have said that we are not proposing to do that' (House of Commons, 2005b). As the Committee noted: 'it does not seem to concern the Secretary of State that the principle of charging people to use the existing road network is precisely what he wants to see Local Authorities bring into force'; it concluded: 'the Government must not duck its responsibility for charging on the most congested stretches of the strategic road network, which is under its direct control'.

Darling was also keen to minimise the government's involvement in providing the technology infrastructure: 'we need to move away from the idea that Government is going to define and specify [the] technology . . . instead we need to start thinking of road pricing "piggy-backing" on systems already being offered by the market' (IPPR, 2005). Although the public sector record on procuring large projects is poor, many saw this as a failure to recognise the need for government to lead in specifying the key requirements; as Christian Wolmar put it: 'this is the wrong way to go about a complex process' (*Transport Times*, 2006b).

While Darling can be credited with bringing road pricing back on the political agenda, *The Observer* (2006) suggested that he 'had spent too much time keeping transport out of the news to the detriment of making meaningful progress, especially tackling congestion and pollution'.

9 ON TO ALEXANDER

In May 2006, a Cabinet reshuffle saw Darling replaced by Douglas Alexander. In appointing him, Blair stated that he wanted 'to advance the debate on the introduction of a national road-user charging scheme. The successful roll-out of local schemes funded from the Transport Innovation Fund will be critical. I would like you to identify the other key steps for the successful introduction of road-user charging within the next decade' (DfT, 2006c). Alexander responded that he would be 'looking closely at how road pricing could help to better manage capacity on the roads and tackle congestion. This will require greater engagement with the public than we have seen previously' (DfT, 2006d). He also stated that 'a personal priority will be to advance the debate about a national system of road pricing in this country – moving the debate from "why" to "how" we might make a national system work in practice' (DfT, 2006d).

Although the *Daily Mail* (2006) reported that, under Alexander, 'we will go full speed to pay as you drive', his own words showed greater caution about the rate of progress:

> I do not believe it would make sense for us to launch straight into a national scheme. All the work that has been done over the last three years confirms that what we should do is take a measured approach. Through pilots and pathfinders . . . we will develop our understanding, apply proven approaches where action is needed soon, and, importantly, be able to demonstrate what works. (DfT, 2006e)

Indeed, Blair's brief of 'within the next decade' provided no sense of urgency.

Despite this, Alexander's public commitment was not welcomed by parts of the media, with the *Daily Express* (2006) bearing the front page headline 'Spy in the sky on motorists' and with 'War on the motorist' as an inside headline. Although the *Financial Times* (2006a) was more welcoming, it thought progress 'too slow' and 'the timidity is all the more surprising' given the fit between road pricing and other elements of government policy, concluding: 'the case for road pricing is overwhelming but obstacles loom large. All the more need for something that so far has been notably lacking in transport policy: leadership'.

The early months of Alexander's stewardship saw a number of key moves. These included a decision to provide £10 million to support the development of charging technologies, which was followed by an invitation for suggestions for demonstrations and supporting research, setting out the key requirements of a national road pricing system (DfT, 2006f). The Department also sought views on the incorporation into UK law the EU Directive on inter-operability (DfT, 2006g) and published a series of guidance notes, *Introduction to Modelling and Appraisal for Road Pricing* (DfT, 2006h).

Most notable was a proposal to introduce a new Road Transport Bill in the 2006/07 parliamentary session, to reform the charging provisions of the 2000 Transport Act, to include provisions to ensure inter-operability between schemes and consistency with a national framework as well as, possibly, to permit charging on trunk roads. These proposals were included in a letter to Jack Straw, Leader of the House of Commons, leaked to the *Sunday Times* (2006). The possibility of charging on trunk roads was included as 'we are considering pilots on the trunk road network as an important stage towards national road pricing', and the letter noted that the government's position on charging was 'very different' from that at the time of the 2000 Transport Act, and that the Bill would 'help pave the way for a national road pricing scheme', although that was only 'in the medium to long term'. The idea of including trunk roads in pilot schemes did indeed represent a major change, given Darling's dogged refusal to consider it.

This apparent change in approach might have been encouraged by the confirmation that traffic congestion really was going to get worse, rather than better as intended in the Ten Year Plan (DfT, 2006i), heralded by *The Times* (2006b) as: 'Promise to keep traffic moving is scrapped as jams get worse'.

However, former adviser to Prescott on the Ten Year Plan, David Begg, told the *Financial Times* (2006b) that although Alexander is 'less cautious' than Darling 'I don't think there's any indication they want to go faster'.

10 DEVOLUTION

As Westminster's full transport powers are limited to England, unless introduced as a tax measure by the Treasury (as had been the intention for the Lorry Road User Charge), a 'national scheme' will require agreement with both the Scottish Parliament and the Welsh Assembly (whether Northern Ireland should be included, given its common border with the Republic of Ireland, is open to question). Without such agreement, offsetting reductions in Vehicle Excise Duty and/or fuel duty would be infeasible.

11 CONSENSUS

During the first nine years of the Blair government, there were four transport secretaries, each of whom had different views on road pricing. In the last seven years of the previous Conservative governments, there were four, again with a range of views. Yet, a national road pricing scheme cannot be developed and implemented within a single Parliament, and certainly not within the typical transport secretary's term of office. Progress requires acceptance both within and between the main Westminster parties, as well as those of Scotland and Wales.

Alistair Carmichael, the Westminster Liberal Democrat shadow transport secretary, saw some form of road pricing as an 'inevitability', which the Lib-Dems have championed for many years (*Transport Times*, 2006c). Despite that championing, Lib Dems campaigned against proposed charging schemes in Bristol and Edinburgh.

Although road pricing might seem a natural Conservative policy, the Tories have been equivocal over recent years. Although Chris Grayling, the shadow transport secretary, had no problems with charges for new capacity, he had reservations about a national charging scheme (ibid.). However, Gummer's Quality of Life policy review might provide for change, and *The Guardian* (2006) was of the view that the principle had the 'outline support' of both Tory and Liberal Democrat party leaders, David Cameron and Sir Menzies Campbell.

Despite the rejection of the planned Edinburgh scheme, the Scottish transport minister, Tavish Scott, described road pricing as the 'best option on the table' for tackling congestion (*Evening News*, 2006). The Welsh transport minister, Andrew Davies, saw congestion charging in Wales as 'obviously a possibility' (BBC, 2006c), and the Assembly government's draft Transport Strategy included a commitment to 'work with the UK Government on the development of a UK wide system of road user charging' and to 'set out the statutory framework' to enable the development of local, pilot, charging schemes (Welsh Assembly, 2006).

12 LOCAL CHARGING SCHEMES

The government had pinned its hopes on persuading some local authorities to adopt charging locally, and on those schemes being successful and making a national charging scheme acceptable. It sought to do this with the promise of TIF monies, making it clear that TIF is the 'only show in town' for government funding.

The government sought to stress that the 'pilot' local schemes would not just be congestion charging, but packages of measures, providing carrots as well as sticks. However, a crucial difference between London and the rest of the country is the control of bus services. In London, the mayor (through Transport for London) determines what services are to be provided, sets the fares and takes the revenues; contractors provide the services. Elsewhere, bus services and fares are largely in the hands of the operators. Livingstone was able to soften some of the impacts of the congestion charge through a marked increase in the quality and quantity of bus services. Lacking his powers, it is almost impossible for local authorities in the 'pump priming' areas to do likewise. Although government has argued that quality bus partnerships can do much, they have not been tested, and many doubt their effectiveness.

In addition to being sure that they can deliver improved services as a complement to congestion charges, the pump priming authorities are also concerned about setting fares and the use of fare revenues. One issue is that the deregulated operators could respond to the introduction of charges by increasing their fares, taking additional profits. Although the government has steadfastly refused to consider a return to the pre-1985 Transport Act days, it is probable that local authorities will need to be given much greater control over services, fares and revenues before they agree to introduce the charging schemes that the government is keen to see, a possibility that Transport Minister Gillian Merron alluded to before the Commons Transport Committee (House of Commons, 2006b).

In deciding to seek to involve local authorities in moving towards a national road pricing system, the government may well have to agree to a number of key policy changes it had not planned for. As well as the issues of hypothecation and control of bus services, there is the challenge of managing a mixed system, once a national system is introduced, including charge setting and allocation of revenues; the chosen implementation path makes it very difficult to avoid independent regulation.

The evidence also suggests that the government might not have saved time in pursuing this route. The procedures involved in the implementation of major local authority schemes are such that, without some major change, it could well be very close to the Steering Group's suggestion of a target of 2014 for implementation of a national scheme before they are up and running.

13 NATIONAL CHARGING

Despite the apparent ultimate government target of a national road pricing scheme, there are real questions about the cost efficiency of implementing, operating and enforcing a scheme covering the complete national network, as opposed to a basic national charging scheme covering those parts of the network where regular congestion occurs, complemented by local schemes in major urban areas.

Such a scheme could be implemented in stages, and could avoid the challenges of dealing with revenue neutrality as well as a need to reach agreement with the Scottish Parliament and Welsh Assembly. It would provide a basis for a clear distinction between an environmental charge, collected nationwide through fuel duty, and a congestion charge levied where the network is regularly congested. It could also ease the challenges associated with setting charges and allocating revenues, but not that of ensuring consistency and inter-operability.

14 IN CONCLUSION – WHAT ARE THE PROSPECTS?

The evidence indicates that if the government really wants to move towards national road pricing using local congestion charging schemes as a first step, it will need to relinquish some controls over local and regional government. It must also be willing to accept that charges are set and managed in a manner that is widely accepted as fair; that certainly requires transparency, and most probably independence, allowing more control to slip away from government. Local political, public and business acceptance make this relaxation of controls unavoidable.

It might be seen that, in adopting the local scheme route, the government is seeking to let the local authorities take any criticisms; to be the 'fall guys'; that once any initial problems have been overcome, it will be easier for government to achieve the real goal of national charging. But, through ensuring that there is consistency between local systems (and avoiding criticism for a lack of 'joined up thinking') and providing a direct route from them to a national scheme, it will not be as easy to stand back as might have been thought, or hoped.

While the evidence suggests, quite strongly, that the government believes that a national congestion charging scheme is inevitable, it also suggests that it is no hurry to move to implementation. Slow progress has its political advantages. With worsening congestion, and only modest steps to increase capacity, the willingness to accept charging is likely to grow, making it easier for future governments to take the final step. It also avoids the need for those individuals in control today to have to make difficult and potentially unpopular decisions. The headlines of the popular press and concerns about the votes of the successor to Mondeo Man provide no encouragement to a government that has lost much popular support to make rapid progress with a policy portrayed as a tax on motorists.

Thus, it seems that the prospects for new charging schemes outside London within the remainder of the 2000s decade are very small, close to zero. However, with the right package of finance and powers, a few local authorities may well implement charging schemes in the early 2010s. By then climate change concerns may have made real, possibly substantial, increases in fuel duty more acceptable, causing some changes in the use of cars – and vans and trucks.

In the meantime congestion, and its impacts on the economy, the environment and the quality of life, will get steadily worse, and future generations will ask why action was not taken earlier. It seems inevitable that the answer will be because political leaders were not willing to lead.

A POSTSCRIPT

Since this Chapter was written, 1.8 million people signed a petition on the Number 10 Downing Street website calling on Tony Blair 'to scrap the planned vehicle tracking and road pricing policy' (www.petitions.pm.gov.uk), the Eddington Report commissioned by Gordon Brown concluded that 'without a widespread scheme by 2015 the UK will require very significantly more transport infrastructure' (Eddington, 2006), Blair has retired to be replaced by Gordon Brown, Ruth Kelly became the fifth Transport Secretary since Labour came to power in 1997, and the Government has published its response to Eddington in which it declared that no decision on the introduction of road pricing on inter-urban roads would be made, yet, describing it as 'a decision for the future' (DfT, 2007).

The prospects for national road pricing remain distant, and local authorities' enthusiasm for local schemes has waned, despite the prospects for TIF funding, with only one Business Plan submitted by the July 2007 target. Further information on the last year of the Blair government and the first months of Brown is in Richards, 2008.

NOTE

1. The Mondeo is Ford's middle-range car; Mondeo Man typifies middle England, whose votes New Labour saw as being critical to its future election success.

REFERENCES

BBC (2006a), 'Raise air travel tax, report says', BBC News, 7 August.
BBC (2006b), 'Climate change fight "moral duty" ', Interview on Radio 4 'Today' programme, 22 April.
BBC (2006c), 'Road tolls part of transport plan', BBC News, 13 July.
CILT (2006), 'Frustratingly Slow: A Report on the Government's Plans to Deal with Road Congestion', Chartered Institute of Logistics and Transport UK, Corby.
Daily Express (2004), '£1.34 a mile driving tax' and 'Pay-as-you-go motoring will take away our liberty', 15 December.
Daily Express (2006), 'Spy in the sky on motorists', 11 May.
Daily Mail (2006), 'We'll go full speed to pay as you drive', 11 May.
DETR (1998), *A New Deal for Transport: Better for Everyone*, London: Department for the Environment, Transport and the Regions.
DETR (2000), *Transport 2010, The 10 Year Plan*, London: Department for the Environment, Transport and the Regions.
DfT (2003a), News Release 20030085, Department for Transport, London, 9 July.
DfT (2003b), *Managing our roads*, London: Department for Transport.
DfT (2004a), *Feasibility Study of Road Pricing in the UK*, London: Department for Transport.
DfT (2004b), *Road User Charging Feasibility Study*, Implementation Workstream, Report on Implementation Feasibility of DfT Road Pricing Policy Scenarios and Proposed Business Architecture, Deloitte Consulting, London: Department for Transport.

DfT (2004c), *The Future of Transport: A Network for 2030*, London: Department for Transport.
DfT (2005), 'Tackling congestion – next step', Press Release, Department for Transport, London, 28 November.
DfT (2006a), *Transport Innovation Fund: Guidance January 2006*, London: Department for Transport.
DfT (2006b), *TIF Pump Priming Round 2: Criteria*, London: Department for Transport.
DfT (2006c), Text of appointment letter – Secretary of State for Transport, Department for Transport, London, May.
DfT (2006d), Text of letter from Douglas Alexander to the Prime Minister, Department for Transport, London, 26 May.
DfT (2006e), Speech by Douglas Alexander in York, Department for Transport, London, 10 May.
DfT (2006f), *Road Pricing Demonstrations and Research*, London: Department for Transport.
DfT (2006g), EC Directive 2004/52 on the interoperability of electronic toll collection systems – proposed legislation to transpose to UK law, Department for Transport, London.
DfT (2006h), *Introduction to Modelling and Appraisal for Road Pricing*, TAG Unit 2.12 (and Unit 3.12), London: Department for Transport.
DfT (2006i), Spending review 2004 PSA 4, Department for Transport, Department for Transport, London, July.
DfT (2007), *Towards a Sustainable Transport System: Supporting Economic Growth in a Low Carbon World*, Department for Transport, London.
DTp (1995a), *London Congestion Charging Research Programme*, Department of Transport, London: HMSO (Note: a report of the Principal Findings is also available from HMSO, and a summary of the Final Report is available in a series of six papers published in *Traffic Engineering and Control*, see Richards et al., 1996).
DTp (1995b), 'Young Publishes London Congestion Charging Study', Press Release, Department of Transport, London.
Eddington, R. (2006), *The Eddington Transport Study: The Case for Action*, London: HM Treasury, HMSO.
Evening News (2006), 'Transport minister backs Scottish road pricing scheme', 20 April, http://news.scotsman.com/roadtolls/Transport-Minister-backs-Scottish-road. 2768717.jp.
Evening Standard (2004), 'Pay every time you drive your car' and 'Transport is still a mess Mr Darling', 26 November; http://www.encylopedia.com/doc/1G1-12535611.html.
Financial Times (2004), 'National news politics and policy: sensitive issue swept under the carpet', 14 October, http://search.ft.com/ftArticle?queryText=Sensitive+issue+swept+under+the+carpet&aje=false&id=041014000936&ct=0.
Financial Times (2006a), 'Leader: Time to give road pricing a green light', 22 May, http://search.ft.com/ftArticle?queryText=Time+to+give+road+pricing+a+green+light&y=2&aje=true&x=11&id=060522000503&ct=0&nclick_check=1.
Financial Times (2006b), 'National news: Transport secretary in driving seat on road-pricing proposals', 9 August, http://search.ft.com/ftArticle?queryText=Transport+secretary+is+driving+seat+on+road-pricing+proposlas&aje=false&id=060809000592&ct=0.

GLA (2006), 'CO_2 emissions and congestion charging', Mayor's Press Release, Greater London Authority, 12 July.
Glaister, Stephen (ed.) (1991), 'Transport Options for London', Greater London Group, Greater London Papers 18, London School of Economics. London.
GLC (1974), *Supplementary Licensing*, London: Greater London Council.
GLC (1979), *Area Control*, London: Greater London Council.
Guardian, The (2006), 'Q&A: National road charging scheme', 7 August, http://www. guardian.co.uk/enviroment/2006/aug/07/travelsenviromentalimpact. transportintheuk.
Hansard (2005), Column 173, House of Commons, 5 July.
Hansard (2006), Column 763, House of Commons, 4 July.
House of Commons (2003), *Urban Charging Scheme*, First Report of Session 2002–03, House of Commons Transport Committee, London: The Stationery Office.
House of Commons (2005a), *Departmental Annual Report 2005*, Fourth Report of Session 2005–06, House of Commons Transport Committee, London: The Stationery Office.
House of Commons (2005b), *Road Pricing: The Next Steps*, Seventh Report of Session 2004–05, House of Commons Transport Committee, London: The Stationery Office.
House of Commons (2006a), *Reducing Carbon Emissions from Transport*, Ninth Report of Session 2005–06, House of Commons Environmental Audit Committee, London: The Stationery Office.
House of Commons (2006b), Uncorrected Transcript of Oral Evidence, House of Commons Transport Committee, London, 28 June.
IPPR (2005), 'Road user charging: building a consensus', Keynote speech by Alistair Darling, Institute for Public Policy and Research, London, 26 October.
Liberal Democrats (2006), 'Fairer, simpler, greener', Policy paper 75, Liberal Democrats, London.
LTT (2005), 'Road use charging: don't wait for the big bang', *Local Transport Today*, 20 January.
LTT (2006a), 'Is climate change moving to the top of the Tories' transport agenda?', *Local Transport Today*, 27 July.
LTT (2006b), 'Are the Dutch finally ready to launch nationwide road pricing?', *Local Transport Today*, 18 May.
Lyons, G. and K. Chatterjee (eds) (2002), *Transport Lessons from the Fuel Tax Protests of 2000*, Aldershot: Ashgate.
McKinnon, A. (2004), 'Lorry Road User Charging: A Review of the UK Government's Proposals', Heriot Watt University, Edinburgh, http://www.sml. hw.ac.uk/logistics/Iruc.html.
McKinnon A. and D. McClelland (2004), 'Taxing Trucks: An alternative method of road user charging', Heriot Watt University, Edinburgh, http://www.sml.hw. ac.uk/logistics/lorryroaduser 2.pdf.
May, A. D. and K. Gardner (1990), 'Transport policy for London in 2021: the case for an integrated approach', *Transportation*, **16** (3).
Ministry of Transport (1964), *Road Pricing: The Economic and Technical Possibilities*, London: HMSO.
Ministry of Transport (1967), *Better Use of Town Roads*, London: HMSO.
NCE (2006), 'Can the UK ever achieve an integrated transport system?', *New Civil Engineer*, 6 July.

Richards, M. G. (2005), *Congestion Charging in London: The Policy and the Politics*, Basingstoke: Palgrave Macmillan.
Richards, M. G. (2008), 'Congestion charging: an idea whose time has come – but not yet, at least not in England', presented at the 87th Annual Meeting of the Transportation Research Board, Washington, DC and awaiting publication in Transporation Research Record, TRB, Washington, DC.
Richards, M. G., C. Gilliam and J. Larkinsan (1996), 'The London Congestion Charging Research Programme', a series of six papers in *Traffic Engineering and Control*, **37** (2), pp. 66–71
SDC (2005), 'Climate Change Programme Review: SDC Submission', Sustainable Development Commission, London.
Sunday Times (2006), 'Motorists face black boxes in cars for road pricing', 6 August, http://www.timesonline.co.uk/tol/news/uk/article601405.ece.
The Economist (2003), 'Fuming: how the transport minister tried to undermine the congestion charge', 13 February.
The Economist (2004), 'Stop–go', 22 July, http://www.economist.com/world/britain/displaystory.cfm?story-id=El_NJOTGJR.
The Observer (2006), 'Darling has a desk full of problems', 7 May.
The Times (1999), 8 July.
The Times (2006a), 'Road toll money may be used to cut council tax', 24 April, http://business.timesonline.co.uk/tol/business/money/consumer_affairs/article708728.ece.
The Times (2006b), Promise to keep traffic moving is scrapped as jams get worse', 1 August, http://www.timesonline.co.uk/tol/news/article69503.ece.
Transport Times (2005), 'U-turn on road user charging cash', 15 December.
Transport Times (2006a), 'TIF – the way ahead', 16 June.
Transport Times (2006b), 'PPP lesson for new boy Alexander' (Christian Wolmar), 19 May.
Transport Times (2006c), 'The third way', 16 June.
Treasury (2002), *Modernising the Taxation of the Haulage Industry: Progress report one*, London: Treasury.
Welsh Assembly (2006), *Wales Transport Strategy*, Cardiff: Welsh Assembly Government.

7. Design tools for road pricing cordons

Anthony D. May, S.P. Shepherd, A. Sumalee and A. Koh

1 INTRODUCTION

In Europe and Asia, most proposals for urban road pricing involve the use of cordon or area charging, in which one or more boundaries are drawn, with charges to cross the boundary (using cordon schemes as in Singapore and Stockholm) or to drive within it (using area schemes as in London). Despite over 40 years of research into such schemes, there is little technical advice on where best to place such boundaries. Most designs are based on a mix of professional and political judgement, with little or no assessment of whether alternative locations would be more effective.

In practice, the performance of any road pricing cordon or boundary will be affected by the combined effects of a reduction in traffic entering the area and an increase in traffic bypassing it. While congestion will be reduced within the area, it might well be aggravated outside it. Since these conflicting impacts will depend on both the topology of the road network and the pattern of demand for its use, it is difficult to offer general advice on cordon location. All that is known is that the benefits of road pricing, usually measured in terms of welfare economic impacts, are critically dependent on the choice of cordon (May et al., 2002).

Section 2 briefly reviews this evidence and our understanding of the approaches which professionals adopt to cordon design. We then report on two promising methods which have been developed to improve the design process. The first of these (Section 3) uses an application of genetic algorithms to represent design options and to highlight those which are most effective. The second (Section 4) provides a short-cut method which is analytically less complex and involves the planner directly in the design process. Both have been shown to provide two- to threefold improvements in performance over judgemental designs. They are not, however, intended to supplant the need for professional and political judgement; rather they are offered as design tools which will help to focus such judgement on those designs which are likely to be technically the most effective. Section 5 discusses future developments.

While many of the principles considered will apply to area charging, the analysis has focused on cordon schemes, which are more common and also simpler to analyse.

2 PAST APPROACHES TO CORDON DESIGN

Evidence from Model-based Studies

Between 1992 and 1995, consultants working for the UK Department for Transport (DfT), conducted one of the most comprehensive studies ever undertaken of the potential for road pricing, and the relative performance of a range of road pricing designs (Richards et al., 1996). The study illustrates well the flexibility of cordon charging and the extent to which design options influence scheme performance. It demonstrated that, while the concept of cordon pricing is simple, its application offers a wide range of options. Those tested in the study (Figure 7.1) included a single cordon around central London (the innermost ring in the figure);

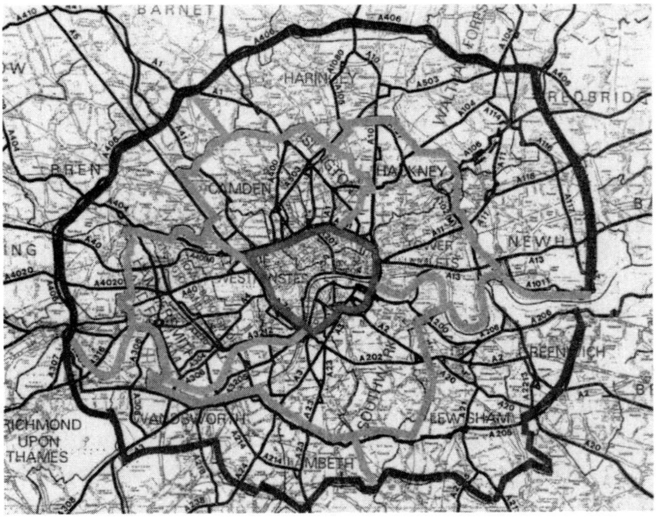

Source: May et al. (1996).

Figure 7.1 The design of the London congestion charging scheme with three cordons and screenlines

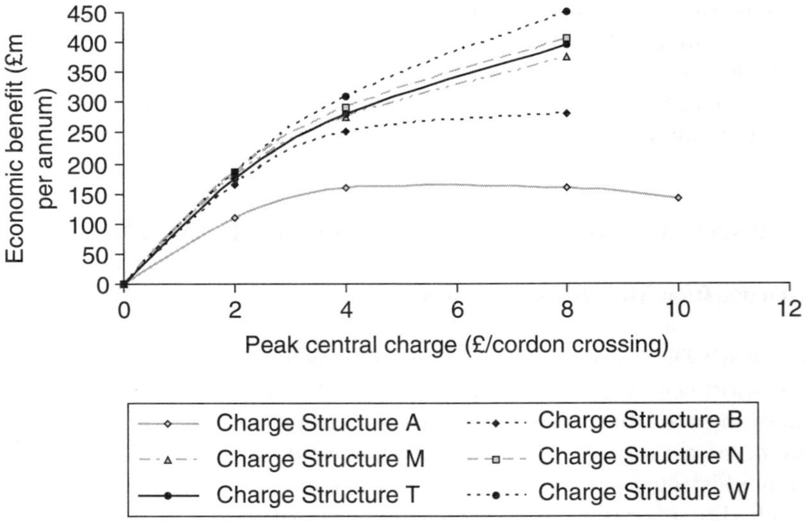

Source: May et al. (1996).

Figure 7.2 Central and inner bi-directional cordon charging for different charging structures, 1991: economic benefits (£m per annum)

a second and third cordon in inner London; the addition of radial screen lines to charge orbital movements; charges either inbound, outbound or both; charges varying by time of day; for the more complex schemes, variations in the ratio of charges between cordons; and, for all of these, variations in the level of charge. In all, some 45 separate options were tested.

Figure 7.2 summarises the impact of 19 options, representing six charging structures and four charge levels, on social welfare benefit. A simple, single cordon around central London performed least well, and reached an optimum level of performance at around £5 per crossing. This is broadly representative of the scheme subsequently implemented (TfL, 2006). A second cordon in inner London added around 50 per cent to the social welfare benefits, before taking account of the additional operating costs. Bi-directional charging on these cordons increased the benefits further, and produced results which were similar to those from three cordons with inbound charges. The best-performing option, with three cordons and four screen lines and bi-directional charging, had benefits at the levels of charge shown of up to three times greater than those from the single cordon. Moreover, there was clear evidence that benefits would have been even higher at higher levels of charge (May et al., 1996). Even allowing for the higher cost of operation, this most complex scheme had a net benefit three to four times greater than the simple single central London cordon (Richards et al., 1996).

The reasons for these differences can be traced back to three principal causes. First, a single cordon intercepts fewer journeys, and thus excludes many which contribute significantly to congestion. Second, it imposes the same charge on all journeys which cross it, thus over-restraining short journeys and undercharging long ones. It is the over-restraint of some journeys which leads to the economic benefit falling at higher charges. Third, and most importantly, it allows many journeys to escape the charge by rerouting around the cordon. The worst congestion in a city is often to be found just outside the central area, and the impact of a single cordon will be to relieve this congestion to the extent that radial journeys are reduced, but to aggravate it through traffic diversion. The more complex schemes, and particularly the screen lines, avoid this, and hence increase the benefits from congestion relief.

Evidence from Interviews

Like other studies before it, the London congestion charging study provided little detail of the basis on which the selected cordons had been designed. Two early inputs to the research described below were therefore a desk study designed to infer the design factors which had been considered (Shepherd et al., 2001), and an interview survey of practitioners.

The survey of practitioners focused on six UK local authorities which were active partners in the UK Charging Development Partnership, and were at differing stages in the development of road pricing proposals. It involved an initial questionnaire and a subsequent in-depth interview with a responsible transport planner which sought further information on key aspects of their proposed scheme. Full details are given in Sumalee (2001). Both questionnaire and interview were structured to cover in turn the context of the proposal, the objectives of the scheme, and the detailed design process.

It became clear that the context and objectives generally had little impact on detailed design. Most respondents suggested that their schemes were being planned both to reduce congestion and environmental impact directly, and to generate revenue to finance other policy instruments. Road pricing was not seen as a solution in its own right, and therefore its design was not critically influenced by its objectives. The key elements in the design process were to avoid adverse impacts and to gain public acceptance; practical considerations were generally less important.

In summary, the most frequent approach was to:

- focus solely on the city centre, together with any major generators on its fringes;
- place the cordon within the city centre ring road if one existed, or alternatively try to avoid charging routes which would allow drivers to bypass the area;
- minimise the crossing points by using existing boundaries to movement, and keep the cordon as simple as possible;
- check that all significant bypass routes were within the city's jurisdiction;
- ensure that on-street parking control extended beyond the cordon;
- use a simple charge structure with uniform charges for all crossing points; and
- keep the charge at a level sufficiently low to be acceptable.

The considerations on location were seen as separate from those on the charge level. While the surveys identified a clear preference for a simple approach, they also demonstrated that respondents were aware of the potential benefits of a more complex scheme. Typical among the responses was:

> [A]t this stage we are trying to find a system that is just good enough to make this scheme work, rather than trying to find an optimal design which may not be possible to implement; but of course there is the possibility of extending the scheme to a more complex system subject to the success of the initial scheme.

The development of design tools described below needs to be seen in this context; they may not be applied initially, but they should help in identifying subsequent extensions of the first schemes.

3 OPTIMAL CORDON DESIGN BASED ON GENETIC ALGORITHMS

As discussed in the previous section, design of charging cordon schemes has primarily relied on professional judgement, which may well fail to identify the best-performing scheme. This section presents a computational method based on the concept of genetic algorithms (GAs) for directly optimising the charging cordon design so as to maximise the scheme benefits. It summarises the method, named GA-AS, as developed in Sumalee (2004a) and the further developments of the method to handle design constraints (Sumalee, 2003) and multiple objectives (Sumalee, 2004b).

The GA-AS Method for Optimal Cordon Design

The problem of charging cordon design is very complex, since it involves an interaction between the scheme design by the planner and the possible responses of travellers. For a given cordon design, road travellers will respond to the changes in generalised travel costs by switching routes or modes or deciding not to travel. These possible responses must be taken into account when designing and evaluating the scheme benefits. Typically, these possible responses are assumed to follow the Wardrop user equilibrium (UE) condition (Wardrop, 1952).

By including this equilibrium condition into the optimisation problem, the problem can be categorised as a Mathematical Program with Equilibrium Constraint (MPEC), which is one of the most challenging optimisation problems. In addition, the topological requirement for the charged links to form a closed cordon imposes further complexity. This combination precludes the application of a conventional gradient-based optimisation algorithm.

Instead, the GA concept has been adopted. GA is an artificial intelligence exhaustive searching technique. It is a stochastic algorithm whose search methods model the natural phenomena of genetic inheritance and Darwinian strife for survival. The basic idea of the GA approach is to code the decision variables of the problem as a finite string, called a 'chromosome', and calculate the 'fitness' (objective function value) of each string. Chromosomes with a high fitness level have a higher probability of survival. The surviving chromosomes then reproduce and form the chromosomes for the next generation through the 'crossover' and 'mutation' operators.

In this framework, the travellers' responses to the scheme have been calculated by SATURN (Van Vliet, 1982), which is a steady-state equilibrium assignment model that predicts route choice and traffic flows on a road network, based on the generalised costs of travel, and takes account of delays due to capacity constraints. The output from SATURN gives the equilibrium flows, which can then be used to evaluate the performance of different scheme designs. GA will act as a planner in this framework to improve the scheme designs so as to maximise a given objective.

For the optimal cordon problems discussed in this chapter, each chromosome in GA represents a uniform charge level for a specific charging cordon. Thus, to apply GA to the charging cordon design problem, we need to develop a chromosome scheme which represents a closed cordon and preserves its formation, even after the genetic operators (that is, mutation and crossover) are applied. Based on Sumalee (2004a), the concept of a 'branch-tree' was used to encode a closed cordon into a chromosome format. The branch-tree is simply a mathematical representation of the links which form a closed cordon. The crossover operation is defined as an exchange between a valid pair of sub-branches from mated chromosomes which will automatically ensure the formation of a charging cordon for the new chromosome. The mutation is based on the branching-in and -out operations as applied to a branch-tree to reduce or widen the coverage of a particular part of that cordon. The detail of the branch-tree structure and its associated crossover and mutation operators can be found in Sumalee (ibid.). The method developed is named GA-AS. Figure 7.3 shows the overall framework of the algorithm. The algorithm is also able to optimise the location of a double cordon scheme.

GA-AS can be extended to deal with outcome constraints (for example, minimum desired revenue level or minimum reduction in total travel time) on the optimal cordon design problem (CON-GAAS). The mechanism used to handle the constraints is to impose a penalty on the fitness value. The algorithm generates potential solutions without considering the constraints, and then penalises those violating the constraints by decreasing the fitness of the chromosomes. In this way, a constrained problem is transformed to an unconstrained problem by associating a scalar penalty with all constraint violations. Sumalee (2003) applied the method of 'dynamic

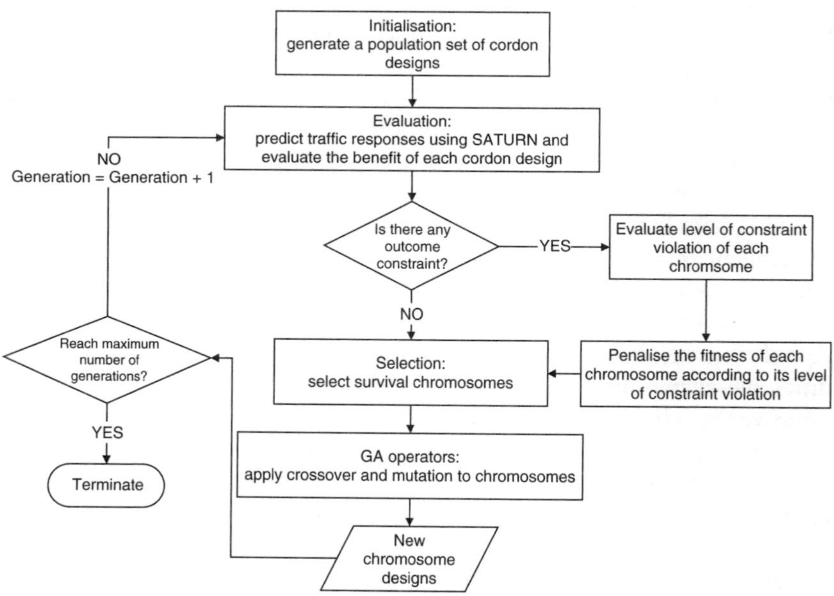

Figure 7.3 Flow-chart of GA-based algorithm for optimal charging cordon design with and without outcome constraints

self adaptive penalty' proposed by Richardson et al. (1989) to the constrained optimal charging cordon design problem.

Similarly, GA-AS can also be extended to handle the problem associated with multiple objectives. Sumalee (2004b) applied the non-sorted dominated GA (NSGA-II) as originally proposed by Deb et al. (2000) to the cordon design problem. The solution to this design problem with multiple objectives will be in the form of a set of non-dominated solutions, also referred to as a 'Pareto front'. The non-dominated solutions represent the set of solutions for which one objective cannot be improved without detracting from the other. This algorithm will try to simultaneously push and spread the solution populations to and along the Pareto front. Once the well-spread Pareto front is enumerated, the planner can then investigate the trade-off between different objectives for different designs.

Tests with the Edinburgh Network

GA-AS was tested with a SATURN model of the Edinburgh road network with the objective of maximising the social welfare benefit. The following tests were conducted:

1. Optimise uniform tolls for three pre-specified judgemental single cordon schemes (see Figure 7.4).
2. Optimise the location of a closed charging cordon with a uniform toll (OPC).
3. Optimise the uniform toll for the 15 links with highest marginal social cost (top-15).
4. Optimise the location of double charging cordons with uniform tolls (D-OPC).

The summary of the test results is shown in Table 7.1. The tests were conducted with a SATURN model operating in 'buffer' mode which represents delays on links rather than at junctions. This simplification was adopted to enable the method to be tested reliably. The values of time and vehicle running cost adopted were 7.63 and 5.27 pence per minute, respectively, with a generalised cost elasticity of –0.57; all these values were based on earlier research. The operating and implementation cost of a toll point were assumed to be £100, based on earlier analysis for London.

From test 1, out of the three judgemental cordons as defined in Figure 7.4, the Inner 2 cordon performs best with a net social welfare improvement of £3.99 thousand per peak hour. In earlier research, an incomplete judgemental outer cordon had been identified with a somewhat higher net benefit of

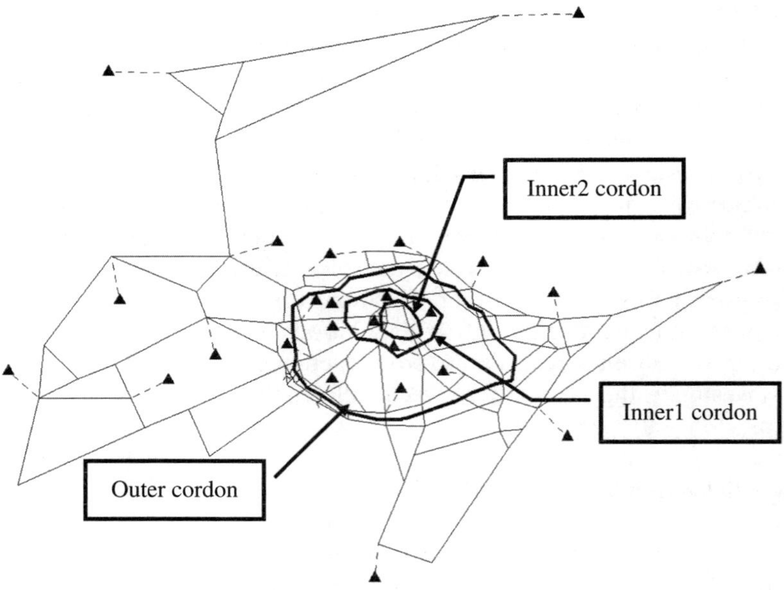

Figure 7.4 Edinburgh network with three judgemental cordons

Table 7.1 Comparison of the performance of different charging regimes for the Edinburgh network

Charging regime	Optimal toll	No. of toll points	Gross social welfare benefit (£k/hour)	Net social welfare benefit (£k/hour)	% of benefit compared to OPC
Inner 1 cordon	£0.50	9	3.00	2.10	−71
Inner 2 cordon	£0.75	7	4.69	3.99	−45
Outer cordon	£0.75	20	3.96	1.96	−73
Top-15	£0.75	15	10.71	9.21	+28
OPC	£1.50	13	8.51	7.21	–
D-OPC	£1.25	38	19.08	15.28	+112

£4.57 thousand per peak hour. This has not been used for comparison, since the focus of GA-AS is on complete cordons.

In test 2, GA-AS found an optimal charging cordon (OPC) as shown in Figure 7.5. OPC is larger than either of the inner judgemental cordons and extends further to the west where congestion is more serious. The net benefit

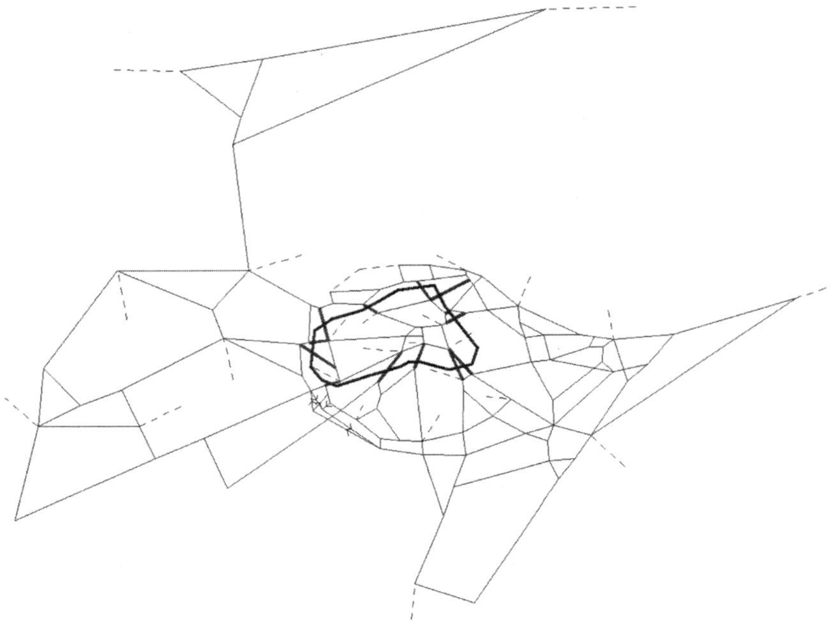

Figure 7.5 Location of the optimal cordon OPC

generated by OPC is £7.21 thousand per peak hour which is 80 per cent higher than the benefit produced by the Inner 2 cordon and over three times the benefit of the other two judgemental cordons. This result clearly indicates the potential loss of scheme benefit by relying on professional judgement.

For the third test, the marginal cost tolls for all links in the network were calculated by running the system optimum assignment (Sheffi, 1985). We then selected the 15 links with the highest level of marginal cost toll (see Figure 7.6). This combination of tolled links was used as an approximation to the best combination of tolled links in the network. The top-15 charging system (with an optimal uniform toll of £0.75) generated a net benefit of £9.21 thousand per peak hour which is around 30 per cent higher than that generated by the OPC cordon. Thus the requirement for a continuous cordon may itself be a serious constraint on optimal design; however, such isolated charging points may be particularly difficult to explain to users.

GA-AS was used in the fourth test to find an optimal double-cordon scheme (D-OPC) (Figure 7.7). The inner cordon is slightly wider than the City Council's proposal, but the outer cordon is very different, crossing the outer ring road in two locations to charge approach corridors contributing to congestion. The optimal uniform toll found for D-OPC is £1.25. The

Figure 7.6 Locations of the top-15 links with the highest marginal costs, numbered in rank order

benefit generated by D-OPC is £15.28 thousand per peak hour, which is more than double the benefit from the OPC.

4 A SHORT-CUT METHOD BASED ON SELECT LINK ANALYSIS

While the GA approach described above has been shown to be capable of generating significantly improved cordon designs, it is analytically complex, and has yet to be tested on the road networks of other cities or with other models. The DfT was interested in providing guidance to local authorities on road pricing design (DfT, 2006), and commissioned work on a short-cut method which could be applied more rapidly on a wider range of network models. The aim of the short-cut approach was to improve on the judgemental designs by using some theoretical modelling while reducing the number of simulations required by the GA-based approach. As such it was not designed to find an optimal closed cordon design but it was expected to improve performance over the judgemental approach.

Figure 7.7 Location of the optimal double cordon D-OPC

Previous work on the GA approach had included as a benchmark the system optimal or 'first-best' solution whereby all links were tolled to give the system optimal or maximum welfare gain (Sheffi, 1985). It was this which led to the top-15 test in Table 7.1 which, as we have seen, outperformed the optimal cordon by almost 30 per cent and achieved almost 30 per cent of the first-best solution (Table 7.2).

Further investigations showed that the percentage of first-best benefits achieved versus the number of links tolled at the system optimal level (added in order of decreasing charges) formed a curve as shown in Figure 7.8. Similar curves were also found for networks in Cambridge, Leeds and York, which have many more links and many more origin–destination pairs than the Edinburgh network discussed here. These curves proved useful in defining how many links (the top-X links) should be used in the design process. The number of links used should be manageable while still achieving a significant proportion of the first-best benefits. In general, less than 10 per cent of the links are required to achieve around 60–70 per cent of the first-best benefits.

As these limited point charges on the highest-charged links from the system optimum could outperform the closed cordons, it was thought that

Table 7.2 Relative performance of OPC and SLA cordons

Cordon	Optimal toll	Cost of implementation /operation per peak hour (£k)	Gross total benefit per peak hour (£k)	Gross total benefit compared to first-best (%)	Net benefit per peak hour (£k)	Flow crossing top 15 links	Proportion of total flow on top-15 links (%)	Proportion of top-15 gross benefits
Top-15 links	£0.75	1.50	10.71	28.8	9.21	66,759	100	100
OPC	£1.50	1.30	8.51	22.9	7.21	34,389	51.5	43.7
SLA-single	£0.75	1.60	7.94	21.3	6.34	51,278	76.7	40.8
First-best condition	na	35.00	37.19	100.0	2.19	All	100	N/A

Note: Notional figures based on 350 links being charged.

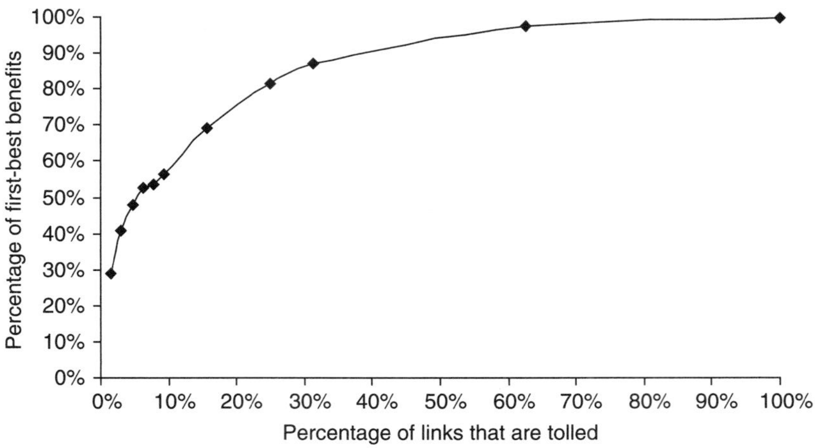

Figure 7.8 Percentage of first-best benefits versus percentage of links tolled: Edinburgh

using these links may prove beneficial in designing a new closed cordon. Figure 7.9 shows the position of the top-15 'highest tolled' links for the Edinburgh network. As can be imagined it would be difficult to find a closed cordon that is not unduly complex which passes through all these links. However, it was realised that it was not essential to include the top links in the cordon; instead the cordon charge should be imposed on the principal path flows through these high-cost links. The higher the proportion of high-cost flows covered by the cordon the higher the potential benefits of that cordon location. This led to the idea of using a select link analysis to aid the cordon design process.

Select link analysis (SLA) is available in many commercial packages such as EMME/2 (Inro Consultants, 1999) or TRANPLAN (Caliper Corp, 2004) and is an easy to use option within SATURN. Basically, SLA shows the paths used by all flows through a set of links. First an SLA is performed for the top-X links (in this case 15, as shown in Figure 7.9). Next, the path information is used to aid in the design of a new cordon on screen, trying to capture as much of the flow from the SLA as possible. A further SLA is performed on the new cordon and this is cross-matched with the top-X links to determine the proportion of flow covered. Cordons with a high proportion of flow covered are taken to the next stage which is to run the simulation for various charge levels to optimise the uniform charge around the cordon.

This heuristic process can be summarised as follows:

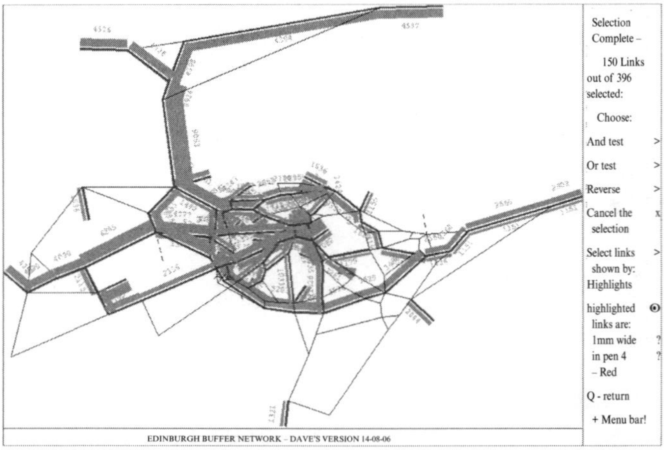

Figure 7.9 Top-15 links and bandwidth from SLA in SATURN

1. Compute the system optimum and calculate first-best benefits.
2. Sort the system optimal charges in descending order and apply those charges using an increasing number of links, creating a graph of relative benefits versus number of links charged.
3. Use the graph from Step 2 to select a subset of top-X links and produce a visual output of these links.
4. Carry out a select link analysis with these links to show, using bandwidths, where the flows come from/go to through these top-X links.
5. Draw a cordon or set of cordons either on-line or off-line which 'catch' a high proportion of the flows from the top-X links.
6. Optimise the charge level for the cordon by plotting benefits for a set of uniform charge levels.
7. Repeat 5–7 until a satisfactory cordon design is achieved.

An example of the single cordon produced by the above approach is shown in Figure 7.10. As shown in Table 7.2, this single SLA cordon achieved 93 per cent of the gross benefits from the GA-optimised single cordon OPC. Thus the method has delivered gains in welfare which are comparable to those produced by a much more time-consuming and complex approach.

This chapter has demonstrated the approach for a small network in Edinburgh. More recent work for the DfT (Shepherd et al., 2006) has proved the method to be effective for much larger networks in Cambridge, York and Leeds. However, some limitations were found to arise from the differing ways in which buffer networks had been specified. Improved approaches are currently being developed. As well as reducing analysis time, the SLA

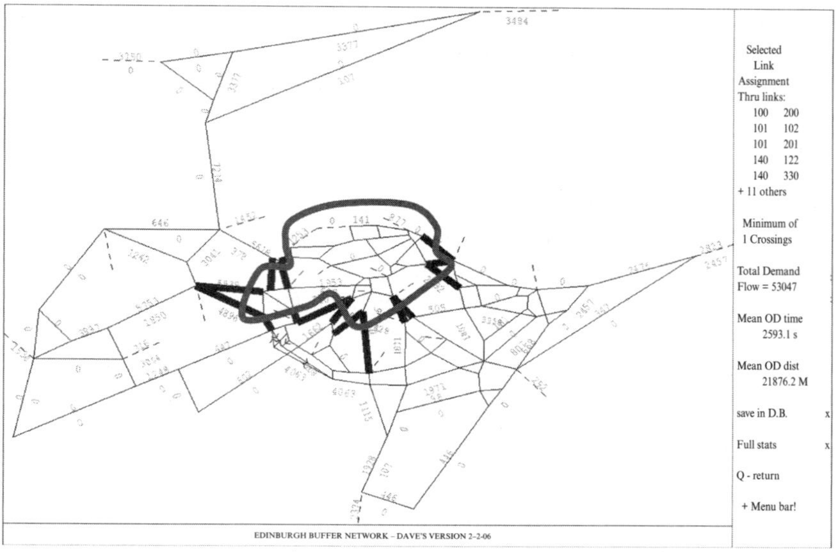

Figure 7.10 Single cordon developed using the SLA approach

approach has the advantage over the GA optimisation approach of involving the planner directly in drawing potential cordons, thus enabling local knowledge to be applied.

5 FUTURE DEVELOPMENTS

The research described above has led to a GA method which can identify the theoretically best-performing cordon for a specified objective in a given city. Such cordons can double or treble the benefits of a cordon located solely on the basis of professional judgement. However, it must be emphasised that these theoretical techniques are not a replacement for professional and political judgement. Instead, they should be used to identify cordons that are worthy of consideration. Where these cordons are found wanting on political grounds, the method can be used to identify constrained optimal designs, or as a benchmark to estimate the benefit lost by adopting a more politically acceptable design.

At present, this method has still to be tested on a wider set of networks and models, and is too complex to be easily transferred into practice. As an alternative, a short-cut SLA method has been developed, which has been found to identify cordons which achieve a large proportion of the benefits of an optimal cordon. The method is much easier to use, and has the advantage of involving the planner directly in the cordon design process. It has highlighted some weaknesses in current practice in network modelling which need to be remedied if these models are to be used successfully in road pricing design.

Both of these methods are being developed further. Future work on the SLA method will look at alternative measures for selecting the 'high-cost' links. This will help get around the problem of solving the system optimum in a SATURN simulation network where, due to the approach used, there is no analytical solution. Work to date has shown that a simple measure of 'total delay' on a link is a good proxy for the high-cost links as defined by the highest system optimal tolls for the Edinburgh network. Further work is required to prove this concept for other networks. If successful, this approach will be applied to simulation networks to improve cordon designs. However, it will no longer be possible to calculate the first-best benchmark.

Both the SLA and the GA approaches need to be enhanced to take into account multiple user classes and multiple time periods. This will enable them to be tested on realistic networks in the UK and elsewhere. In the meantime we continue our research into optimal tolling, extending the work to consider both tolling and investment in capacity. The theoretical approach is based around the work of Lawphongpanich and Hearn (2004), which uses a constraint-cutting algorithm to solve the second-best optimal toll problem for a given set of links. This method has been selected as it avoids the pitfalls of other methods reported in Shepherd and Sumalee (2004) whereby a change in path set causes a discontinuity which disturbs the optimisation procedure. While this approach is being adopted to solve the joint second-best problem of investment in capacity and charge levels for predefined links it will not be able to solve the cordon location problem. For this we intend to extend the GA approach to include capacity investments under typical planning constraints.

REFERENCES

Caliper Corporation (2004), TransCAD, Transportation GIS Software, Overview, Newton, MA, www.caliper.com.

Deb, K., S. Agrawal, A. Pratap and T. Meyarivan (2000), 'A fast elitist non-dominated sorting genetic algorithm for multi-objective optimization: NSGA-II', *Proceedings of the Parallel Problem Solving from Nature VI (PPSN-VI)*.

Department for Transport (DfT) (2006), 'Designing efficient local road pricing schemes', TAG Unit 3.12.1 (draft), DfT, London.

Inro Consultants (1999), *EMME/2 Users Manual Release* 9, Montreal: Corp.

Lawphongpanich, S. and D.W. Hearn (2004), 'An MPEC approach to second-best toll pricing', *Mathematical Programming B*, **101**(1).

May, A.D., D. Coombe and T. Travers (1996), 'The London Congestion Charging Research Programme: 5. Assessment of the impacts', *Traffic Engineering and Control*, **37**(6), 403–9.

May, A.D., R. Liu, S.P. Shepherd and A. Sumalee (2002), 'The impact of cordon design on the performance of road pricing schemes', *Transport Policy*, **9**.

Richards, M., C. Gilliam and J. Larkinson (1996), 'The London Congestion Charging Research Programme: 6. The findings', *Traffic Engineering and Control*, **37**(7/8), 436–40.

Richardson, J.T., M.R. Palmer, G. Liepins and M. Hilliard (1989), 'Some guidelines for genetic algorithms with penalty functions', *Proceedings of the 3rd International Conference on Genetic Algorithms*, eds J. Schaffer and Morgan Kaufmann, CA.

Sheffi, Y. (1985), *Urban Transportation Networks: Equilibrium Analysis with Mathematical Programming Methods*, Englewood Cliffs, NJ: Prentice-Hall.

Shepherd, S.P., A. Koh and A.D. May (2006), 'Investigating a select link analysis approach to cordon design: Executive Summary', Report to Department for Transport, June, unpublished.

Shepherd, S.P., A.D. May and D.S. Milne (2001), 'The design of optimal road pricing cordons', *Proceedings of the 9th World Conference on Transport Research*, Seoul, Korea.

Shepherd, S.P. and A. Sumalee (2004), 'A genetic algorithm based approach to optimal toll level and location problems', *Networks and Spatial Economics*, **4**(2), 161–79.

Sumalee, A. (2001), 'Analysing the design criteria of charging cordons', ITS Working paper no. 560, Institute for Transport Studies, University of Leeds.

Sumalee, A. (2003), 'Optimal toll ring design with spatial equity impact constraint: an evolutionary approach', *Journal of Eastern Asia Society for Transportation Studies*, **5**, 1813–28.

Sumalee, A. (2004a), 'Optimal road user charging cordon design: a heuristic optimisation approach', *Computer-Aided Civil and Infrastructure Engineering*, **19**, 377–92.

Sumalee, A. (2004b), 'Optimal road pricing scheme design', PhD thesis, University of Leeds.

Transport for London (TfL) (2006), *Central London Congestion Charging: Impacts Monitoring*, London: TfL.

Van Vliet, D. (1982), 'SATURN: a modern assignment model', *Traffic Engineering and Control*, **23**, 578–81.

Wardrop, J. (1952), 'Some theoretical aspects of road traffic research', *Proceedings of the Institution of Civil Engineers*, **1**, 325–62.

PART II

London

8. The London Congestion Charging Scheme, 2003–2006

Georgina Santos*

1 INTRODUCTION

In this chapter the London Congestion Charging Scheme (LCCS) is assessed. A brief overview of how the scheme works is given in Section 2 and a summary of its impacts on traffic in Section 3. With information and data from Transport for London demand elasticities are estimated in Section 4 and the area marginal congestion costs in Section 5. Finally, a description and potential impacts of the Western Extension are presented in Section 6. Section 7 concludes.

2 HOW THE LONDON CONGESTION CHARGING SCHEME WORKS

The LCCS, designed and managed by Transport for London (TfL), was implemented on 17 February 2003 and is essentially an area licensing scheme. It was designed and managed by Transport for London (TfL), and is essentially an area licensing scheme. All vehicles entering, leaving, driving or parking on a public road inside the zone between 7:00 am and 6:30 pm Monday to Friday, excluding public holidays, must pay a congestion charge. This was initially £5, but on 4 July 2005 it was increased to £8. Traffic signs make it clear where the limits of the charging zone are.

Figure 8.1 shows the limit of the area, the inner ring road, which runs along Euston Road, Pentonville Road, City Road, Old Street, Commercial Street, Tower Bridge Road, New Kent Road, Kennington Lane, Vauxhall Bridge Road, Park Lane, Edgware Road and Marylebone Road. No charge is made for driving on the inner ring road itself. The charging area is relatively small. It covers only 21 sq km (8.4 sq miles), representing 1.3 per cent of the total 1,579 sq km (617 sq miles) of Greater London.

A number of payment options are available, as the charge can be paid for a day, a week, a month and a year, up to 90 days in advance. The methods

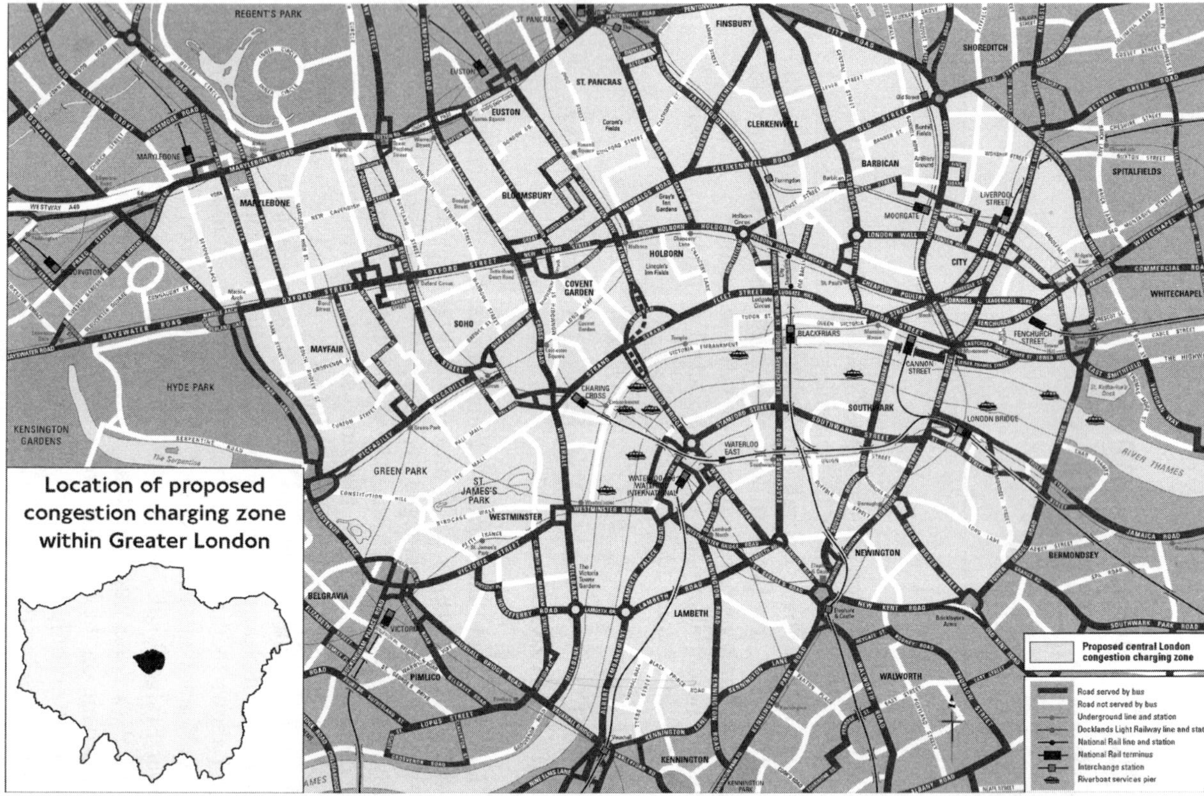

Source: www.london.gov.uk/approot/mayor/congest/pdf/zone_map.pdf.

Figure 8.1 Map of the London congestion charging zone

of payment are online, in person at selected shops, petrol stations and car parks, by post, by telephone, by mobile phone, and at BT internet kiosks. The charge can also be paid on the day or on the day after until midnight. A 'next day payment', however, costs £10 (instead of £8) and can only be paid via the call centre or via the internet.

There are a number of exemptions and 100 per cent discounts in place, which apply to two-wheelers, emergency vehicles, vehicles used by or for disabled people, public buses, licensed London taxis and mini-cabs, some military vehicles, alternative fuel vehicles (with stringent emission savings), and roadside assistance and recovery vehicles. Finally, vehicles registered to residents of the charging zone are entitled to a 90 per cent discount when at least a week's worth of congestion charge is bought.

Enforcement is undertaken with Automatic Number Plate Recognition (ANPR). There are 203 camera sites located at every entry and exit to the charging zone and also inside the zone. After a manual check, violators are tracked through the Driver and Vehicle Licensing Agency and issued a Penalty Charge Notice (PCN) of £100. As of September 2006 the PCN of £100 is reduced to £50 if paid within 14 days, and increased to £150 if not paid within 28 days.

3 IMPACTS ON TRAFFIC

The main objective of the LCCS was to reduce congestion by reducing the volume of traffic entering and circulating within the charging zone during charging hours. In this sense, the scheme is a success. Table 8.1 shows the percentage change in vehicle counts entering the charging zone.

Table 8.1 Percentage change of vehicles entering the charging zone during charging hours

Vehicle type	Total change 2003 vs 2002	Total change 2004 vs 2003	Total change 2005 vs 2004
Cars	−33	−1	−3
Taxis	17	−1	0
Buses and coaches	23	8	−6
Vans	−11	−1	−3
Lorries and other	−11	−5	−4
Pedal cycles	19	8	−3
Motorcycles	12	−3	−12

Source: TfL (2006, Table 2.1, p. 22).

During 2003 and 2004, average reductions in congestion within the charging zone were 30 per cent when compared to pre-charging levels (TfL, 2005a, p. 14). Congestion is defined by TfL as 'the difference between the average network travel rate and the uncongested (free-flow) network travel rate in minutes per vehicle-kilometre' (TfL, 2003a, Table 3.1, p. 46). Using the uncongested network travel rate of 1.9 min. per km (approximately 32 km per hour) from TfL (2003a, p. 52), and pre- and post-charging average travel rates of 4.2 and 3.5 min. per km, respectively, it can be seen that congestion decreased from 2.3 to 1.6 min. per km (TfL, 2005a, p. 15). Most of these reductions in travel times were the result of reduced queuing 'time at junctions, rather than increases in driving speeds' (ibid., p. 13).

The overall picture for congestion in 2005 was more mixed. Typical delays in 2005 were 1.7 min. per km, rather than 1.6 min. per km and the consequent reduction in congestion was 22 per cent, rather than 30 per cent (TfL, 2006, p. 41).

The reason TfL gives for this 'increase' in congestion is the network adjustments that were made in London in 2004 and 2005. They argue that some roads have increased priority for buses, taxis and/or cyclists, thus reducing the available road space for other vehicles (ibid., p. 41). As a consequence, while some vehicle types may be enjoying higher speeds, others are facing lower average speeds.

Both in 2004 and 2005, there was lower congestion on the inner ring road and on the main radial routes approaching the zone, when compared to 2002 (ibid., p. 41). On the other hand, congestion on main roads in inner London (outside the charging zone) in 2005 was on average higher than in any previous year, with delays of 1.5 min. per km, in contrast to 1.3 min. per km in 2002 (TfL, 2006, p. 41).

4 DEMAND ELASTICITIES

The formula to compute the generalised cost elasticity of demand for trips is:

$$\eta = \Delta q/\Delta GC \cdot GC/q,$$

where q is number of trips by the relevant mode, and GC is generalised cost of a trip by the relevant mode: car, taxi, van or lorry.

The GC per km is defined as:

$$GC \text{ per km} = VOC \text{ per km} + VOT \text{ per km} + \text{congestion charge per km} - \text{reliability benefits},$$

where *VOT* is value of time and *VOC* is vehicle operating cost. The reliability benefits are a percentage of the time benefits due to higher speeds.

VOT is usually given in pounds per hour. Using speed this can be expressed in pounds per km. Both *VOT* and *VOC* were taken from TAG Unit 3.5.6 (Department for Transport, 2004), which provides recommendations on the values of time and vehicle operating costs for use in economic appraisals of transport projects in Great Britain.

Values of Time

The working, commuting and other values of time per person were all converted to 2003 prices and values and adjusted to London prices. Using vehicle occupancies and trip purpose, the value of time per vehicle was obtained. The results are presented in Table 8.2.

The speeds pre- and post-charging inside the charging zone are reported to be 14 and 17 km per hour, respectively (TfL, 2003a,b, 2004a, 2005a). However, the average speed in Greater London is higher. Most trips that start or finish inside the charging zone have an origin or destination outside the charging zone. Average speed has always been and still is higher the further the vehicle is from the charging zone. Average speed in the rest of London is assumed to be 26.65 km per hour. This assumption is based on a weighted average of the average speeds in TfL (2003c, Table 3.2, p. 17).

Using the vehicle-km from TfL (2005a, Figure 15, p. 29) and the total traffic counts in the charging zone provided by TfL, the average number of kilometres travelled by the different vehicle types inside the zone on a typical weekday can be computed. These are presented in Table 8.3, together with average trip lengths.

Using average speeds outside and inside the charging zone, whole average trip length and average distance travelled inside the zone by each vehicle type, the average time costs per vehicle type in pence per km can be estimated. These are presented in Table 8.4.

Table 8.2 Average value of time per vehicle type (pence per hour, 2003 prices and values)

Car	Taxi	Van	Lorry
1,096	2,309	1,367	1,145

Note: Non-working values of time are the same for all modes.

Sources: Tables 1, 2 and 3, TAG Unit 3.5.6 (Department for Transport, 2004) and HM Treasury GDP deflator series.

Table 8.3 Average distance (km) driven by cars, taxis, vans and lorries

	Cars	Taxis	Vans	Lorries
Average trip length	11.6[a]	8.4[a]	12.0[b]	28.0[b]
Average daily km driven	23.2	n.a.	24.0	56.0
Average km driven inside the charging zone during charging hours	2.0	2.30	2.56	2.22

Sources: a. TfL (2003c), Table 3.6, http://www.tfl.gov.uk/tfl/pdfdocs/ltr/london-travel-report-2003.pdf; b. LSDP (2005), http://www.tfl.gov.uk/tfl/downloads/pdf/Freight-Plan-sc-app-bc.pdf, p. 12.

Table 8.4 Average time costs per vehicle type (pence per km, 2003 prices and values)

	Car	Taxi	Van	Lorry
Pre-charging	44.3	97.4	56.2	47.1
Post-charging	43.1	93.4	54.4	45.6

Source: See text.

Vehicle Operating Costs

The vehicle operating costs have two components: the fuel and the non-fuel costs.

Following the TAG Unit 3.5.6 (Department for Transport, 2004), the fuel consumption was estimated using a function of the form:

$$L = a + bV + cV^2,$$

where L is consumption in litres per km, V is average link speed in km per hour (although in this study this was used as average speed in km per hour), and a, b and c are parameters defined for each vehicle category. The different values were uprated to 2003 values and prices as well as 2003 fuel efficiency.

The components of the non-fuel vehicle operating costs include oil, tyres, maintenance, depreciation and vehicle capital saving (only for vehicles in working time). Again, following the TAG Unit 3.5.6, the non-fuel elements of VOC were combined in a formula of the following form:

$$C = a_1 + b_1/V,$$

where C is cost in pence per kilometre travelled, V is average link speed in kilometres per hour (although in this study the average area speed per hour

Table 8.5 Average vehicle operating costs per vehicle type (pence per km, 2003 prices)

	Car		Taxi		Van		Lorry	
	Pre	Post	Pre	Post	Pre	Post	Pre	Post
Fuel	7.6	7.6	5.9	5.9	9.1	9.1	25.8	24.6
Non-fuel	4.4	4.4	4.7	4.7	7.5	7.5	14.4	14.1
Total	12.0	11.9	10.7	10.6	16.6	16.5	40.2	38.7

Notes:

1. The perceived cost of non-fuel VOC of working vehicles (that is, work car and van) is the cost perceived by businesses and is therefore equal to the resource cost. The perceived cost of non-work cars on the other hand is the cost perceived by the consumer and is therefore equal to the market price, which includes all indirect taxation (such as VAT) (TAG Unit 3.5.6, Point 1.1.6, Department for Transport, 2004).
2. Non-fuel VOCs are assumed to remain constant in real terms over the forecast period. This assumption is made because the main elements which make up non-fuel VOCs are subject to less volatility than fuel VOCs (TAG Unit 3.5.6, Point 1.3.17, Department for Transport, 2004).
3. 'Pre' refers to pre-charging (lower average speeds) and 'post' refers to post-charging (higher average speeds).

Sources: Tables 10, 11, 13, 14 and 15, TAG Unit 3.5.6 (Department for Transport, 2004) and HM Treasury GDP deflator series.

was used instead), a_1 is a parameter for distance-related costs defined for each vehicle category, and b_1 is a parameter for vehicle capital saving defined for each vehicle category. Table 8.5 presents the VOC for each vehicle type.

Generalised costs

With all the information above, the GC for cars, taxis, vans and lorries was computed before and after the charge. Dodgson et al. (2002) argue that reliability benefits are worth approximately 25 per cent of time benefits, and this was the value assumed for the increase in reliability. The GC of each vehicle type was the time cost per km plus the fuel and non-fuel VOC per km minus 25 per cent of the value of the time saved. Taxis of course are exempt from the charge, so the congestion charge per km was computed and added into the calculation of GC only for cars, vans and lorries.

In order to estimate the congestion charge per km, the £5 charge was divided by the average distance travelled by each vehicle type multiplied by 2, as any vehicle going in would need to come out, and the £5 is a charge per day. The GC pre- and post-charging is presented in Table 8.6.

The GC of cars and working vehicles increased after charging was introduced because the charge of £5 was not compensated by the time savings,

Table 8.6 Pre- and post-charging GC (pence per km, 2003 prices)

	Cars	Taxis	Vans	Lorries
Pre-charging	56.4	108.0	72.9	86.3
Post-charging	76.4	103.0	90.8	104.3

Sources: Tables 8.4 and 8.5.

even when including the reliability benefits of 25 per cent. The GC of taxis on the other hand decreased because taxis do not pay the charge, and the average speed inside the charging zone increased, thus reducing time costs.

GC Elasticities

The percentage change in the number of trips made is presented in Table 8.7.

Table 8.8 shows the final numbers input in the elasticity formula presented at the beginning of this section.

The GC elasticities of demand for trips are presented in Table 8.9.

For all four elasticities computed in Table 8.9 it should be borne in mind that they are point elasticities, using the quantities and costs pre-charging as the base point; and that they all correspond to the change in GC when the £5 charge was first introduced. The changes correspond to the period 2002 vs 2003.

A number of important observations also need to be made:

1. The GC elasticity of demand for trips by car is –0.96. This elasticity is fairly high because London has a reasonably good public transport system and many car users switched to the bus. Some also switched to taxis, and this is discussed further, below.
2. The GC elasticity of demand for trips by taxi is very high: –2.67. This can be explained by the substitution effect between cars and taxis. The cross-price elasticity of demand for trips by taxi with respect to the GC of trips by car is approximately 0.35.[1] Unfortunately, when calculating this cross-price elasticity, the GC of trips by taxi cannot be held constant, as it also changes because the average speed in the charging zone is higher. Most of the increase in demand for trips by taxi can be explained by the increase in the GC of trips by car, rather than by the 5 per cent decrease in the GC of trips by taxi. The own-price elasticity of demand for trips by taxi in Table 8.9 is capturing the substitution effect resulting from the increase in the GC of trips by car.
3. Being commercial vehicles, vans and lorries have less flexibility to change route or time of travel, and not having close substitutes, the own-price elasticity is lower.

Table 8.7 Vehicle counts pre- and post-charging

	Cars	Taxis	Vans	Lorries
Spring 2002 total	386,752	113,007	113,267	31,585
Spring 2003 total	255,256	127,133	98,542	27,953
Changes (%)	−34	+13	−13	−11

Source: Transport for London, data provided on request.

Table 8.8 Values used to compute GC elasticities

	Cars	Taxis	Vans	Lorries
Δq	−131,496	14,126	−14,725	−3,632
ΔGC	20.0	−5.0	17.9	17.9
GC	56.4	108.0	72.8	86.27
q	386,752	113,007	113,267	31,585

Sources: Tables 8.1–7.

Table 8.9 GC elasticities of demand for trips

Cars	Taxis	Vans	Lorries
−0.96	−2.67	−0.53	−0.56

Source: Table 8.8 values.

Congestion Charge Elasticity

In July 2005 the charge was increased from £5 to £8; using that increase, the point and arc elasticities can be computed. Since taxis do not pay the congestion charge, the exercise cannot be carried out for them. Table 8.10 shows the values necessary to compute the congestion charge elasticities.

After the charge increase, the number of lorries increased. This is difficult to interpret in relation to the charge increase itself. However, changes made to the fleet schemes, which took place at roughly the same time as the charge increase, may provide an explanation. Following the introduction of discounts for vehicles on the fleet scheme, users of the scheme increased by 8 per cent (TfL, 2006, p. 7). The additional 10 per cent surcharge for vehicles on the 'fleet' was removed and a £1 discount was introduced, thus resulting in a charge of £7 per charging day per fleet-vehicle.

The arc and point congestion charge elasticities computed on the basis of the increase from £5 to £8 are presented in Table 8.11.

The congestion charge elasticities of demand for trips, computed for the £3 increase, are very low. One explanation for this is that the main changes took place with the initial introduction of the congestion charge. By the time the increase was implemented, most traffic had already settled and those paying £5 were actually willing to pay £8 as well. As a result, the reduction was positive but small in the case of cars, even smaller in the case of vans and negative in the case of lorries, although this last effect was probably the consequence of the changes to the fleet-scheme.

5 AREA MARGINAL CONGESTION COSTS

The marginal congestion cost (MCC) per passenger car unit (PCU)[2] can be computed with the following formula, standard in the road pricing literature:

$$MCC = e_{sq} \cdot b \,/\, s(q),$$

where b is value of time in pence per PCU per hour, s is speed in km per hour, dependent on q, the traffic volume[3] in the area, in PCUs per hour, and e_{sq} is the elasticity of speed with respect to traffic volume.

The information needed is then:

1. value of time in pence per PCU per hour;
2. speed in the charging zone (which is 14 and 17 km for pre- and post-charging respectively); and
3. number of PCUs (which approximates traffic volume).

Table 8.10 Values used to compute congestion charge elasticities

	Cars	Vans	Lorries
Δq	−5,933	−211	221
Δcharge	3	3	3
Charge	5	5	5
Q	244,085	94,175	25,060

Sources: Tables 8.1–8.9.

Table 8.11 Congestion charge elasticites of demand for trips computed on the basis of an increase from £5 to £8

	Cars	Vans	Lorries
Point elasticity	−0.041	−0.004	0.015
Arc elasticity	−0.053	−0.005	0.019

Source: Table 8.10.

In order to compute the value of time in pence per PCU per hour the average value of time per vehicle type per hour is needed first. The average values of time per vehicle type in pence per hour for cars, taxis, vans and lorries were presented in Table 8.2. The value of time for buses, motorcycles and pedal cycles can be estimated in exactly the same way, using TAG Unit 3.5.6 (Department for Transport, 2004) and uprating the numbers to London 2003 prices and values. Table 8.12 presents these values.

The value of time per PCU can be computed by multiplying the VOT per vehicle type by the standard PCU rating for each vehicle type and by its share in total traffic and then adding them all up. Table 8.13 presents vehicle counts, PCUs and shares.

Table 8.14 presents the VOT per PCU.

In 2002 the average value of time in the charging zone was £15.9 per PCU. In 2003 the average VOT per PCU had increased to £18.3 in real terms. The reason for this was the change in traffic composition due to the congestion charge.

The VOT, speeds and changes in traffic volume can be input into the formula to estimate the area MCC. It should be noted that the MCC is not estimated per link but for the whole charging zone. Therefore, it is an area MCC. Rather than flow, total PCU counts are used.

The elasticity of speed with respect to traffic volume is the percentage change in speed over the percentage change in traffic volume, computed as a point elasticity with 2002 as the base year. The change in speed was 21.5 per cent (change from 14 to 17 km per hour) and the absolute value of the change in traffic volume was 15.5 per cent (change from 844,229 to 712,954 PCUs, as indicated in Table 8.13, below). The elasticity of speed with respect to traffic volume was then 1.38. The MCC can now be computed:

$$\text{MCC pre-charging} = 1.38 * 15.93 / 14 = 1.57 = 157 \text{ pence per PCU/km},$$

$$\text{MCC post-charging} = 1.38 * 18.29 / 17 = 1.48 = 148 \text{ pence per PCU/km}.$$

Table 8.12 Average values of time for buses and two-wheelers (pence per hour, 2003 prices)

Vehicle type	Value of time	
Bus driver	1,150	
Bus passengers (assumed to be 23)	14,220	
Bus total		15,370
Pedal cycle		727
Motorcycle		890

Source: See text.

Table 8.13 Vehicle counts pre- and post-charging (and relevant PCU values)

Vehicle type	Vehicles		PCUs		Share of PCUs	
	2002	2003	2002	2003	2002	2003
Cars	386,752	255,256	386,752	255,256	0.46	0.36
Taxis	113,007	127,133	113,007	127,133	0.13	0.18
Vans	113,267	98,542	169,901	147,813	0.20	0.21
Buses	26,472	32,296	66,180	80,740	0.08	0.11
Lorries	31,585	27,953	78,963	69,883	0.09	0.10
Pedal cycles	25,181	28,329	5,036	5,666	0.01	0.01
Motorcycles	48,780	52,926	24,390	26,463	0.03	0.04
Total	745,044	622,435	844,229	712,954	1	1

Source: Transport for London, data provided on request.

Table 8.14 VOT per PCU (£ per PCU per hour, 2003 prices)

Year	VOT per PCU
2002	15.9
2003	18.3

Source: Own calculations explained above.

These values can now be used to estimate the area MCC per vehicle type, presented in Table 8.15.

A congestion charge that approximated the MCC would need to be different for each vehicle type. If the area MCC per km from Table 8.15 is multiplied by the average number of km driven inside the zone, a rough approximation of an efficient charge can be obtained. Table 8.16 compares the £5 charge with the area MCC and the actual and efficient km driven inside the charging zone.

With the £5 charge cars were overcharged while vans and lorries were undercharged. With an £8 charge cars are even more overcharged but vans and lorries are paying a charge closer to their MCC. Under the fleet-vehicle scheme, where goods vehicles are entitled to a £1 discount, the £7 paid by vans is not too different from the MCC they impose. Lorries, on the other hand, are still undercharged.

Even with the simple ANPR enforcement system, it would probably not be administratively expensive to have different congestion charges (for example, three different levels) for different vehicle types. The reason why

Table 8.15 Area MCC by vehicle type (pence per km, 2003 prices)

Vehicle type	MCC pre-charging	MCC post-charging
Cars and taxis	157.0	148.0
Vans	235.5	222.0
Buses	392.5	370.0
Lorries	392.5	370.0
Pedal cycles	31.4	29.6
Motorcycles	78.5	74.0

Note: These figures are simply computed as the MCC in pence per PCU/km multiplied by the respective PCU ratings for each vehicle type.

Table 8.16 Area MCC and the £5 charge

Vehicle type	Efficient km £5 charge	Actual km £5 charge	Efficient charge with actual km
Cars	3.38	2.00	£2.96
Vans	2.25	2.74	£6.08
Lorries	1.35	2.50	£9.27

Source: See text.

this has not happened can mainly be found in the lobbying from the haulage industry that opposed a congestion charge altogether, and especially a higher charge for goods vehicles. There had originally been proposals for a £15 charge for bigger goods vehicles but this encountered much opposition.

6 THE WESTERN EXTENSION

On 19 February 2007 the charging zone was extended to the West, to include all of the Royal Borough of Kensington and Chelsea and also areas of the City of Westminster that are not covered by the original scheme. Figure 8.2 shows the old and new charging zones. The limits are Grosvenor Road, Chelsea Embankment, the southbound route of the Earl's Court one way system, the West Cross route, Scrubs Lane, Harrow Road, the Grand Union Canal and the Great Western Railway line. Harrow Road, however, is the main route for diverting traffic.

As with the original scheme there will be no charge on the boundary route around the extended charging zone. There will also be two free corridors:

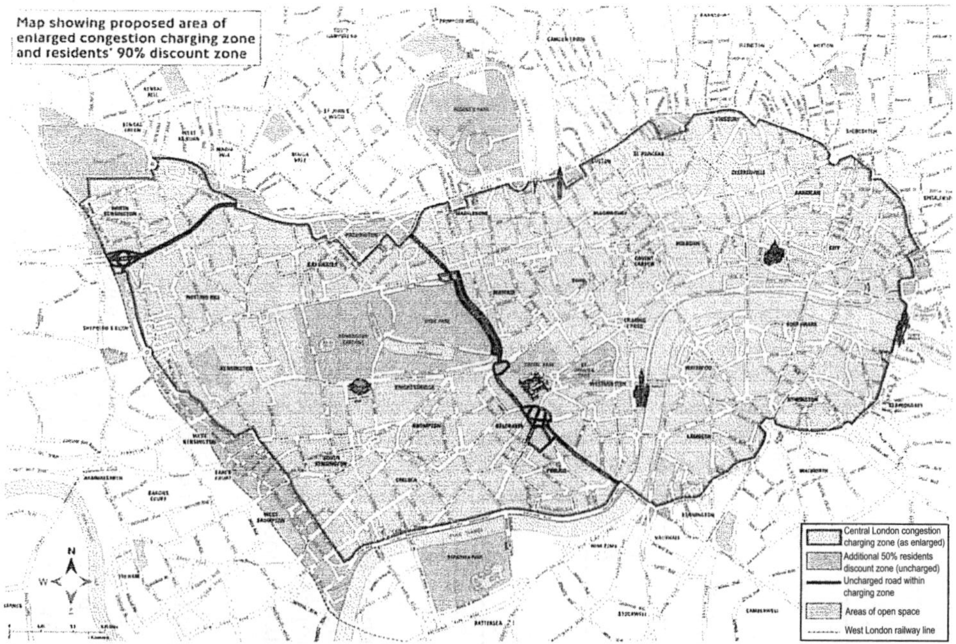

Source: www.london.gov.uk/mayor/congest/docs/zone-map-092005.pdf.

Figure 8.2 Map of the original congestion charging zone and the western extension

one north to south through the proposed extended zone, which is the boundary of the original charging zone, and another one north-west of the zone, east to west, as the diversion route would have been too long for drivers just wanting to cross that segment of the Westway A40.

The hours of operation will be shortened to finish at 6:00 pm and the same exemptions and discounts will apply, with the addition of an extended residents' discount zone. This zone affects some areas around the original charging zone as well. The extended residents' discount zone in the original charging zone came into force in December 2005. The areas where the discount (will) apply are the shaded areas just outside the limit of the zone in Figure 8.2.

6.1 Impacts of the Western extension

The impacts on traffic in the extension will not be as pronounced as in the original charging zone. Indeed almost two-thirds of the traffic currently entering the extension will not be affected by the charge (TfL, 2005b, p. 72, point 6.4.12). The reason for this is that 9 per cent of that traffic are vehicles which are exempt or discounted (residents, disabled, alternative fuel vehicles), 19 per cent are taxis and buses, and 36 per cent are vehicles that go through or come from the original charging zone and have already paid the charge (ibid., p. 72, point 6.4.12). Furthermore, since residents within the extension will be entitled to a 90 per cent discount, they may be attracted on to the roads. By paying the discounted charge they will be able to drive not only in the extension but also in the original charging zone. Some residents who currently do not drive may start driving, including those who initially made alternative arrangements after the original scheme was first introduced (ibid., p. 72, points 6.4.11 and 6.4.12). It should also be mentioned that there is a greater proportion of car travel by residents in the extension than there is in the original zone and therefore a higher proportion of households will be able to take advantage of a residents' discount. The number of cars registered for a resident discount could increase by more than 150 per cent with the extension (TfL, 2004b, p. 3).

Notwithstanding all that, a reduction in vehicle-kilometres of between 10 and 14 per cent within the extension is expected and average speeds are projected to increase by between 10 and 14 per cent (TfL, 2005b, p. 72, point 6.4.10).

The extension, however, will also cause an increase in vehicle-kilometres in the original charging zone of roughly 2 per cent, mainly because, as explained above, residents will be priced on to the roads. As a result of this, average speeds in the original charging zone are expected to decrease by 2 per cent (ibid., p. 74, points 6.4.17 and 6.4.19).

Since the end time will be put forward to 6:00 pm, inbound traffic to the enlarged zone between 6.00 pm and 6.30 pm will increase to pre-charging levels.

7. CONCLUSIONS

The only aim of the LCCS was 'to reduce traffic congestion in and around the charging zone' (TfL, 2004a, p. 7). This aim has indeed been achieved. The London congestion charge is not a first-best (Pigouvian) charge that internalises the marginal congestion cost exactly. It is rather a practical, unsophisticated charge, equal for all vehicle types, despite their different congestive effects. It does not vary in time or location, except for the fact that it applies in a specific area between 7:00 am and 6:30 pm.

The generalized cost elasticities of demand in the first year of operation varied from –0.53 to –0.96, according to vehicle type. The estimated GC elasticity of demand for trips by taxi is very high because it picks up the transfer from private car to taxi.

When the charges of £5 and £8 are compared with the area MCC it can be seen that cars are overcharged. Heavy goods vehicles were/are undercharged with either, and LGVs were undercharged when they paid £5 but are now probably paying just about the MCC they impose, if not slightly more.

The extension to the west will reduce traffic although the effect will be relatively smaller when compared to the effects experienced in the original zone. The reason for this is that two-thirds of vehicles entering the extension would be unaffected, and some residents of the extension would be priced on to the roads.

Although expensive to operate, the LCCS has reduced traffic and increased speeds. The investment in public buses in London prior to the implementation of the scheme should not be overlooked by any other town or city considering the introduction of any kind of congestion charging.

NOTES

* The author is grateful to Transport for London for provision of data and to the Rees Jeffrey's Fund for financial support. Any errors are the author's sole responsibility.

1. This is computed as the percentage change in trips by taxi divided by the percentage change in the GC of trips by car, using the numbers from Table 8.8.
2. PCU is a measure of the relative disruption that different vehicle types impose on the network. A car has a PCU rating of 1, a van has a PCU rating of 1.5, a lorry has a PCU rating of 2.5 or 3, a bus has a PCU rating of 2.5, a motorcycle has a PCU rating of 0.5,

and a pedal cycle has a PCU rating of 0.2. In the US, passenger car equivalents (PCEs) are used instead. The meaning, however, is the same.

3. Since it is an area rather than a link that we are considering, we need to talk about traffic volume rather than traffic flow.

REFERENCES

Department for Transport (2004), *Transport Analysis Guidance* (Webtag), Values of Time and Operating Costs (TAG Unit 3.5.6), December, http://www.webtag.org.uk/webdocuments/3_Expert/5_Economy_Objective/3.5.6.htm.

Dodgson, J., J. Young, and J. van der Veer (2002), *Paying for Road Use, Technical Report*, Report to the Commission for Integrated Transport, National Economic Research Associates (NERA), London, February, http://www.cfit.gov.uk/research/pfru/pdf/pfru-tech.pdf.

London Sustainable Distribution Partnership (LSDP) Freight Plan Working Group (2005), *Strategic Choices for Freight in London, Pre-read for London Sustainable Distribution Partnership meeting. Appendix B: Freight and Servicing Patterns, trends and forecasts*. http://www.tfl.gov.uk/tfl/downloads/pdf/Freight-Plan-sc-app-bc.pdf.

Transport for London (2003a), *Impacts Monitoring Programme: First Annual Report*, Transport for London, London, June, http://www.tfl.gov.uk/tfl/cclondon/cc_monitoring-1st-report.shtml.

Transport for London (2003b), *Congestion Charging: Six Months On*, Transport for London, London, October, http://www.tfl.gov.uk/tfl/downloads/pdf/congestion-charging/cc-6monthson.pdf.

Transport for London (2003c), *London Travel Report 2003*, http://www.tfl.gov.uk/tfl/pdfdocs/ltr/london-travel-report-2003.pdf.

Transport for London (2004a), *Congestion Charging Central London – Impacts Monitoring: Second Annual Report*, April, http://www.tfl.gov.uk/tfl/cclondon/cc_monitoring-2nd-report.shtml.

Transport for London (2004b), *Report to the Mayor – Annex E: Economic Assessment*, July, http://www.tfl.gov.uk/tfl/cclondon/cc_report_mayor 2005.shtml.

Transport for London (2005a), *Congestion Charging Central London – Impacts Monitoring: Third Annual Report*, April, http://www.tfl.gov.uk/tfl/cclondon/pdfs/ThirdAnnualReportFinal.pdf.

Transport for London (2005b), *Proposed Western Extension of the Central London Congestion Charging Scheme*, Report to the Mayor following consultation with stakeholders, businesses, other organisations and the public, September, http://www.tfl.gov.uk/tfl/cc-ex/tfl-report.shtml.

Transport for London (2006), *Central London Congestion Charging – Impacts Monitoring: Fourth Annual Report*, June, http://www.tfl.gov.uk/tfl/cclondon/pdfs/FourthAnnualReportFinal.pdf.

9. The Big Smoke: congestion charging and the environment

David Banister

1 INTRODUCTION

> Like most other large cities around the world, London experiences high levels of air pollution. This is not a new phenomenon. London has long been popularly referred to as the 'Big Smoke'. (Mayor of London, 2002)

Most of the debate over congestion charging in London has been focused on the reductions in traffic and the savings in travel time, with little attention being paid to the environmental issues. One of the main benefits from the congestion charging scheme has been the improvements in air quality, reductions in noise and accidents in the central area. The reductions in the amount of traffic, particularly in central London, have been one of the main aims of the mayor's Transport Strategy (GLA, 2001). This is seen as the best means to reduce environmental pollution, the use of carbon-based fuels and to improve air quality by tackling the problem at source. This has been part of the rationale behind the heavy investment in the public transport network, congestion charging and the use of appropriate planning and other demand management methods to facilitate such a change. In addition to promoting behavioural change, there is also pressure to improve the quality of the vehicle fleet operating in London to reduce emissions and fuel use through technological innovation.

Traditionally, London has had a very different pattern of travel and modal shares from other parts of Great Britain (Table 9.1). Londoners travel less than the average Briton (about 80 per cent of the national figure), they walk more (+7 per cent), and they make much more use of local bus services (+20 per cent) and other public transport (+96 per cent – this is mainly underground, and local rail). They make substantially less use of the car, both as a driver (60 per cent of the national average) and as a passenger (70 per cent of the national average). In addition, 36 per cent of Londoners do not own a car, even though income levels are higher than elsewhere in Great Britain.

Table 9.1 Travel in distance (km) per person per year (1999–2001)

	Walk	Car driver	Car passenger	Other private	Local bus	Other public	Total (weighted average)
London	349	3,402	2,268	174	523	1,919	8,723
Great Britain	327	5,668	3,270	327	436	981	10,900

Source: DfT (2006).

This chapter brings together the evidence of the impacts of congestion charging on the environment in the central London zone. It was difficult to carry out such an analysis before implementation, as the exact impacts of the scheme on traffic volumes, speeds, and stop and start driving were unclear. But there is now two years of after data that allows such analysis, even though these changes also include improvements in the vehicle stock. The chapter also looks at the likely impacts of the western extension of the London congestion charging zone, and the proposals for a low emissions zone (LEZ) covering the whole of Greater London. It is concluded that there is substantial potential for improvement in the environmental quality of London arising from the implementation of innovative charging systems and that these benefits should be included as part of the scheme assessment. The levels of all charges should also relate to the emissions profiles of the individual vehicles.

2 THE LONDON CONGESTION CHARGING ZONE

In the original debate over the implementation of congestion charging, the environmental issues were not seen as being an important consideration. Yet some of the main benefits from congestion charging have been environmental. In the original Road Congestion Charging Options for London (ROCOL) (GOL, 2000) report that reviewed the options for congestion charging in London, there were very few mentions of the environment. There was one short paragraph under the benefits which said 'there could be environmental and amenity benefits from the reduced traffic volumes but much would depend on how the benefits of reduced traffic were allocated. Emissions of CO_2 [carbon dioxide] would be reduced, but there would be no significant improvement in local air quality' (Para 5.10.6). The main conclusion on the environmental effects of congestion charging from the option adopted was that there would be no significant impact on nitrogen dioxide (NO_2) or particulate matter

(PM_{10}) concentrations, and that up to 3 per cent reduction in traffic-based CO_2 might be possible.

The headline figures immediately after implementation (2003) are that there have been significant reductions of 16 per cent in both nitrogen oxides (NO_x) and PM_{10} emissions (on the annual average day). About 75 per cent of this was due to the lower traffic volumes and more even travel, but the remaining 25 per cent was accounted for by improvements in vehicle technology. The effects of stop–start driving on emissions seem to have been underestimated, and the effects of reduced queuing times at junctions have also contributed to these benefits. Beevers and Carslaw (2005) suggest that estimation of these benefits in advance of the implementation of congestion charging was difficult, as they were speed related, but now there is more data on traffic, vehicle mix, speeds and emissions. Outside the congestion charging zone on the inner ring road, there was some modest (about 1 per cent) increase in these two pollutants. Reductions in CO_2 emissions are estimated at 20 per cent, and this can be directly attributed to the lower levels of fuel use (−19 per cent) in the congestion charging zone (TfL, 2005a).

In the second year of congestion charging, emissions have increased marginally for NO_x (+0.6 per cent), but decreased for PM_{10} (−0.7 per cent). But when the recent trends are compared with levels of NO_x concentrations since 1998, there is a general decline across the whole of London, and this is also reflected in the congestion charging zone (running annual mean concentrations $\mu g.m^{-3}$). There is no evidence of an identifiable 'congestion charging impact' for NO_x (TfL, 2005b, p. 10). This conclusion contrasts with the increase in NO_2 levels, particularly along the Marylebone Road, of between 8–9 per cent, and the background levels are much higher (50 per cent) than the target level of $40\mu g.m^{-3}$. For PM_{10} concentrations, the picture is better, with levels close to the national air quality objectives (for exceedence days).

In the latest monitoring report (TfL, 2006b), these figures have been revised with greater reliability being placed on them. The early figures were only for one year (2003), and reflected the average day over the whole year, whether or not the charging regime was in operation. They covered only the major roads, with tailpipe measurements related to an old database for vehicles in London. The revised figures (Table 9.2) have speed flow data for all roads, with a new national vehicle stock database and information on particulates for tyres and brakes, as well as the tailpipes.

For NO_x emissions, the reductions in traffic volume gave about 10 per cent of the savings (2002–03), while traffic speed increases accounted for about 49 per cent of the savings and the vehicle stock change the remaining 41 per cent. For PM_{10}, the respective proportions were 5, 35 and 60 per cent, and for CO_2 emissions, they were 51, 45 and 4 per cent. These figures are all less

Table 9.2 Emissions data for vehicles in the London charging zone (% change)

	2002–2003				2003–2004
	Traffic volume change	Traffic speed change	Vehicle stock change	Total change	Technological change
NO_x	−1.4	−6.5	−5.5	−13.4	−5.2
PM_{10}	−0.8	−5.5	−9.2	−15.5	−6.9
CO_2	−8.4	−7.3	−0.7	−16.4	−0.9

Note: NO_x and PM_{10} changes are based on measurement, but the CO_2 change is calculated from fuel savings.

Source: Based on TfL (2006b, Table 6.3).

than those originally estimated using the poorer quality data. There seem to be further reductions in emissions (2003–04) resulting mainly from further improvements in vehicle technologies (as volume levels and speeds have not changed substantially since the first year of the scheme).

Noise monitoring has been carried out, but only on a limited set of sites (50 sites over four years, with three within the congestion charging zone and two on the inner ring road or approaching it). Basically, no change has been identified and so it is not considered in this chapter. London does have an ambient noise strategy (GLA, 2004a), where the intention is to reduce noise through better management of the transport system, and better planning and design of buildings, and a series of London noise maps are being produced to identify noisy and quiet areas.

3 THE WESTERN EXTENSION

The congestion charging zone doubled in size in February 2007 with its extension to the west of the central area of London (Figures 9.1 and 9.2). This was confirmed by the Variation Order issued on 26 September 2005 by the mayor of London. There will be more residents with discounts and traffic levels may also increase in the existing zone, as those with discounts make more use of their cars. In the consultation, there were some 100,000 responses, mainly negative, and in polls carried out by the London Borough of Westminster, 66 per cent of residents and 84 per cent of business were against the extension. But the mayor confirmed the draft order, giving little away except to reduce the end of the charging period. The extended scheme

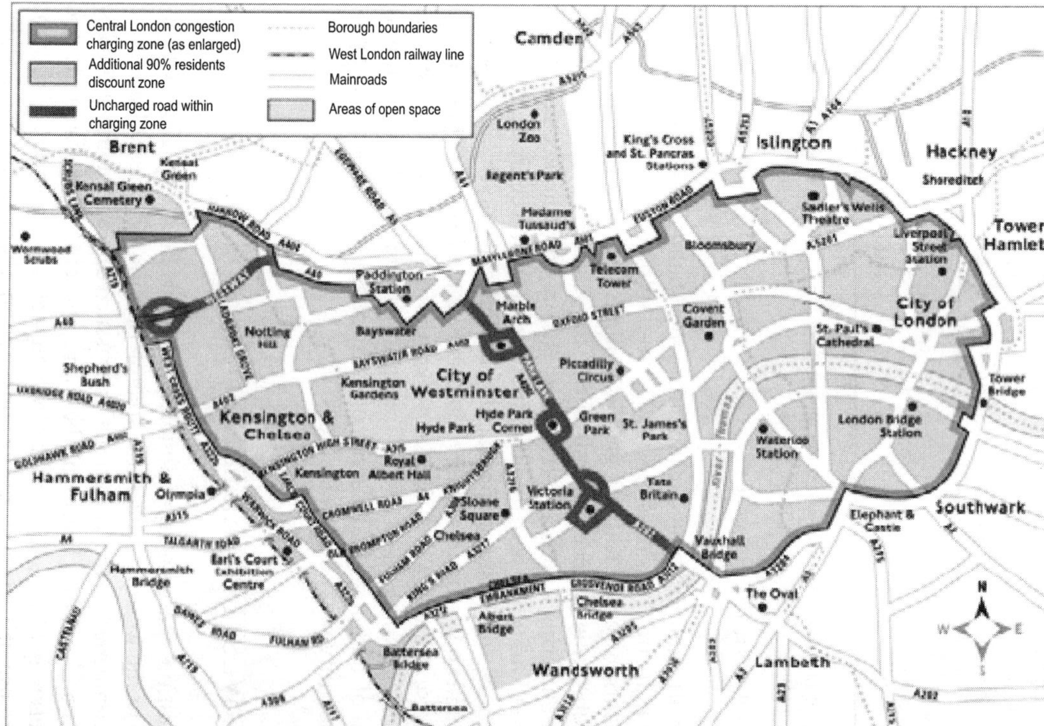

Source: TfL (2006b, Figure 11.2).

Figure 9.1 *The old and new congestion charging zones in London*

Note: The low emissions zone covers all of Greater London shown above.

Source: TfL (2006b, Figure 11.3).

Figure 9.2 Map comparing extended London congestion charging area with Greater London

will operate from 07:00 to 18:00 on weekdays. The net revenues from the extension will be limited (about £10 million per annum), but traffic levels are expected to be cut by between 5 and 10 per cent and speeds will increase by about 20 per cent.

In terms of the environment, less than 1 per cent of the responses to the public consultation related to this topic. But the environmental case for the western extension should be strong and the benefits of improved air quality should reflect the same orders of magnitude as those already experienced in the existing congestion charging zone. Nevertheless, there were many supporters and objectors to the expected environmental impacts (TfL, 2005a, Annex B, Theme 18 Environmental Impacts, pp. 335–40), and Table 9.3 summarises some of the main points raised.

The responses (over 700) on environmental issues are both positive and negative, with a ratio of positive (25 per cent) to negative (75 per cent)

Table 9.3 Responses to environmental issues raised in the consultation on the western extension of the congestion charging zone

Issue: Proposals will improve air quality pollution	
Greenpeace Lancaster Youth Centre Royal Brompton and Harefield Hospital 162 Members of the Public	Response: 'Comparable reductions are expected from a Western Extension, though there would be a small increase in the existing zone. On the boundary, there has been effectively no overall change in emissions from traffic as a result of the scheme'.
Issue: Proposals will reduce air quality/pollution	
RB Kensington and Chelsea Abingdon Ward Councillor Earls Court Neighbourhood Associations Kensington Red Route Action Group Lillington and Longmoore Gardens Residents' Association Morpeth Mansions and Ashley Gardens Residents' Association North Fulham NDC Partnership West London Residents' Association Paddington Residents' Active Concerns on Transport 399 Members of the Public	Response: 'TfL's projections of traffic changes on the western boundary route as a result of charging indicate some increases in traffic, balanced by reductions on radial routes into and out of the extension area. Experience of the existing zone and the Inner Ring Road, where comparable changes in traffic conditions have occurred, suggests that the net impact of charging on air quality would be negligible'.
Issue: Proposals will reduce air quality/pollution – diverted traffic	
Corporation of London Councillors for Tachbrook Ward Councillor for Vincent Square Ward Friends of the Earth West London Lillington and Longmoore Gardens Residents' Association	Response: 'No overall increase in flows along the Inner Ring Road is expected. While some new traffic movements may be attracted to such a "free" route, this would be offset by an overall reduction in vehicles travelling to and from a Western Extended charging zone'.
Issue: Infrastructure/street clutter	
RB Kensington and Chelsea Paddington Business Improvement District	Response: 'TfL has met with the affected Boroughs to discuss the infrastructure that

Table 9.3 (continued)

158 Members of the Public	would be associated with the proposed extension. If the Mayor decided to confirm the Variation Order TfL would continue to liaise with the affected Boroughs and have regard to their streetscape guidance'.

Note: There is a statutory requirement to give a response to issues raised in the public consultation.

Source: Based on TfL (2005b).

reflecting the expectations that people (and organisations) are more likely to be against any proposal than for it. The representations come from local authorities and several local associations, but there is little comment from business on the environmental issues. The overwhelming number of responses came from individuals (719 out of 738). The experience from the existing congestion charging zone has been used in the responses from TfL to the concerns from (and support for) the consultation on the western extension, and the recommendation from TfL to the mayor was that 'no change to the Variation Order is appropriate in light of the representations made on the environment' (TfL, 2005b).

It seems that even with the experience of the original congestion charging zone, there has been little weight placed on the environmental benefits of the western extension. Even the responses made by TfL to the comments raised in the consultation seem cautious. The interpretation must be that there is still some uncertainty over the environmental benefits, yet it is the same issues that feature in the next section covering the introduction of the LEZ in London. There seems to be a strong emphasis on the congestion reduction potential of the western extension rather than the combined power of increasing efficiency and contributing to environmental improvement.

4 LOW EMISSIONS ZONE

There are now plans to create an LEZ to cover the whole of Greater London (by 2008), so that EU air quality objectives can be reached by the target date of 2010.[1] This covers all the area shown in Figure 9.2. As noted in the quotation at the beginning of this chapter, the mayor has a strong desire to improve health and quality of life for those people living and working in London (GLA, 2002; TfL, 2006a) The plans do not include cars, but would cover lorries, buses and coaches. Road transport contributes 47 per cent of both PM_{10} and NO_x emissions in London in 2005, and some 1,600 premature deaths are caused by poor air quality, with PM_{10} and NO_x being the main contributors (GLA Economics, 2005). It is estimated that cars account for 39 per cent of all transport-related NO_x emissions and 33 per cent of PM_{10} emissions in London, but they were excluded from the scheme because of the difficulties in and the costs of enforcement (TfL, 2004).

The scheme would operate at all times and it would use a similar technology to that operated in the existing congestion charging zone, which uses automatic number plate recognition technology that matches up heavy goods vehicles (HGVs over 3.5 tonnes) and light goods vehicles (LGVs) with a database. The set-up costs for the basic scheme (HGVs only) would be £9.3 million, with operating costs of £6.4 million per annum and revenues of £3.9 million per annum (Watkiss et al., 2003). The combined scheme (HGVs and LGVs) would be £10.4 million, with operating costs of £7 million per annum and revenues of £4.3 million per annum. So the LEZ would not generate revenues, but it would have substantial health benefits estimated at £26 million in the first year (Watkiss et al., 2003). These benefits would increase as the emission restrictions become tighter to £40 million in 2010.

The emissions criteria would apply to about 37 per cent of the most polluting HGVs entering London. These vehicles account for 34 per cent of all NO_x on London roads and 25 per cent of all PM_{10} in 2005 (ibid.). No buses would be affected as they will already conform to the emission criteria by 2005, but about 56 per cent of all coaches (5,800) entering London would be affected. All HGVs (and LGVs), buses and coaches would have to pay a charge (£200 per day) to enter any part of London, if they exceeded the defined initial levels of pollution for PM_{10} (2008) and for tighter levels of PM_{10} and NO_x (2010).[2] The estimated costs to industry of compliance would be about £65 million (a mid-estimate of the £37 – £95 million range calculated in Watkiss et al.).

Taxis are responsible for 24 per cent of PM_{10} and 12 per cent of NO_x emissions from transport in central London. The taxi emissions strategy aims to reduce emissions by 37 per cent from the fleet of 20,000 London taxis (to meet Euro III standards by July 2008). Since April 2005 an environmental surcharge (20p per trip) has been made on all taxi journeys to fund the investment needed (GLA, 2004b). Some companies are already taking action, as can be seen from Box 9.1. Taxis are not being included in the initial LEZ proposals for 2008, but there is an option to include them in the second phase from 2010 when the Euro IV standards will be imposed.

BOX 9.1 TAXI EMISSIONS STRATEGY

Radio taxis – operate in London and run 2,500 cabs and 80 executive cars, producing 24,000 tCO_2 per year. They have adopted a carbon offset programme for £120,000 per annum (2006) to fund renewable energy projects overseas in Bulgaria and Sri Lanka (80%) and forestry projects in the UK and Germany (20%). Radio taxis claim to be the world's first carbon-neutral taxi company, and this policy has added £1.2 million to its revenues since it started in January 2005.

Taxis – all taxis to be Euro III for PM_{10} and NOx by mid-2008.

In addition to imposing increasingly strict constraints on emissions, the intention is to encourage a wider use of alternative fuels in all forms of commercial transport to again reduce levels of local and global emissions. London now has one of the cleanest bus fleets in the world, as all new buses match Euro III standards, with 93 per cent also having particulate traps that reduce emissions of PM_{10}, CO and HC by 90 per cent (meeting Euro IV standards). Innovations in new fuels for buses include the use of hydrogen and hybrid technologies (Box 9.2).

BOX 9.2 INNOVATIONS IN NEW FUELS FOR BUSES

Hybrid buses – these have a 336v battery pack to provide power via a 120kW electric motor. The battery is recharged by a 1.9 litre diesel Euro IV engine and additional power is provided by regenerative braking. These hybrid buses reduce NO_x emissions by 89%, CO emissions by 83%, fuel use by 40%, CO_2 emissions by 38% and perceived noise by 30% (from 78dB to 74dB).

Hydrogen buses – hydrogen is stored on the roof of the bus and used to power a fuel cell to produce power (electricity) and water vapour. There is no pollution, but the sourcing of the hydrogen can produce pollution if carbon based energy sources are used. There are three such buses operating in London as part of the nine-city EU CUTE project (http://www.fuel-cell-bus-club.com).

The LEZ tackles the local emission problems (NO_x and PM_{10}) but not the global issues (CO_2), as the intention is not to reduce the levels of traffic, but to improve local air quality. Congestion charging tackles both the local and global air quality problems as it cleans up vehicles and reduces the number of cars in central London.

5 TRANSPORT IN LONDON

The environment features high on the list of priorities in terms of encouraging people to switch from their cars to rail, bus and green modes of transport. Perhaps one of the greatest achievements of the mayor over the last five years has been the renaissance of the bus in London, partly triggered by congestion charging and the substantial investment in new vehicles. Buses have shown a remarkable comeback, with an increase of 44 per cent in patronage over five years (2000–05), while elsewhere in Great Britain it has declined by 5 per cent over the same period. Over 40 per cent of all bus trips made in Great Britain are in London (Table 9.4). In London there has been heavy investment in new buses, bringing the total fleet to 8,000 buses – there are now 4,500 new buses in London, of which 3,000 are replacements. The level of expenditure per head of population on buses is two to three times the levels elsewhere in the country. Congestion charging has helped improve bus speeds and reliability. It should be noted that only 25 per cent of all bus passenger trips involve the central area, where congestion charging is in operation, and 29 per cent of bus passenger trips are wholly in the suburban zones.

There has been a substantial increase in cycling in London (2000–05), albeit from low initial levels. The 50 per cent increase (TfL forecast was 22 per cent) reflects investment in cycling schemes (now £14 million per year), and the availability of a cycling network of 420 kilometres. Investment is also taking place in walking schemes (£6 million per year), including new

Table 9.4 London's dominance of the bus market

	Trips (m) 1999/2000	Trips (m) 2004/05	Bus km run (m) 2004/05
London	1,296 (30%)	1,872 (41%)	470 (18%)
Rest of Great Britain	2,982 (70%)	2,827 (59%)	2,144 (72%)
Overall	4,278	4,609	2,614

Source: http://www.dft.gov.uk/stellent/groups/dft_transstats/documents/page/dft_transstats_610507.pdf.

pedestrian crossings, wider pavements and better street lighting. Part of these initiatives has been funded out of the new revenues from congestion charging, which have also been used for improving safety. The number of people killed or seriously injured on London's roads in 2004/05 fell by 19.3 per cent on the previous year (5,164 to 4,169, including a drop in fatalities from 286 to 216).

The range of transport innovations mentioned here offer substantial potential for improving air quality in London, and together with considerable investment in cleaner public modes of transport and networks for walkers and cyclists mean that the 'Big Smoke' is now a distant memory. The air quality issue is seen as a central element of the London Plan, and transport has been targeted as one of the main areas for improvement. But it seems strange that the same issues have not featured in the congestion charging debate which has consistently focused on the congestion reduction potential, rather than its contribution to a better quality environment in London.

6 TWO PROPOSALS

London is one of the few great cities of the world that is actively tackling the environmental problems created by traffic.[3] A range of measures have been introduced, ranging from the congestion charging scheme and the LEZ to heavy investment in clean public transport and green modes of transport, and a reassessment about the best use of available road space. However, the prominence given to the environmental costs has been limited, and it is not clear why this has been the case. One explanation could be that the data and the methods were not sufficiently robust to analyse the environmental implications of radical transport policies. But we now have substantial *ex post* data, which suggests that air quality has improved considerably inside the congestion charging zone and that similar results would be expected in the western extension. Certainly, the LEZ for the whole of Greater London has substantial air quality and health benefits.

As part of the thinking about implementation of the new wider congestion charging zone, two proposals should be considered. First, discounts should be given to all clean vehicles, including the already exempt vehicles such as hybrid vehicles, electric vehicles and those using alternative fuels (Table 9.5). The charge should be related to the pollution profile of the vehicle. That information is already available on the database in terms of the registration details of the vehicle. To some extent the differential rate of the VED (Vehicle Excise Duty) reflects the CO_2 emission levels and fuel

Table 9.5 Congestion charging: discounts

2004 – all the discounts are for 100% except for residents	Registrations	Usage on a typical day
Blue badge – disabled drivers	119,600	8,000
Residents – 90% discount	29,200	18,000
Alternative fuelled vehicles	4,500	}
Vehicles with 9 or more seats	11,500	} 2,500
Other	1,300	}

Note: 'Other' includes: some public service vehicles, breakdown vehicles, some National Health Service employees' and patients' vehicles. There are also exempt vehicles – buses, taxis, motorcycles, bicycles and emergency service vehicles – these account for 7,300 vehicles, most of which are used every day in the charging zone. There has been an increase of 21% in the low emission vehicles registered from March 2004 (4,554) to March 2005 (5,489) (Mayor of London, 2006, p. 39).

Source: http://www.tfl.gov.uk/tfl/downloads/pdf/congestion-charging/Final-consolidated-scheme-order-with-vos 2004-4and5.pdf.

types, but charges for entering the extended congestion charging zone could be directly related to the CO_2 emission levels (and fuel type) to reflect more accurately the environmental pollution created.

At present a super clean small car (for example, Citroen C1 or Toyota Aygo) with a CO_2 emission level of 109 g/km is charged as much as a more polluting vehicle with a CO_2 emission level of 3–4 times as much (for example, Audi A8 and the Range Rover V8 both emit about 355 g/km). Even the Toyota Prius, which is an exempt vehicle (hybrid) has a CO_2 emission level of 104 g/km. There is concern over the quality of air in London. In the consultation on the air quality strategy, 71 per cent of respondents believed that air pollution was a problem, and 43 per cent thought that it had got worse over the last five years, when in fact it had got better (GLA, 2002).

The proposal would relate the congestion charge directly to the CO_2 emissions profile of the vehicle, with a zero charge for vehicles below 120 g/km, rising to £4 for those within the range 120–150 g/km, and £8 for those between 150 and 165 g/km. For those cars in the 165–185 g/km band, the charge would be £12, for those in the 185–225 g/km band the charge would be £16, and for those cars in excess of 225 g/km, the charge would be £20.[4] These charges could apply to all vehicles. The CO_2 information is available on the national DVLA (Driver and Vehicle Licensing Agency) for all vehicles registered after March 2001 (this includes 43 per cent of the total car stock). Of the vehicles registered since 2001[5], only 3 per cent would qualify for the zero rate for the congestion charge, with 28.5 and 23 per cent coming into the next two bands, and 16, 22 and 7.5 per cent in the highest three bands. This distribution of values would bring the average price of the congestion charge to about £10, or 25 per cent above the current levels. To reflect other emissions, these charges may have to be adjusted for the fuel used in the vehicle. Currently the national car stock is 20 per cent diesel and 80 per cent petrol, but since 2001, 43 per cent of all vehicle sales have been diesel (DfT, 2006, Tables 5 and 6). It is likely that more revenue would be raised than is the case currently (about £121 million in 2005–06), but over half the vehicles would pay no more than at present.

If environmental and congestion objectives are to gain in importance, the discounts to the residents in the existing and extended congestion charging zones should be reduced, with incentives being given to them (and others) to drive more fuel-efficient and 'cleaner' vehicles in the zones. Even exempt vehicles like taxis could be given clear incentives to reduce their emissions so that the more-polluting vehicles would be charged for entry into the extended congestion charging zone. At present, taxis produce over 20 per cent of CO_2 emissions from public transport, but only account for 3.6 per cent of public transport trips (Mayor of London, 2006, p. 23). This would lead to both environmental and congestion benefits.

The second proposal is that the environmental effects of transport should be fully reflected in the cost–benefit analysis (CBA) on all congestion charging schemes, as the potential benefits of such improvements are undervalued. Reductions in CO_2 emissions can now be valued, as a market has been created in the EU with the Emissions Trading Scheme (ETS), which has been in operation since January 2005 (covering 12,000 large industrial plants in the EU). The second round of negotiations will include transport (to run from 2008 to 2012). At present the price of a tonne of CO_2 is about €20 (8 February 2008, http://www.carbonpool.eu/), and this means that some of the environmental costs can be included in the CBA. The price has been volatile since figures were released on levels of carbon emissions from those industries within the scheme, where it was found that target levels for reductions had been set too low. But stability has returned to the ETS market; firms are now pricing in their CO_2 allowances, and it is affecting their long-term decisions (EC DG Environment, 2005). Similarly, there are estimates of the hospital costs and premature deaths in terms of lost output that can be attributed to poor air quality. In London, these figures are estimated to be 1,600 premature deaths and 1,500 respiratory hospital admissions per year that are attributable to air pollution (GLA Economics, 2005).

Air quality is improving in London (Table 9.6). The level of transport-related CO_2 emissions is about $1tCO_2$ per person in London, with the corresponding figure for other parts of England over double that level

Table 9.6 Emission levels in London and England in 2003 (tonnes)

	Transport CO_2 emissions	Motor vehicles NO_x emissions	Motor vehicles PM_{10} emissions
London	7,367,728	30,618	1,618
England (including London)	96,189,687	384,331	21,097

Source: GLA Economics (2005).

(2.1 tCO_2 per person). Levels have increased by only 1 per cent as compared with the increase nationally of 11 per cent (1993–2003). This reduction is due to a reduction in emissions from cars and taxis (–8 per cent as compared with an increase in England of 4 per cent), and a further reduction in underground emissions, but an increase of 26 per cent in bus emissions[6] (ibid.).

Even though the overall levels of air pollution are lower than those found elsewhere in England, the concentration of London pollution within a relatively small area (1,580 sq km) causes problems. The levels are 3–6 times as high as those found elsewhere in England (Table 9.7). Figures can be placed on the transport CO_2 emissions through the ETS carbon market, initially through the market rate created for other industries but eventually to the transport sector itself. For local air pollutants (NO_x and PM_{10}), values can be obtained from the EU BeTa database on the costs which assess the implications of air quality on health, including respiratory diseases, hospital admissions, and premature death (for more details see the EU database benefits tables (BeTa): http://europa.eu.int/comm/environment/enveco/air/betaec02aforprinting.pdf). The health costs are likely to be higher in London where the concentration levels are high (Woodcock et al., 2007).

GLA Economics (2005) have used the figures from AEA Technology Environment (2004) and the Government Economic Service (2002) to give estimates of the average damage costs attributed to the three major transport pollutants (CO_2, NO_x and PM_{10}) in 2003 prices £ per tonne. The values derived all have uncertainty attached to them, but they provide a basis for some estimates of the monetary costs of emissions. For carbon the value used is £70 per tonne (equivalent to £19 or €27 per tonne of CO_2, somewhat higher than the current value of £15 or €20). For NO_x and PM_{10} the values derived are £500 per tonne and £25,000 per tonne, respectively.[7] The GLA Economics (2005) study then uses these values to calculate the damage costs associated with transport emissions in London (Tables 9.6 and 9.7). These estimated costs for PM_{10}, NO_x and CO_2 range from

Table 9.7 *Concentrations of pollutants in London and other parts of England*

2003	Area sq km	Population	2003 GVA £m	NO_x tonnes	PM_{10} tonnes
London	1,580	7,387,868	155,069	30,618	1,618
Other regions	128,293	42,467,872	659,506	353,713	19,479
England	129,873	49,855,740	814,575	384,331	21,097
London %	1.2%	14.8%	19.0%	8.0%	7.7%

Concentrations	Grams/sq km		Grams/capita		Grams/£1m GVA	
	NO_x	PM_{10}	NO_x	PM_{10}	NO_x	PM_{10}
London	19,378,481	1,024,051	4.14	0.22	197.44	10.43
Other regions	2,757,071	151,832	8.32	0.46	536.33	29.54
England	2,959,283	162,443	7.71	0.42	471.82	25.90
Ratio	6.5x	6.3x	0.54x	0.53x	0.42x	0.40x

Note: The ratio is the London/England values.

Source: Based on data in GLA Economics (2005).

£167 million to £305 million per annum, but they need to be set against the total transport gross value added (GVA) of £10.4 billion in 2003 (at 2001 prices).

There seems to be a considerable potential for including environmental factors in the appraisal of innovative transport schemes in London. For both the original congestion charging scheme and the western extension, it has been assumed that the environmental impacts are small scale, and it is only with the London-wide LEZ that the environmental issues have come to prominence. But even here it is the local air quality factors that have featured and not the carbon emissions associated with the reduction in traffic. It is only when the local and global emissions are combined that the real benefits can be observed in terms of cleaner air, better quality local environment (less traffic), and potentially considerable health benefits.

NOTES

1. Consultation ran from 30 January 2006 to 24 April 2006.
2. The LEZ would define emission standards that certain categories of vehicle would have to comply with in order to travel in London without charge. The standards would be based on euro standards (see Appendix 9A1). These are emission standards that vehicles must be manufactured to by a certain date. TfL proposes the following emission standards for the proposed LEZ (TfL, 2006a):

   - For 2008, a standard of Euro III for small particulates (PM_{10}) for HGVs, coaches and buses.
   - For 2010, a standard of Euro IV for small particulates (PM_{10}). In the event that NO_x certification capability is available, a standard of Euro IV for PM_{10} and NO_x. This would apply to HGVs, coaches and buses, with an option of extending the scheme to LGVs and taxis.

3. LEZs have been operating in Sweden since 1996 in Gothenburg, Malmö and Stockholm, with Lund joining in 2002. Tokyo introduced an LEZ in 2003 and Berlin has also recently introduced an LEZ in the central city area.
4. Since this chapter was written, the Mayor has announced that from 27th October 2008 he will introduce three levels of charging – vehicles with under 120g CO_2/km will pay no charge to enter the congestion charging zone, vehicles between 120–225g CO_2/km will pay £8 and those over 225g Co_2/km will pay a daily charge of £25.
5. These estimates assume that the stock of vehicles in central London reflect the car stock nationally for 2005 (DfT, 2006).
6. The headline figure for buses reflects the increase of 43 per cent in bus passenger-km, with passengers per bus increasing by 7 per cent and vehicles-km by 23 per cent. There is a reduction of 15 per cent in emissions per passenger-km, but the net increase is 26 per cent. On the underground, there is a 21 per cent reduction in emissions per passenger-km, but this must be balanced against an increase of 25 per cent in passenger-km, giving a net figure of 10 per cent.
7. These are midpoints of ranges – NO_x ranges from £173 to £1,098 per tonne and PM_{10} from £9,492 to £59,230 per tonne discounted at 1.5 per cent as advised by the Treasury. The range used for carbon emissions is £35 to £140 per tonne. The GLA's economy–environment model provides emission factors for NO_x, PM_{10} and CO_2 per kilometre travelled by

various types of vehicles. Total vehicle-kilometres are then used to estimate city-wide emissions: www.london.gov.uk/mayor/economic_unit/docs/Enviroseemodelfinalreport.pdf.

REFERENCES

AEA Technology Environment (2004), 'An Evaluation of the Air quality Strategy', Final Report to Defra, ED50232, December, http://www.defra.gov.uk/environment/airquality/strategy/evaluation/pdf/exec-summary.pdf.

ApSimon, H. (2005), 'The air over London', in J. Hunt (ed.), *London's Environment: Prospects for a Sustainable World City*, London: Imperial College Press, pp. 83–98.

Beevers, S.D. and D.C. Carslaw (2005), 'The impact of congestion charging on vehicle emissions in London', *Atmospheric Environment*, **39**(1), 1–5.

Department for Transport (DfT) (2006), *Transport Statistics Bulletin: Vehicle Licensing Statistics 2005*, National Statistics, SB(06)25, May.

European Commission DG for the Environment (2005), 'Review of the EU Emissions Trading Scheme: Survey Highlights', Report produced for DG Environment by McKinsey and Company and Ecofys, November, http://ec.europa.eu/environment/climat/pdf/highlights_ets_en.pdf.

Government Economic Service (2002), 'Estimating the Social Costs of Carbon Emissions', London, WP140.

Government Office for London (GOL) (2000), 'Road Charging Options for London', Independent Working Group of Transport Professionals Report for GOL, The ROCOL Report, The Stationery Office, London, March.

Greater London Authority (GLA) (2001), 'The Mayor's Transport Strategy', GLA, London, July.

Greater London Authority (GLA) (2002), 'Cleaning London's Air: The Mayor's Air Quality Strategy', GLA, London.

Greater London Authority (GLA) (2004a), 'Sounder City: The Mayor's Ambient Noise Strategy', GLA, London.

Greater London Authority (GLA) (2004b), 'Cleaner, greener taxi fleet for London', Press Release 490, December.

Greater London Authority Economics (2005), 'The Environmental Effectiveness of London: Comparing London with other English Regions', GLA Economics, London, June.

Mayor of London (2006), 'Environment Report 2005', Transport for London Group Transport Planning and Policy, February.

Transport for London (TfL) (2004), 'Low Emissions Zone Feasibility Study', Transport for London Report.

Transport for London (TfL) (2005a), 'Central London Congestion Charging: Impacts Monitoring', 3rd Annual Report, April.

Transport for London (TfL) (2005b), 'Report to the Mayor on the proposed western extension', September, http://www.tfl.gov.uk/tfl/cc-ex/pdfs/reports/Annex-B-18-Environmental-Impacts.pdf.

Transport for London (TfL) (2006a), 'Report to the Mayor on Consultation – Draft Transport and Air Quality Strategy Revisions: London Low Emissions Zone', London, January.

Transport for London (TfL) (2006b), 'Central London Congestion Charging: Impacts Monitoring', 4th Annual Report, June.

Watkiss, P. et al. (17 authors) (2003), 'The London LEZ Feasibility Study', Report for the GLA by AEA Technology Environment, July.
Woodcock, J., D. Banister, I. Roberts, A. Prentice and P. Edwards (2007), 'Energy and transport', *The Lancet*, **370**, 1078–88.

APPENDIX 9A1

The EU sets standards for vehicle emissions that all new vehicles must comply with – Euro III standards were introduced in October 2000 and Euro IV standards are effective from October 2005 for new and from October 2006 for all vehicle types. They apply to all HGVs, coaches and buses (Table 9A1.1).

Changes in the engine test cycles have been introduced in the Euro III standard (2000). The old steady-state engine test cycle ECE R-49 has been replaced by two cycles: the European Stationary Cycle (ESC) and the European Transient Cycle (ETC). Smoke opacity is measured on the European Load Response (ELR) test. For the type approval of new vehicles with diesel engines according to the Euro III standard (2000), manufacturers had the choice between either of these tests. For type approval according to the Euro IV (2005) and later limit values and for Enhanced Environmentally friendly Vehicles (EEVs), emissions have to be determined on both the ETC and the ESC/ELR tests.

Emission standards for diesel engines that are tested on the ETC test cycle, as well as for heavy-duty gas engines, are summarised in Table 9A1.2.

Table 9A1.1 EU emission standards for HD diesel engines, g/kWh (smoke in m^{-1})

Tier	Date	Test	CO	HC	NOx	PM	Smoke
Euro I	1992, <85 kW	ECE R-49	4.5	1.1	8.0	0.612	
	1992, >85 kW		4.5	1.1	8.0	0.36	
Euro II	1996.10		4.0	1.1	7.0	0.25	
	1998.10		4.0	1.1	7.0	0.15	
Euro III	1999.10, EEVs only	ESC & ELR	1.5	0.25	2.0	0.02	0.15
	2000.10	ESC & ELR	2.1	0.66	5.0	0.10 0.13*	0.8
Euro IV	2005.10		1.5	0.46	3.5	0.02	0.5
Euro V	2008.10		1.5	0.46	2.0	0.02	0.5

Note: * for engines of less than 0.75 dm^3 swept volume per cylinder and a rated power speed of more than 3,000 min^{-1}.

Table 9A1.2 Emission standards for diesel and gas engines, ETC Test, g/kWh

Tier	Date	Test	CO	NM HC	CH_4[a]	NOx	PM[b]
Euro III	1999.10, EEVs only	ETC	3.0	0.40	0.65	2.0	0.02
	2000.10	ETC	5.45	0.78	1.6	5.0	0.16
							0.21[c]
Euro IV	2005.10		4.0	0.55	1.1	3.5	0.03
Euro V	2008.10		4.0	0.55	1.1	2.0	0.03

Notes:
a. For natural gas engines only.
b. Not applicable for gas fuelled engines at the year 2000 and 2006 stages.
c. For engines of less than 0.75 dm^3 swept volume per cylinder and a rated power speed of more than 3000 min^{-1}.

Table 9A1.3 Emission durability periods

Period[a]	Vehicle Category[b]
100,000 km or 5 years	N1 and M2
200,000 km or 6 years	N2
	N3 ≤ 16 ton
	M3 Class I, Class II, Class A, and Class B ≤ 7.5 ton
500,000 km or 7 years	N3 > 16 ton
	M3 Class III, and Class B > 7.5 ton

Notes:
a. km or year period, whichever is the sooner.
b. N1 are goods vehicles <3.5 tons, N2 are between 3.5 and 12 tons, and N3 are over 12 tons. M1 are for passenger travel with <8 seats, M2 have >8 seats and a weight <5 tons, and M3 have >8 seats and a weight >5 tons. Mass designations (in tons) are 'maximum technical permissible mass'. The classes refer to different weight categories.

Emission Durability

Effective October 2005 for new type approvals and October 2006 for all type approvals, manufacturers should demonstrate that engines comply with the emission limit values for useful life periods which depend on the vehicle category, as shown in Table 9A1.3.

Effective October 2005 for new type approvals and October 2006 for all type approvals, type approvals also require confirmation of the correct operation of the emission control devices during the normal life of the vehicle under normal conditions of use ('conformity of in-service vehicles properly maintained and used'), http://www.dieselnet.com/standards/eu/hd.php.

APPENDIX 9A2

There are nine main air pollutants in the UK Sustainability Strategy, and these are confirmed by the EU's legally binding targets to ensure that emissions are reduced to levels 'at which no or minimal effects on human health is likely to occur'. Targets are set at levels or concentrations covering different time periods (Table (9A2.1).

Note that NO_x is a more complex pollutant in that the oxidised nitrogen is emitted mainly as nitric oxide (NO), with only a small fraction as nitrogen dioxide (NO_2). But NO also combines with ozone (O_3) to form more NO_2. This means that it is very difficult to reduce levels of NO_2 (ApSimon, 2005, p. 86).

PM_{10} are small particles of carbon (soot) emitted mainly from diesel engines with a diameter of less than 10μ. They affect cardiovascular and respiratory systems, and contribute to premature deaths.

Table 9A2.1 Air pollutants in the UK Sustainability Strategy

Pollutant		Target
Seven pollutants need to be addressed at the local level		
Nitrogen dioxide	NO_2	200 $\mu g.m^{-3}$ (105ppb) 1 hour mean (exceed <18 times/year) by end 2005 40 $\mu g.m^{-3}$ (21ppb) annual mean by end 2005
Particulate matter	PM_{10}	50 $\mu g.m^{-3}$ daily mean (exceed <35 times/year) by end 2004 40 $\mu g.m^{-3}$ annual mean by end 2004
Sulphur dioxide	SO_2	350 $\mu g.m^{-3}$ (132ppb) 1 hour mean (exceed <24 times/year) by end 2004 125 $\mu g.m^{-3}$ (47ppb) daily mean (exceed <3 time/year) by end 2004 266 $\mu g.m^{-3}$ (100ppb) 15 minute (exceed <35 times/year) by end 2005
Carbon monoxide	CO	11.6 $\mu g.m^{-3}$ (10ppm) 8 hour mean by end 2003
Benzene		16.25 $\mu g.m^{-3}$ (5ppb) annual mean by end 2003
1,3 -Butadine		2.25 $\mu g.m^{-3}$ (1ppb) annual mean by end 2003
Lead	Pb	0.5 $\mu g.m^{-3}$ annual mean by end 2005 and then halved again by 2008
Two pollutants need to be addressed at the national and EU levels		
Ozone		
Polycyclic aromatic hydrocarbons		

Note: ppb/ppm: parts per billion/million.

Source: Based on ApSimon (2005, Table 1).

10. The effects of the London Congestion Charging Scheme on ambient air quality

Kenny Ho and David Maddison

1 INTRODUCTION

Congestion charging is a form of road pricing which involves defining area, such as the central business district or a greater metropolitan area, and charging motorists as they enter that area. Such a pricing scheme has been introduced in several international cities, such as those in Singapore, Norway and Sweden. The main objective of these and related schemes is to reduce traffic congestion by raising travel costs. Such a scheme, however, also brings an indirect benefit by improving ambient air quality. Our main focus in this chapter is to evaluate the impact of the London Congestion Charging Scheme (LCCS) on the ambient air quality in the capital.

Motor vehicle emissions contribute to five major pollutants: volatile organic compounds (VOCs), ozone (O_3), carbon monoxide (CO), nitrogen dioxide (NO_2) and particulate matter (PM_{10}). We use the concentration of PM_{10} as our air pollution index because clinically it is one of the most important pollutants in vehicle exhaust emissions. Many researchers have linked it to various health problems, especially respiratory illnesses such as asthma and chronic bronchitis. Long-term exposure to fine particles can cause premature death from heart and lung disease, and even lung cancer. It has been estimated that PM_{10} concentrations are associated with approximately 8,100 deaths and 10,500 hospital admissions nationally each year. The UK government recommends that air pollution standards should be based on PM_{10} concentrations.

Two studies have recently considered the link between particulate air pollution and health impacts in London. Maddison (2005) determined that particulate air pollution was responsible for 10.8 per cent of respiratory hospital admissions. Note that particulate matter is here being used as a general indicator for all air pollutants. The same study also considered cardiovascular admissions to be linked to particulate air pollution, but the study found that these hospital admissions were not additional and would have occurred anyway at some point in the future. Atkinson et al. (1999) also studied the effect of particulate matter on mortality in London. The extent of life lost per case of premature mortality is not known but is unlikely to be comparable with the extent of life lost through traffic accidents.

A number of authors have drawn attention to the possible complementarities that exist between measures intended to deal with the problem of traffic congestion, such as road pricing and raised parking fees, and the measures required to tackle the environmental problems posed by road transport, in particular the problem of urban air pollution. These are quite distinct from measures intended to operate directly on vehicle emissions such as minimal fuel efficiency requirements and emissions standards, which in some contexts have been found either ineffective in tackling rising levels of air pollution or too costly in relation to the benefits obtained (for example, Krupnick and Portney, 1991; Harrington, 1997). Crawford and Smith (1995) discuss briefly the relationship between traffic speeds and vehicle emissions and also provide a more general analysis of the role of a variety of fiscal instruments in attaining air quality objectives in transport.

Hall (1995) was one of the first to draw attention to the complementarities between measures intended to reduce traffic congestion and the effect of such policies on air quality objectives. Her paper also drew attention to some of the difficulties inherent in making a simple link between measures intended to reduce traffic congestion and emissions, including the substantial divergence between the predictions of transport models and direct observation. Daniel and Bekka (2000) recognise congestion charging as a way of simultaneously solving the problem of traffic congestion and air pollution. However, they point out that congestion charging does not directly internalise the cost of emissions, which varies across vehicles, and should therefore be accompanied by additional policy instruments such as emission taxes or standards. Their research in Delaware, Newark and Wilmington indicates that after the introduction of congestion charges, travel time on highways decreased by about 22 per cent and average speed increased from 37.8 mph to 48.3 mph. In addition, emissions of hydrocarbons (HC), carbon monoxide (CO) and NO_x fell by 3.4 to 10.5 per cent, depending on the elasticity of travel demand.

Proost and Van Dender (2001) construct a model for the city of Brussels that integrates the demand for urban transport as a function of the price of competing modes with a model that calculates not only changes in the speed and volume of traffic, but also the change in noise levels, accidents and emissions of key air pollutants. This model can be used to explore the linkages between policies intended to deal with traffic congestion and their effects on environmental objectives. In a series of simulations the authors find that environmental policies (for example, an emission tax), although resulting in a substantial reduction in urban air pollution, have little effect on the congestion externality. On the other hand, transport policies (for example, a congestion fee) do cause a large shift in urban traffic and bring significant environmental benefits through the impact on traffic demand.

Modelling the relationship between policies intended to reduce traffic flow and the ambient concentration of air pollution is very difficult. It is clear that the primary relationship between vehicle-kilometres travelled and emissions is positive. At the same time, however, it is known that vehicle emissions vary widely according to the level of maintenance. It is also known that vehicle emissions vary between vehicle types and also with vehicle speed, and that vehicle emissions depend on occupancy rates. Therefore even if one is able to model the change in vehicle flows and modal switch between private and public transport of various forms, calculating changes in emissions is far from trivial. Even if this were successfully achieved, it would then be necessary to predict changes in concentrations on the basis of the altered level of emissions. One scenario is that if the owners of poorly maintained older cars are dissuaded from entering the charging zone, then the reduction in ambient concentrations may exceed expectations. If, on the other hand, individuals respond by increasing vehicle occupancy rates then expectations regarding the likely improvement in air quality may not be met. If drivers are diverted onto roads that are uncontrolled, then depending on dispersion patterns the effectiveness of the charge will be restricted.

It is also true that emission characteristics may be based on emission rates for average vehicles across different classes and calculated for average driving speeds. Emissions, however, will depend on the distribution of driving speeds. Another important empirical question is the speed with which changes in behaviour can be expected to occur as well as an assessment of the transaction costs involved in paying the charge. The conclusion is not of course that one should not seek to model these benefits when considering the desirability of introducing a scheme involving traffic restraint. Rather it is necessary to conduct some *ex post* assessment in order to determine whether expectations have been met.

Computer-based models may fail to take account of important changes and consider only the most obvious or those effects that are possible to predict on the basis of available data. An *ex post* statistically based approach does not make any assumptions with regard to the manner in which the change in policy produced its effect on air pollution. On the other hand, there is a risk of inadvertently attributing a change in the level of air pollution to a particular change in policy when it should more correctly be attributed to contemporaneous events. None the less, *ex post* assessments have obvious importance to the question of whether the congestion charging scheme should be maintained, modified, extended or copied.

Although there have been several well-known experiments involving attempts to restrain traffic involving the use of electronic road pricing (ERP), an area licensing system (ALS), cordon charging and licence plate bans, we are not aware of any published attempt to produce an *ex post* assessment of the effects on ambient air quality of such schemes. The main objective of this chapter is to evaluate whether congestion charging in London has had a beneficial effect, by examining the data from air pollution monitors for evidence of a structural break associated with the implementation of the scheme. Data on PM_{10} concentrations are collected from monitors inside and outside the charging zone covering the period before and after the scheme was introduced. Controlling for autonomous trends, the analysis tests for the presence of a structural break around the time when the scheme was introduced.

2 THE LONDON CONGESTION CHARGING SCHEME

The LCCS was launched on 17 February 2003 by the mayor of London, Ken Livingstone. Although London is not the first city to adopt the congestion charging system, it is the largest city that is currently employing such a system. If successful in reducing traffic congestion, other busy cities around the world are likely to follow suit. London has the worst traffic congestion problem in Europe. Research by Transport for London (TfL) indicates that drivers in central London spend 50 per cent of their travel time in queues, and this costs the city £2–4 million every week in terms of lost time. Under the LCCS, motorists wishing to enter central London on Monday to Friday are subject to a congestion charge of £5 per day (£8 from 4 July 2005). There is, however, exemption from the charge for certain classes of road user, including the owners of low emission vehicles, and rebates for those living inside the congestion-charging zone. The money from congestion charging will be spent on the expansion and development of the traffic network in London, especially the public transport system. Several short-term (for example, bus network improvements) and long-term (for example, expansion of the Underground network and access) programmes are planned and will be funded by the revenue from the congestion charge. Our analysis might pick up the consequences of early investments as well as the 'pure' effect of congestion charging.

The primary objective of the LCCS is to create an efficient and reliable transport system by reducing traffic congestion and encouraging the use of public transport. Although the scheme was subject to opposition it has succeeded in reducing the number of vehicles in central London and also in dramatically reducing travel time. Measurements of congestion within the charging zone indicate a reduction in congestion averaging 30 per cent since the start of congestion charging, which is at the top end of TfL's range of prior expectations. The proportion of time that drivers spend stationary or moving slowly in queues in the charging zone has been reduced by up to 30 per cent.

Congestion charging has succeeded in reducing the volume of traffic in central London. There are observed reductions of 18 per cent in traffic (vehicles with four or more wheels) entering the zone, and 15 per cent in traffic circulation within the zone. On the other hand, congestion charging also changes the traffic pattern in central London. The number of cars and other potentially chargeable vehicles operating within the charging zone during charging hours has been reduced by 34 and 25 per cent, respectively. In contrast, the number of licensed taxis and buses/coaches increased by 22 and 21 per cent, respectively, during that time inside the zone. Again, these changes in traffic patterns are towards the top end of TfL's prior expectations.

In addition to the increase in the number of buses circulating in the zone, there is an 18 per cent increase in the number of passengers on each bus. This can be explained by the improvement in bus service reliability, a 30 per cent fall in additional waiting time and 60 per cent fall in disruption of the bus journey. However, there has been no systematic net change in National Rail and Underground services.

An important secondary goal of the scheme was to reduce air pollution in central London. In the view of some commentators this was even more important than the value of time savings, given the current evidence linking air pollution and a range of adverse health effects. By reducing the amount of traffic in and around the charging zone, congestion charging was expected to contribute to improving the general air quality in the zone. The reduction in the volume of traffic implies that less fuel will be consumed and fewer polluting emissions will be produced. Moreover, faster vehicle speeds and reduced congestion also mean that vehicles are generating fewer emissions per unit distance travelled. However, several local councils, including Westminster, sought to block the plan, arguing among other things that the consequence of the scheme would be to increase the levels of air pollution immediately outside the congestion charging zone. In this they received support from a number of neighbouring councils such as Chelsea and Kensington. Moreover, air pollution can move from these perimeter areas into the charging zone, offsetting the benefit from the scheme. Our empirical analysis will shed some light upon this possibility.

3 DATA AND MODEL SPECIFICATION

Daily data on particulate air pollution is taken from the UK's Automated Urban and Rural Network, covering the period from 1 January 1997 to 31 May 2005. These concentrations are automatically measured on an hourly basis. A number of monitors are located in Greater London although only three of these (Bloomsbury, Marylebone and Westminster) are located within the congestion charging zone. The remaining nine monitors are all at varying distances outside the congestion charging zone but within Greater London. The monitor in Shaftesbury Avenue (operated by Camden Council, see Camden Council, 2003) is located inside the zone, but not included in the network. However, its data can be transformed and compared with other data by multiplying each observation by 1.3 to account for differences in the method of capture. A summary description of these measured PM_{10} data is presented in Table 10.1.

Monitors are occasionally affected by occurrences such as road works. Such activities could affect the results if they occur concurrently with the implementation of the congestion charging scheme. However, checks revealed that none of the monitors was affected by unusual activities taking place nearby. Nevertheless, there was a serious technical problem which affected the PM_{10} monitoring equipment in Bloomsbury from June

Table 10.1 Summary description of PM_{10}

Monitor	Site type	Location	Obs.	Mean	Std dev.	Min	Max
Shaftesbury	Unknown	Inside	1691	35.21	12.34	8	110
Bloomsbury	Urban	Inside	2570	24.20	9.56	7	85
Marylebone	Kerbside	Inside	2787	36.89	13.69	9	139
Westminster	Urban	Inside	2430	22.96	11.75	4	134
Kensington	Urban	Outside	2992	21.61	9.49	5	89
A3	Roadside	Outside	2808	23.66	10.85	5	88
Bexley	Suburban	Outside	2945	20.47	10.26	3	141
Camden	Kerbside	Outside	2941	28.30	11.02	8	98
Haringey	Urban	Outside	2975	22.77	9.54	6	89
Brent	Urban	Outside	2968	19.37	9.03	4	99
Hillingdon	Suburban	Outside	2945	22.15	10.26	5	114
Eltham	Suburban	Outside	2833	19.32	8.86	5	94
Sutton	Suburban	Outside	1852	27.06	11.53	8	133

2002 to May 2003, which meant that PM_{10} data had to be deleted from that period.

A dummy variable is created (*Charge*) that takes the value unity onwards from the point when congestion charging came into force. The statistical significance of this variable indicates the effect that the policy has had on air pollution concentrations as measured at particular locations. The variable *Day* represents a time drift of the observation period from 1 January 1997 to 31 May 2005.

Since the congestion charging operates only during weekdays from Monday to Friday, we believe that there will be a significant difference in the level of PM_{10} inside the charging zone between weekdays and the weekend. To capture this effect, six dummy variables are created denoting different days of the week (*Mo*n, *Tue* and so on).

The methodology employed uses multiple regression analysis to examine the introduction of the LCCS while holding other known influences on air pollution constant. While conceptually straightforward, a number of factors require particular attention. The first of these is that pollution levels are subject to a range of long-run trends due to the effects of variations in the price of fuel, fluctuations in the rate of economic growth, technological improvements in the emission characteristics of the vehicle fleet and innovations such as the introduction of low sulphur diesel. These could perhaps be dealt with by including non-linear time trends. Such an approach, however, is not very attractive, making it more difficult to identify the contribution of congestion charging to pollution levels.

The approach taken here is to identify these trends by using nonparametric smoothing techniques on air pollution records from the centres of other major cities in England after London. These are Birmingham, Leeds, Liverpool, Manchester, Sheffield, Newcastle and Southampton. None of these cities has a congestion charging scheme. Specifically, the approach involves combining the records from these six different cities and then subjecting the resulting average to locally weighted smoothing with a bandwidth equal to a period of 365 days. The effect of this is to capture the long-run trends while averaging out the effect of seasonal cycles and meteorological conditions.

Sine and *Cosine* terms were included to capture the seasonal effects. In order to capture day of the week effects, weekday dummy variables are included in the regression. To control for the meteorological influences variables such as maximum temperature (*Maxtemp*), minimum temperature (*Mintemp*), grass temperature (*Grass*), average wind speed (*Speed*), maximum wind speed (*Gust*), precipitation (*Rain*) and sunshine (*Sun*) are included. The lag and higher order of these meteorological variables are also employed. Data on meteorological conditions are obtained from the UK

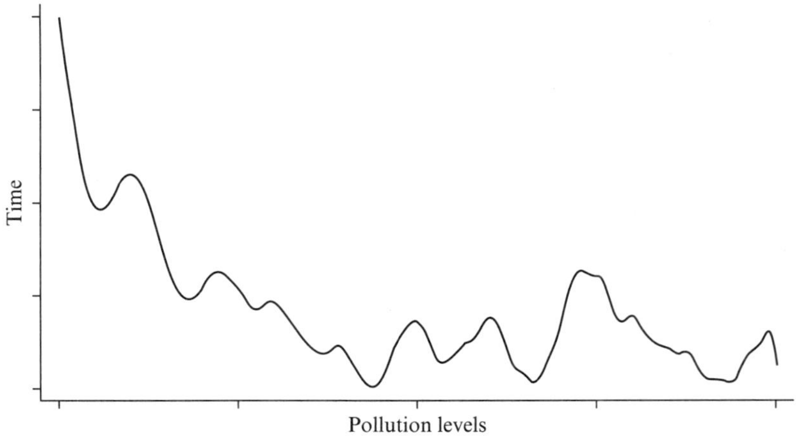

Figure 10.1 Autonomous trends in air pollution concentrations

Table 10.2 Summary description of meteorological data

Variable	Mean	Std dev.	Min	Max
Maxtemp	15.48005	6.285264	–1.7	37.9
Mintemp	7.877872	5.117104	–5.5	21.1
Rain	1.723462	3.70041	0	45.6
Sun	4.681543	4.090332	0	15.5
Grass	5.052522	5.778419	–12.1	19.6
Speed	6.860746	3.339278	0.1	24.8
Gust	21.86262	7.40882	4	64

Meteorological Office and refer to the weather station located at Heathrow Airport. A summary description of these data is shown in Table 10.2.

The variable *Charge* is also interacted with the time trend *Day* in order to allow for an evolving impact. This allows for the possibility that the full impact of changes in the price of entering inner London do not have an immediate effect. The use of both a dummy variable and a dummy variable interacted with a time trend also allows for the possibility of a step change up or down as well as a long-run trend. One possibility is that the immediate effect of congestion charging was to increase emissions before drivers gradually learned to adapt to the presence of the charging zone.

The variation in air pollution concentrations recorded by different monitors is allowed to differ between monitors. Meteorological changes are believed to have a similar effect on all monitors at the same time. Moreover, other non-modelled factors are likely to affect all monitors in a small area. This suggests that the residuals are likely to be correlated across panels. To the extent that air pollution persists, it is also likely that these errors are autocorrelated. Autocorrelation might also be expected if there are temporary factors affecting particular monitors such as building work, in which case the autocorrelation parameters might differ between monitors. Finally a variable *Inside* is included in the model, which takes the value unity if the monitor is located in inner London.

4 RESULTS AND DISCUSSION

Column 2 in Table 10.3 displays the impact of the congestion charging scheme on the time trend of PM_{10}. To recap, '*Day*' and '*Day* × *Inside*' represent the time trend of PM_{10} before the introduction of the scheme, while '*Charge*' and '*Day* × *Charge*' and '*Charge* × *Inside*' and '*Day* × *Charge* × *Inside*' represent the change of the trend after the scheme. We can identify two clear structural breaks from this table. For outer London, the slope of the trend is steeper after the introduction of the charge; while for the area inside the charging zone, there is a change in the sign in the trend, showing that PM_{10} concentration actually started to decline after the scheme.

Table 10.3 The estimated regression model

Variable	Parameter	*t*-statistic
Day	0.0002163	0.67
Day × *Inside*	0.0011481	1.36
Day × *Day* × *Inside*	3.10E-07	0.85
Inside	1.63072	4.04
Charge	–2.221761	–0.97
Day × *Charge*	0.0015706	1.88
Charge × *Inside*	12.03041	4.11
Day × *Charge* × *Inside*	–0.0054089	–4.07
Trend	0.1726677	2.73
Lag PM_{10}	0.6031405	71.05
Sine	0.726654	3.31
Cosine	1.217447	4.72
Mon	0.0385141	0.09
Tue	0.0683119	0.16
Wed	–0.4271761	–0.98
Thur	–2.687056	–6.19
Fri	–3.574453	–8.26

Table 10.3 (continued)

Variable	Parameter	*t*-statistic
Sat	0.8371978	1.91
Mon × Charge	–0.9439561	–1.11
Tue × Charge	–0.4428538	–0.52
Wed × Charge	–0.2206232	–0.26
Thur × Charge	–0.1989431	–0.23
Fri × Charge	0.1027872	0.12
Sat × Charge	–0.3447631	–0.41
Mon × Charge × Inside	–1.199698	–2.03
Tue × Charge × Inside	0.3566049	0.61
Wed × Charge × Inside	–0.735366	–1.27
Thur × Charge × Inside	–3.061041	–5.72
Fri × Charge × Inside	–2.655122	–4.60
Sat × Charge × Inside	1.199115	2.06
Maxtemp	–0.15492	–1.37
Mintemp	0.7273886	4.86
Rain	–0.1261692	–2.35
Sunshine	0.1147224	1.41
Grass	–0.6287749	–8.79
Speed	–1.165218	–6.89
Gust	–0.5042266	–5.72
Lag Maxtemp	–0.5439012	–8.12
Lag Mintemp	0.0389091	0.45
Lag Rain	–0.1079095	–3.56
Lag Sunshine	0.2349638	6.01
Lag Grass	0.0764026	1.32
Lag Speed	0.4585072	6.13
Lag Gust	–0.0153446	–0.51
Maxtemp × Maxtemp	0.0285417	8.58
Mintemp × Mintemp	–0.0177985	–1.98
Rain × Rain	0.0050877	2.29
Sunshine × Sunshine	–0.0302604	–4.74
Grass × Grass	0.0058314	1.05
Speed × Speed	0.0497803	5.69
Gust × Gust	0.0066554	4.19
Constant	15.14831	6.69
R-squared	0.6246	

Note: Dependent variable = PM_{10t} – autonomous trend; method = two-step generalised least squares with heteroskedastic panels, cross-sectional correlation and panel specific first order correlation.

Figure 10.2 below provides a visual representation of the structural break occurring on 17 February 2003. The results indicate that the congestion charge saved about 3.13 micrograms (10.58 per cent) of PM_{10} inside the zone. In contrast, the results reveal that there was an increase in the PM_{10} concentration outside the charging zone, amounting to 2.42 micrograms (10.72 per cent). This is particularly serious at the perimeter of the zone. For instance, the monitor in Marylebone Road indicates an average increase of 4.92 micrograms (13.35 per cent) in PM_{10} concentration after the introduction of congestion charging. This is consistent with the fact that individuals previously contemplating a journey through central London are now more likely to skirt around the edge of the congestion charging zone.

Results described in this section indicate that the LCCS led to structural breaks in the trend of PM_{10} concentration in London. There was a reduction inside the charging zone after the introduction of the scheme, showing that its secondary objective, namely improving the ambient environmental quality, was attained. However, the increase outside the charging zone (especially the perimeter area) implies that air pollution is worsening in

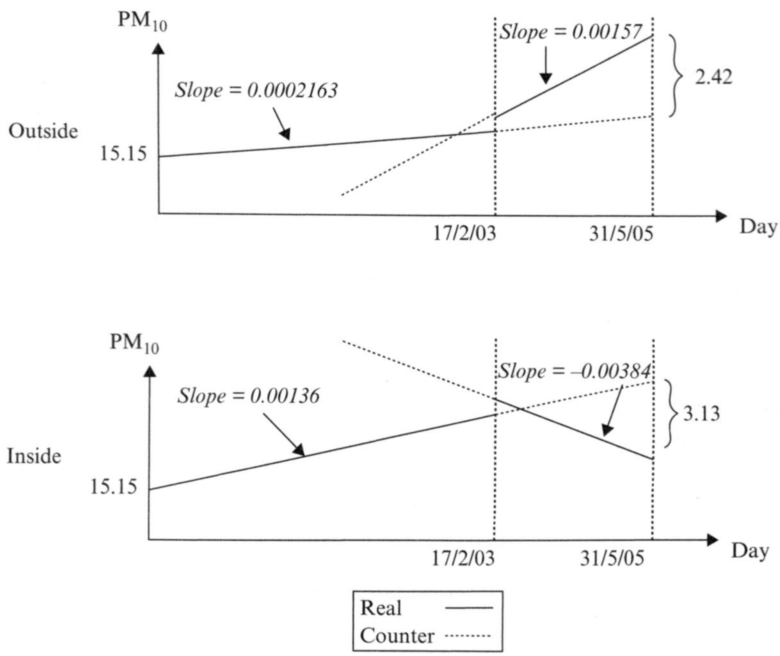

Figure 10.2 The structural break associated with congestion charging

those areas. This highlights the fact that the LCCS is indeed disadvantageous for households located on the perimeter of the charging zone. This finding will increase pressure either for the scheme to be abandoned or alternatively for the zone to be extended, with neither outcome seeming more likely than the other.

Recently, a proposal involving the possible western extension of the charging zone has been submitted to the mayor of London. The proposed new boundary in the south would follow Grosvenor Road and Chelsea Embankment; in the west, the southbound route of the Earls Court one way system, the West Cross route and Scrubs Lane; and in the north, Harrow Road in part, and then the Grand Union Canal and the Great Western Railway line.

It is proposed that the congestion charge for the extended scheme would be the same as for the existing central zone and that there would be only one charge from the whole combined area. All discounts and exemptions would be in line with those prevailing for the central London scheme, with a 90 per cent discount from the congestion charge for all residents in the defined discount area, including the extended zone. Since the improvement to environmental quality within the current charging zone is recognised, it is anticipated that the reduced traffic congestion in the western extension could bring comparable results. Another possibility is to create a second zone surrounding the first in which the fee for driving is set lower than for the inner zone. This might have the consequence of limiting the reduction in air quality at the perimeter of the zone but is clearly harder to implement. It also means that many more people will be located on the perimeter of a congestion charging zone. The final possibility is that some kind of compensation will have to be paid by the residents within the zone on the basis of the reduced environmental quality experienced by those living in neighbouring boroughs. It will of course also be extremely interesting to see the extent to which the benefits of living inside the zone and the disadvantages of living on the perimeter become capitalised into land prices.

We now turn to deal with two important limitations of this chapter. The first one regards the unusual meteorological conditions in 2003 and the other regards the technical problem in the Bloomsbury monitoring equipment from June 2002 to May 2003.

Air quality measurement of PM_{10} is very susceptible to the influence of 'secondary' or 'imported' pollution from elsewhere. This is particularly important during periods of unusual weather. A major limitation of this chapter is the unusual weather conditions in 2003, which distort the accuracy of the evaluation. In 2003, the UK had an exceptionally long and hot summer, causing a number of pollution episodes and an increase in average pollution levels over the year. This also led to particularly high levels of

PM_{10}. Hence, we believe that part of the benefit from the congestion charging scheme was 'offset' by this abnormally high air pollution.

Another limitation concerns the reliability of the data collected in Bloomsbury. During June 2002 and May 2003, there were problems with the PM_{10} monitoring equipment there. This means that PM_{10} data recorded during that period had to be deleted from the database. We have investigated this point by deleting these observations and find that it does marginally reduce the impact of congestion charging on PM_{10} concentration inside the charging zone, although the overall effect is still highly significant statistically.

5 CONCLUSION

London's recently adopted congestion charge has resulted in a sizeable change in ambient air quality. By performing a multiple regression analysis on time-series panel data recording PM_{10} concentrations, our results have shown a distinct structural break in PM_{10} trend both inside and outside the zone following the implementation of the LCCS. Air pollution is relieved within the charging zone (by 10.58 per cent) but worsened outside the zone (by 10.72 per cent), especially in the perimeter area. We cannot, however, attribute these changes wholly to congestion charging because of the related measures paid for by the charging revenue.

There are two ways in which these findings might be improved by future researchers. An obvious possibility is to extend the time period covered by the analysis: we could collect data only up to 31 May 2005, when this chapter was being prepared. An extension in the data collection period would not only increase the reliability of the estimation results, but also lessen the distortion caused by the unusual weather influence in 2003.

Second, since the congestion charge was increased from £5 to £8 on 4 July 2005, another possible improvement would be to create a new dummy variable reflecting the increase in the charge and to evaluate the impact of this increment. Air pollution level might then be compared over three different periods: before 17 February 2003, 17 February 2003 to 3 July 2005, and after 4 July 2005.

REFERENCES

Atkinson, R.W., H.R. Anderson, D.P. Strachan, J.M. Bland, S.A. Bremner and A. Ponce de Leon (1999), 'Short-term associations between outdoor air pollution and visits to accident and emergency departments in London for respiratory complaints', *European Respiratory Journal*, **13**: 257–65.

Atkinson, R.W., H.R. Anderson, J, Sunyer, J.A. Michelabaccini, J.M. Vonk, A. Boumghar, F. Forastiere, B. Forsberg, G. Touloumi, J. Schwartz and K. Katsauyanni (2001), 'Acute effects of particulate air pollution on respiratory admissions: air pollution and health: a European approach', *American Journal of Respiratory and Critical Care Medicine*, **164** (10): 1860–66.

Crawford, I. and S. Smith (1995), 'Fiscal instruments for air pollution abatement in road transport', *Journal of Transport Economics and Policy*, **29**: 33–51.

Daniel, J. and K. Bekka (2000), 'The environmental impact of highway Congestion Pricing', *Journal of Urban Economics*, **47**: 180–215.

Hall, J. (1995), 'The role of transport control measures in jointly reducing congestion and air pollution', *Journal of Transport Economics and Policy*, **29**: 93–103.

Harrington, W. (1997), 'Fuel economy and motor vehicle emissions', *Journal of Environmental Economics and Management*, **33**: 240–52.

Krupnick, A. and P. Portney (1991), 'Controlling urban air pollution: a benefit cost assessment', *Science*, **252**: 522–8.

Maddison, D. (2005), 'Air pollution and hospital admissions: an ARMAX modelling approach', *Journal of Environmental Economics and Management*, **49** (1), 116–31.

Proost, S. and K. Van Dender (2001), 'The welfare impacts of alternative policies to address atmospheric pollution in urban road transport', *Regional Science and Urban Economics*, **31**: 383–411.

Reports

Camden Council (2003), 'Air Quality in Camden', 2003 Annual Review, London.

Websites

Air Quality Archive, http://www.airquality.co.uk.
Camden Council, http://www.camden.gov.uk.
Congestion Charging at BBC, http://www.bbc.co.uk/london/congestion.
Congestion Charging Home, http://www.cclondon.com.
London Air Quality Network, http://www.londonair.org.uk.
Met Office, http://www.met-office.gov.uk.
Transport for London, http://www.tfl.gov.uk.

11. Transferring London congestion charging to US cities: how might the likelihood of successful transfer be increased?

Shin Lee

1 INTRODUCTION

The London Congestion Charging Scheme (LCCS) has largely been celebrated as a success. It has prompted numerous cities in and outside of Britain to consider implementing congestion charging. In what sense is it a success? The official review on the scheme's performance for the first two years reported a 30 per cent reduction in congestion, a 15 per cent reduction in traffic circulating in the zone (60,000 fewer car movements), and a 60 per cent decrease in bus operation disruption caused by traffic delays, among other positive figures (TfL, 2005). However, some Londoners – for example, those who had to transfer from car to bus and some retailers – are less happy, as occasional newspaper articles suggest. About a year after its introduction, research at Imperial College London suggested that congestion charging has had a significant effect on the sales in a major retail establishment studied (*Local Transport Today*, 2004). Prud'homme and Bocarejo (2005) claimed that the scheme was an economic loss, if a political success. Peak-hour congestion in the charge zone, measured in terms of excess travel time per kilometre beyond what is expected during uncongested hours, was 1.8 minutes per km in 2005 – still down from the pre-charge equivalent of 2.3 minutes but slightly up from the immediate post-charge level of 1.6 minutes (TfL, 2006). This offered the opponents of the scheme a ground for instant criticism, albeit seemingly short-lived.

Nevertheless, there is little doubt that London's 'success' has contributed in large part to the fact that more cities, not just in the UK but also in the rest of the world, are now considering congestion charging ever more than before. Within the United States, varying degrees of progress towards road pricing have been made recently in San Francisco,

Seattle (featured in this volume), New York City, Boston and the State of Oregon. San Francisco County and New York City are more or less openly discussing a scheme similar to that of London (Chan, 2005; Bowman, 2006).

Congestion charging is a long-standing idea and a much discussed policy device. Its effect on aggregate travel behaviour is reasonably predictable, although the substitution to different modes and time periods is subject to considerable uncertainty – depending on the price elasticity of demand to travel by car, there will be fewer drivers on the road with congestion charging. This is well known to policy analysts and policy makers alike. Equally well known is the difficulty associated with its implementation, which is why not so many other major world cities have implemented it. It is also this well-known implementation-related challenge that really made the LCCS so widely celebrated. Thus, in considering ways to enhance the likelihood of successful transfer of the scheme to other places, it will be useful to pay particular attention to how in London the policy device has been made acceptable to the public. At the same time, we need to identify the conditions under which desired outcomes can be brought about. In other words, when a policy such as congestion charging is considered, we need to think about success in terms of both successful implementation and successful outcome.

This chapter considers transference of the LCCS to US cities. Its primary objectives are: (i) to identify and apply a framework of analysis for policy transfer; (ii) to argue that a successful transfer should involve transference of key aspects of the overall planning process (and not just the scheme) to help overcome the implementation barriers in the UK–US context; and (iii) to examine the key travel characteristics likely to affect the outcome of congestion charging in the US.

Section 2 discusses general frameworks deemed useful for the purpose of this chapter, while Section 3 overviews the implementation process of the London scheme. Section 4 compares the key travel characteristics in the UK and the US which are likely to influence the effectiveness of congestion charging. Section 5 concludes the chapter.

2 POLICY TRANSFER – THE GENERAL FRAMEWORK

This section briefly reviews the literature on policy transfer in an attempt to identify a conceptual framework in which the overall analysis can be placed. In so doing, it emerges that a successful transfer would be ensured only through sufficient attention to contexts. A separate subsection underlines the importance of understanding the decision making and implementation process in particular, for an application to congestion charging.

While policy transfer has been attracting a growing interest as an area of investigation relatively recently, borrowing a novel idea from another country or adopting programmes implemented elsewhere has not been uncommon throughout the history of government policy making. For colonists, transferring a set of policies from their mother country to their new colony was the most natural form of policy making. Post-colonial states and many of the non-Western countries for that matter have invariably been adopting 'Western' policies in the process of modernisation and development, for modernisation almost meant westernisation for them. Along a similar line, some Eastern European post-communist countries adopted the Western European-style tax systems during the process of transition. Policy transfer of a larger scale and more coercive nature has been practised by international agencies in their attempt to help strengthen some of the weak/problematic economies in the world.

The exchange of policy ideas between the US and the UK dates back to the nineteenth century (for example, income tax, social security, unemployment insurance, and so on). However, the 1980s are especially marked by the proliferation of policy transfer, in most cases, from the US to the UK by the Thatcher administration. This appears to have drawn some scholarly attention to this area. Numerous cases such as the British Urban Development Grant programme, which was based on the American Urban Development Grant programme, the British Child Support Agency resonating with the American Child Support Enforcement System (Dolowitz, 2000), and Magnet Schools have been examined in detail. Not surprisingly, many of these represent unsuccessful transfers to varying degrees and for various reasons, the analyses of which constitute a bulk of the literature on policy transfer.

It may be possible to draw more general principles governing the process of policy transfer, and many of these seek explanations for the occurrence or pattern of policy transfer, identification of the key actors in the process, or construction of a broader-embracing theoretical framework (Dolowitz, 2000). The title of a review paper by Dolowitz and Marsh (1996) appears to sum up the nature of the literature – 'Who learns what from whom: a review of the policy transfer literature'.

Studies of immediate use for a prescriptive inquiry such as the current chapter attempts to address are relatively few. Dolowitz (2000) identifies three types of transfer failure, drawing from a number of case studies on the US–UK transfer of social policies. These include: (i) uninformed transfer; (ii) incomplete transfer; and (iii) inappropriate transfer. He shows that these are associated with incomplete knowledge, omission of crucial elements of the policy, and insufficient attention to contexts, respectively. Whereas this typology is not explicitly proposed as a framework for evaluating policy transfer, it is useful for a normative analysis concerning the ways to ensure successful policy transfer.

Noting the gap in the literature more recently, Mossberger and Wolman (2003) analyse policy transfer as a form of policy evaluation and suggest a set of criteria for assessing the probable effectiveness of transfer prior to implementation, along with some general considerations for overcoming undesirable outcomes. While such a framework – what they call 'policy transfer as a form of prospective policy evaluation' – may be seen as a repackaging of the reasons for the failure or success of policy transfer, Mossberger and Wolman add further useful insights and depths to the latter and produce a more explicit model for a prescriptive analysis with a discernible focus.

Mossberger and Wolman's key criteria are placed in a sequence of phases in policy transfer: awareness, assessment and application. Dolowitz's concern about sufficient knowledge is addressed in the first stage 'awareness'. To assess the quality and extent of information, Mossberger and Wolman suggest two criteria: scope of information, which underlines reference to a range of cases where the policy has been adopted, and adequacy and accuracy of information.

Dolowitz's concern about 'sufficient attention to contexts' is largely covered in the 'assessment' stage. But with Mossberger and Wolman, assessment as an activity is given a definitive role. Information that is gathered needs to be processed, paying particular attention to similarities of problems and goals, policy performance and differences in setting. Their call for a systematic assessment of the policy to be transferred by a group of experts is a good one, although this can possibly undermine the very motivation behind policy transfer for saving time and information costs and reducing uncertainty – 'bounded rationality' to use the term coined by Simon (1957). Nevertheless, so-called bounded rationality is in fact severely challenged in many cases whereby policies are emulated with very little assessment. A full assessment can help reduce uncertainty while at the same time it may expend considerable time and resources. If the reductions in uncertainty outweigh the increases in resources, bounded rationality will still be intact.

'Application' refers to whether the information gathered as an outcome of awareness and assessment is actually used in executing policy transfer. There is no direct link to Dolowitz's three types of failures here. However, it is implicit in the way 'incomplete transfer' is illustrated, which refers to deliberate omission of crucial elements of the policy, that the omission does not follow a sound assessment. Therefore, incomplete transfer can be seen as a form of inappropriate application. In any case, the decision to leave out some elements is made at the application stage.

The issue of incomplete transfer is an interesting one and is addressed in both studies. Incomplete transfer was identified as a reason for unsuccessful transfer in the case of the British Child Support Agency, thus engendering a type of failure (Dolowitz, 2000). In highlighting the need for careful assessment of policy performance, Mossberger and Wolman (2003) make a point that 'problems with particular aspects of a policy may not merit total rejection of assessment or even partial transfer' (p. 431). Note that the motivation behind this 'partial transfer' differs from Dolowitz's 'incomplete transfer', whereas the omission of certain elements is deliberate in both concepts. The core message from the quotation itself is actually similar to what Dolowitz refers to as 'negative lessons' which may be drawn from unsuccessful experiences.

The need to understand various contexts is underscored in both studies and is probably the most common reason for unsuccessful transfer of any type of policy. Its importance becomes even more crucial when transfer of congestion charging is considered – a policy well known for the political risk it might involve. It is proposed here that there are two levels of context, which should each be accorded appropriate attention: one relating to the decision-making process and another associated with the likely response of the consumers of the policy in question. One can focus on the product (of planning) – the design of the scheme – without fully acknowledging the process: what barriers were involved in the decision-making process or the implementation process and how these were overcome. These are of particular relevance to the UK–US context where the political structure is relatively comparable (in contrast to that, for example, of Singapore, with its substantially different political system which did not have to go through the complicated decision-making process to put its recent electronic road pricing scheme, or indeed its earlier supplementary licensing policy, into practice). What is transferred (for example, policy device, scheme, policy or planning process) matters in this way, and a complete transfer of a successful scheme may well turn out to be a failure. Understanding the entire process, including decision making at various stages of the process, gaining support from the public and other stakeholders, coming to an agreement on the scheme specifications and overcoming other intervening factors, and 'applying' all this information in the process of transfer, are probably more of a challenge than ensuring complete knowledge on the scheme itself or complete transfer of all the elements of the scheme in dealing with congestion charging in the UK–US context.

Understanding the Contexts: Decision Making, Implementation and Outcome

If the implementation of the original policy does not involve much difficulty, understanding the entire decision-making process may not be necessary to the core of the transfer process. For any policy involving formidable challenges at the implementation stage, however, the decision-making process in a wider sense (beyond the decision made by the key policy maker) should be an essential part of the context that needs to be sufficiently understood for a transfer to be successful. The case of congestion charging perfectly exemplifies this. Public aversion and implementation issues are the key challenges, at least in the Western world. By understanding how the unpopular policy was eventually accepted by the public and the process in which the implementation (detailed specifications) of the scheme was determined, the ways to manage or resolve similar yet subtly different conflicts may be deliberated and devised in advance. If the political context on both sides involved in transfer is well understood, the chance of the borrowing city to enjoy the adopter's advantage can be heightened.

It is also important in the transfer of any type of policy to understand the context of the behaviour change. Policies are intended to change certain behaviour of people who will be subject to them. What was the prevailing behaviour like before the implementation of the policy in question? For example, travel behaviour is immediately linked to the location of jobs, residence and other facilities as well as the relative costs of available travel options. These determine a large part of the policy outcomes. How are these travel and spatial patterns compared between the two places in question? Why did people respond to the policy in the way they did? Answers to these questions will shed light on whether similar effectiveness can be achieved as an outcome of policy transfer.

Clearly, there are cultural, social and economic dimensions to the context. The economic consideration can be of great importance, especially when a city with a vulnerable economy is considered. The issue was not so acute in the case of London, the UK capital and one of the world's greatest cities, with a powerful economy. Likewise, the American cities considered here typically have strong economies and are considering some type of congestion pricing based on the forecast rapid traffic growth, which is a reflection of strong economic growth. The social dimension is undoubtedly a significant one, given the regressive nature of pricing. It will merit a focused investigation on the socio-economic impact of the scheme, such as the spatial distribution of commuters and the behavioural adjustment made by people of different incomes. Cultural and life-style factors clearly affect the behavioural response to policy. Again, this is an area that deserves special attention. These various factors are discussed here only briefly.

3 LONDON CONGESTION CHARGING: DECISION MAKING AND IMPLEMENTATION

This section discusses what is considered to have been essential in putting congestion charging into practice and having it pass the initial political test in London. It concerns the decision-making and implementation process, which may be transferable, in the form of lesson-drawing, to the US context, considering the relative similarities in the political system. The process as experienced in the originating country should be 'assessed' as well as 'applied' to the transferring country. Application allows modifications according to the assessment of similarities and differences in setting (Mossberger and Wolman, 2003).

How Was the LCCS Made 'Acceptable'?

Congestion charging has been known to be unpopular among both politicians and the general public. Although it is supposed that the latter's attitude is usually a direct cause of the former's general position, the London experience demonstrates how the acceptance on the part of politicians is separable from that of the public. In addition, the public acceptability can be further divided into two layers – the public as a group and the public as a composite of groups with different interests. These different layers of acceptability play different roles in the overall process of decision making and implementation and therefore might need to be considered separately.

Politicians are not in favour of congestion charging generally because they assume that the majority of the voters – the car drivers – would resent it. When the mayor of London, Ken Livingstone, nevertheless chose to introduce it if elected, besides his conviction and pro-public transport position, he must have decided that even more citizens (than the number of car drivers who would resent the charges) resent severe congestion. 'Introducing' in this context meant to take forward the ROCOL (Road Charging Options for London) – an independent research group appointed by the Government Office for London – proposal. Under the 1999 Greater London Authority Act, powers to introduce congestion charging were given to the mayor.

Over 80 per cent of the people entering central London were non-drivers before congestion charging. Even considering: (i) that they do not precisely represent the constituencies for Greater London, (ii) the significant proportion of the users of the Underground (who do not directly suffer from congestion as much as road users) and (iii) the question of who really votes among those eligible, it is likely that the opponents of severe congestion outnumber the opponents of the congestion charge.

The situation would be quite different in US cities, which is a highly plausible reason why there has not been any noteworthy political leader championing such a scheme in that country. Even in New York, which has a higher usage of public transport than most other US cities, nearly 40 per cent of the entry to its central business district (CBD) is made by car. In the US as a whole, over 88 per cent of all trips are made by private motor vehicles.

For Livingstone, the prospect of congestion charging, if politically accepted, would bear two fruits – reductions in congestion and associated externalities, and funding for improving public transport. Its impact on the local economy was not a significant issue for the British capital with its unique position in the world economy. As a time-honoured promoter of public transport, Livingstone had compelling reasons to win public support. So, the acceptance of the idea on the part of the mayor, the key decision maker, might have been rather straightforward.

However, the acceptance by the 'public' was not so straightforward. Initially, there was a preliminary survey on the overall view of the public prior to drafting the mayor's Transport Strategy. The survey, Hearing London's Views, which asked the 'key stakeholders' their views on transport issues in general but also on the initial ideas of congestion charging, was sent to local councils, businesses and representatives of road users in July 2000. This informal consultation revealed that 67 per cent of the 400 respondents would support congestion charging if the revenues were used for improving public transport. Interestingly, of these supporters, 45 per cent were car drivers. The initial political test was answered with a largely positive sign and the views represented in this round of consultation influenced the draft Transport Strategy.

It should be noted that Livingstone was in a position to meet the conditions the public as a group required for their approval – recycling the revenue to expand the public transport service. There was indeed a 23 per cent increase in bus provision (TfL, 2005). (However, other British cities might face a challenge in just doing that. The 1985 Transport Act which deregulated the bus industry effectively limited the power of local authorities over the provision of bus services. London was not subject to this act, and here a system of competitive tendering was introduced for planned bundles of routes.)

The next layer of public acceptance – ensuring the support of various stakeholders – was even more complicated and, in effect, significantly costly in terms of the exemptions and discounts that had to be given away as a response to the views received from different stakeholders during the public consultation process. This is not surprising considering the multiplicity of interest groups that comprise the 'public'. It is always easier to resolve conflicts within a person or a single group faced with choices than to resolve interpersonal or inter-group conflicts (Coombs, 1987).

In London, having the proposed scheme accepted required two years of intensive exercise involving several rounds of public consultation, scrutiny reviews by the London Assembly, a court decision, and considerable modifications to the proposed scheme. For a transport project, especially of the scope of the proposed London congestion charging scheme, two years is not a long period but rather considered to be remarkably short. However, it signifies the level of political commitment and exceptional efficiency it commanded and not the simplicity of the process. The entire process was in fact highly intense, complex and demanding. Political commitment played a fundamental role in progressing through various phases of planning without major delay, which is often the case in the transport policy implementation process. The extra efforts to speedily progress the whole process were quite visible, including the choice of a relatively low-technology system with video cameras and flat charges instead of electronic transponders and variable rates, and the relatively quick negotiation processes. The rapidity and timing were considered crucial in overcoming the political hurdle.

How Was the Implementation (Scheme Specifications) Determined?

The formal public consultation began with the publication of the mayor's draft Transport Strategy on 11 January 2001. This round of public consultation gave an opportunity for a larger number of people (compared to Hearing London's Views) – 8,000 – to comment on the congestion charging scheme as proposed in the draft Transport Strategy. This was fed into the final Transport Strategy which was published on 10 July 2001, shortly after which Transport for London made a Scheme Order for central London congestion charging, providing the legal basis for its implementation and detailing the key aspects of the scheme. Then another round of public consultation on the congestion charging scheme itself was held over a two-month period from 23 July to 28 September 2001. Taking into account the views received in this main consultation, TfL suggested modifications to the scheme, primarily to discounts and exemptions. These modifications were then put out for further consultation until 18 January 2002. In February 2002, the mayor confirmed the Scheme Order with modifications.

Banister (2003) makes a very insightful observation that very few raised any issues with respect to the principle of congestion charging in the later consultations and the debate on the principle, which emerged earlier, was addressed in the Transport Strategy between January and July 2001. This may be an indication that education of relatively complex concepts and ideas can be facilitated through extensive public consultation and a well-focused and well-targeted supply of information. It should be noted that the campaign undertaken by an appointed communications consultancy was highly effective throughout the planning process. Getting the message across about the rationale for pricing has often been regarded as a major challenge.

Issues which were raised through consultation include, among others, the extent of exemptions and discounts, the higher charges for lorries, the charging area, the regressive nature of flat charges, and the charging hours. Major modifications involved extended exemptions and discounts, no higher charges for lorries, one charge per vehicle per day and a slight shortening of the weekday charge hours from 7 am–7 pm to 7 am–6:30 pm. The latter was in response to the views represented by theatre owners who feared a loss of potential custom for evening entertainment with the charge in place until 7 pm (to which the mayor responded without hesitation). The Scheme Order published in July 2001 as well as the Report to the Mayor which presents the analysis of representations received during all consultations and recommendations for modifications, are available on TfL's website http://www.tfl.gov.uk/tfl/cclondon/cc_public_consultation.shtml.

The proposal further underwent: the scrutiny committee's review; judicial challenge (by Westminster City Council and the Royal Borough of Kensington and Chelsea); and a final review until its implementation on 17 February 2003. The entire process, involving the series of consultations and modifications which took place over about two years, illustrates how intimately the public was involved in the process of implementation by expressing their views and influencing the specifications of the scheme. In a similar fashion, even the initial design of the scheme was approached with sufficient political sensitivity, such as beginning with a politically acceptable price level with a view to progressive adjustment later if appropriate.

It is important to note that the concessions added as a consequence of this process can and do weaken the effectiveness of the scheme by (i) not being able to discourage driving among those exempted and substantially discounted; (ii) reducing the revenue base which will in turn be reinvested in the improvements of public transport; and (iii) further complicating the structure of intergroup inequity (inherent in pricing schemes in general) by causing cross-subsidy between those who pay and those who do not pay. It has been estimated that about 26 per cent of vehicles operating in the charge zone will have various types and extents of discount, 29 per cent will be exempted from the charge altogether, and only 45 per cent will actually pay the full charge (Banister, 2003). As Banister rightly points out, the drivers with the exemptions and discounts are the real winners (benefiting from improved speeds without paying more). The aforementioned rise in traffic congestion in 2005 (which started to crawl back in the second half of 2004), despite the increase in charge from £5 to £8, has been attributed to the greater number of exempted vehicles, such as taxis and buses, entering the zone (TfL, 2006). Although some hold the view that exemptions and discounts awarded in the central London scheme were excessive, the prevailing wisdom at the time appeared to be that those concessions were necessary for the implementation of the scheme. It is probably true that they were necessary to achieve the required level of efficiency, and it is possible that that level of rapidity was required to achieve political success.

4 UNDERSTANDING THE TRAVEL CHARACTERISTICS: LONDON AND US CITIES

In trying to identify the transport variables which are of greater importance than others in terms of their influence on the performance of the policy, it would be useful to think in terms of why Londoners decided on the scheme they have and responded to congestion charging in the way they did.

The Severity and Characteristics of Traffic Congestion

First, the public approved the idea, subject to conditions (recycling the revenue solely to invest in public transport). This may be surprising but is reasonably explained by the presence and perception of severe congestion. Immediately prior to the implementation of congestion charging, the average peak traffic speed in central London was 8.7 miles per hours (14 km/h). With the possible exception of New York City, there are few cities in the US experiencing anything near this level of congestion in their central areas today.

Indeed, there is documented evidence that the average commuting time in the major US metropolitan areas has been relatively constant, and occasionally falling, despite the constant and strong growth in commuting flows, depending on the area and the type of location – whether central or suburban – at origin and destination (Gordon et al., 1989, 1991; Lee et al., 2007). Categorising commuting trips in this way, which yields four trip types (suburb-to-centre, centre-to-centre, centre-to-suburb and suburb-to-suburb), helps explain how the average commuting times have been more or less stable in large, rapidly suburbanising metropolitan areas. In the New York metropolitan area, commuting trips originating from a suburban area and ending in a suburban area have risen from 48.7 per cent in 1980 to 49.6 per cent in 1990, while the corresponding figures for the Los Angeles metropolitan area are 52.3 per cent and 56.1 per cent respectively (Lee et al., 2007). These suburb-to-suburb commuting times are the least time-consuming commuting of all types although they were slightly more time-consuming in 1990 compared to 10 years before. Much of this increase is accounted for by the increase in non-work trips during the peak hours (Giuliano, 1998; Lee and Gordon, 2006). The most time-consuming trips are those originating from a suburban area and ending in the central area. The increasing dominance of trips within the suburbs (and the corresponding decrease in the most time-consuming trips from suburb to the centre) must be more than offsetting the slight increase in the suburb-to-suburb travel time.

Nevertheless, according to a survey by the US Bureau of Transportation Statistics (2003), more than two of five adults in the US report that traffic congestion is a problem in their communities. Most of the cities considering congestion charging have forecasts which indicate significant and rapid increase in traffic in the near future. Traffic congestion is present and perceived as a current or a future problem by a significant number of people in the US. But this is to a lesser degree and normally attributable to certain stretches of motorways (and probably in certain directions) and not concentrated in or limited to the centre of the city.

Mode Share

Along with the presence of severe traffic congestion, the fact that a relatively small proportion of the daily peak hour entry to the priced zone is made by private automobiles has contributed to the acceptance of the idea by the citizens of London. In 1993, 16.5 per cent of these entries to central London were made by car. This was reduced to 9.9 per cent by 2002, even before the implementation of congestion charging, which must reflect how severe the congestion was prior to the implementation of the policy. Therefore, only about 10 per cent of the travellers were considered for charging – that is, before exemptions and discounts were outlined and determined. There is a stark difference between London and a typical US city in this regard. Even in New York City, where the use of public transport is relatively high by US standards, nearly 40 per cent of the CBD entry was made by car in 1998. In the central cities of Greater Los Angeles (Los Angeles and Long Beach), commuting by public transport was about 6.6 per cent of all commuting. In the suburban counties of the metropolitan area, the corresponding figures range from 1.1 to 2.8 per cent (US Bureau of Economic Analysis, 2000). Table 11.1 shows a national comparison on the modal share.

Table 11.1 Mode shares in the UK and US (%)

Mode of transport	UK[1]	US[2]
Personal-use vehicle	64.43	88.7
Public transportation	9.20	1.2
Bicycle/walking	26.37	0.8
Other*		8.3

Note: * Other includes air transportation (7.2%) and 'other', including 'charter/tour/intercity bus, taxi, limousine, hotel/airport shuttle bus; intercity train and other not elsewhere classified' (1.0%).

Sources: 1. DfT (2006); 2. US Department of Transportation (2004).

The Significance of Non-work Trips

The significance of non-work trips is likely to influence the effectiveness of the policy as non-work trips can often be more easily diverted to different times (from weekdays to weekend or from charged hours to uncharged hours). National statistics show that the percentages of work trips are quite similar in the two countries as indicated in Table 11.2.

What is notable here is that the percentages of work trips are very small in both countries. Studies have shown that non-work trips now account for a significant proportion of peak-hour traffic (Giuliano, 1998; Lee et al., 2007), suggesting that there is scope for pricing the hours of heavy traffic in both societies. The relevance and the extent of increases in non-work trips are treated in detail in Lee and Gordon (see Chapter 17 in this volume).

Spatial Patterns, Density and Physical Size

There were about 60,000 fewer car movements 2 years after the inception of the LCCS. Of these, 20–30 per cent were estimated to have diverted around the zone and 50–60 per cent to have switched to public transport. The rest of those who stopped driving to central London during the charging hours could have: moved to other modes of travel such as car share, bicycle and motorcycle; changed the time of travel (to outside of charging hours); or simply reduced the number of trips. Are the same or similar patterns of change expected in the US? Given the far longer commuting distance, is diversion a practical option? Is public transport really considered an alternative to driving in the US? The extremely small share of public transport (see Table 11.1) may be one of the many indications that it is not.

Table 11.2 Trips by purpose in the UK and US (%)

Purpose	UK[1]	US[2]
Commuting/business	18.3	17.7
Education[a]	11.9	9.8
Family/personal business[b]	39.2	44.6
Leisure	26.5	27.1
Other	4.0	0.8

Notes:
a. Education refers to 'education/escort education' for the UK classification and 'school/church' for the US classification.
b. Family/personal business includes 'shopping, other escort and personal business' for the UK classification.

Sources: 1. DfT (2006); 2. US Bureau of Transportation Statistics (2005).

Several major US cities rely on private automobiles even more than the national statistics suggest.

London's public transport sector was helped in attracting previous car users by the expansion of its service being concurrent with the congestion charge scheme. But an equivalent extent of investment in public transport in US cities is unlikely to have a similar effect. The wide spread of activities over an enormous space in many US cities often makes it extremely difficult, if not impractical, to have the city covered by a well-developed public transport system. Rail is worse in this sense than the more flexible and less costly bus system (but the latter is usually subject to traffic congestion). How the massive costs of rail projects were unmatched by a significant increase in ridership in the Los Angeles metropolitan area has been previously argued and documented (Rubin et al., 1999a, 1999b). It has been acknowledged that low density makes a very unfavourable case for mass transit in general.

5 CONCLUSIONS

This chapter has sought to identify the conditions which would increase the likelihood of a successful transfer of congestion charging from London to US cities. From the literature on policy transfer which is principally concerned with a positive analysis of why and how policy transfer occurs, we identified two relevant models which are applicable to a prescriptive task. First, we reflected on the typology of transfer failure suggested by Dolowitz

(2000) and turned it around to a model for successful transfer. Second, a framework for assessing policy transfer recommended by Mossberger and Wolman (2003) is of direct relevance to the purpose of this chapter. There are both considerable parallels and subtle differences in these two approaches, and the two together provided a conceptual framework for this work.

Of the three types of failure (namely, uninformed transfer, incomplete transfer and transfer with insufficient attention to context), congestion charging is deemed to be most susceptible to a failure resulting from insufficient attention to context. Congestion charging has been discussed for decades and has been under consideration in various contexts, not just in the US but more widely. The primary reason why it has not been more widely adopted is its potential political risk and the challenges associated with its implementation. Particular attention, therefore, should be given to the contexts in general and especially the context in which decision making and implementation had taken place in London. Significant lessons learned from the London experience could be applied in the US where the political and planning arrangements are relatively similar, although appropriate modifications should be made to reflect differences in the political culture which exist between the two countries.

Strong political commitment, which was helped by the institutional framework, several rounds of public consultation which began at the stage of shaping the mayor of London's Transport Strategy, the role of the scrutiny review committee, and the way the city responded to the views of stakeholders all characterise the decision-making and implementation process in the London case.

An assessment of similarities and differences between the two countries – the transferred and the transferring – in the context of various institutional arrangements, decision making and political culture, local economy and travel characteristics is of paramount importance. At the same time, applying the knowledge obtained through the assessment in various stages of decision making is equally important, as Mossberger and Wolman suggest.

In order to predict the likely performance of congestion charging in the US, it is also necessary to assess similarities and differences of problems and goals. These require consideration of similarities and differences in travel characteristics. Problems are considerably different in the sense that traffic congestion in the US does not tend to be contained in the city centre. It is often along major corridors – a stretch of motorway or arterial road in one direction, and so on – connecting suburban communities with the central area or between suburban communities (between job-rich and house-rich communities). This makes London-style area charging less appealing. Furthermore, some corridors on the US urban highways such as SR-91 in Riverside Country, I-15 in San Diego and I-394 in Minneapolis, are currently priced through a variable charge scheme whereby one or two median lanes are reserved for drivers who pay or those carrying two or more additional passengers – 'HOT' (high occupancy toll) lanes. The 1991 Intermodal Surface Transportation Efficiency Act (ISTEA) enables states to apply for congestion pricing pilot projects and, through the funds made available for such projects, encourages the adoption of some form of pricing mechanism. Fourteen states so far have been awarded funding for what is called a 'value pricing pilot program' (US Department of Transportation, 2006).

The extent of spatial dispersion in many US metropolitan areas, and its implications for the geographical distance one is typically bound to cover on a daily basis, often makes it extremely difficult for Americans to use conventional public transport and for governing bodies to provide a high-performance public transport network. This, putting aside the cultural and life-style factors which are frequently cited, renders mode shift ambitious at least as a short-term goal. Drivers who are more sensitive to pecuniary penalties are more likely to be priced off the road. However, the level of mode shift made by Londoners is unlikely in a US context. Pricing in combination with other forms of travel demand management measures will possibly persuade more Americans to leave their cars at home, but this persuasion and the corresponding behavioural response will only occur after a considerable amount of time. Peak-hour charging might be a more practical near-term goal. Lee and Gordon in this volume report that a large and increasing proportion of peak-hour trips are non-work trips. This confirms that there is great scope for manipulating the pattern of travel over the time of day in the US context.

The unit cost of driving is considerably lower in the US compared to Britain, where substantial fuel taxes are already in place. Will this make a modest (or proportional) increase in the cost of driving more acceptable in the US? Possibly, although it may be determined by whether car users are sensitive in their political attitude and behavioural response to the absolute level of the charge or its proportion to their current user costs. But at the same time, there are more low-income car drivers in the US (the predominance – over 90 per cent in many cities – of car drivers and the oft-cited great income inequality together clearly point to this). Even a modest increase in motoring costs may hit these low-income drivers hard. The whole issue of equity consideration is likely to differ in the context of US metropolitan areas in relation to spatial patterns of wealth and poverty specific to each area and the design and location of the pricing scheme – an area which merits further investigation.

There were a number of aspects other than those discussed here in detail which contributed to the successful implementation of the LCCS within a relatively short period. In addition to strong project management, integrating the communications team with the project management team for the purpose of an effective campaign has been acknowledged as a main strength, as has been well demonstrated by the near absence of confusion among drivers from day one. Presenting congestion charging as part of an overall transport strategy, thus being consistent with the longer-term strategy, contributed to general acceptance by the public. These, along with what has been discussed thus far, including a strong political will, well-exercised participatory planning, political sensitivity, delivery of bus improvements and confidence in the local economy, may be listed as conditions for the successful implementation of congestion charging. However, they will by no means comprise sufficient conditions. In the same national context and with similar strategies, Edinburgh's project has fallen by the wayside. Some unexpected political turns, for example, can pose a serious barrier to implementing this politically delicate policy device. The greatest significance of the London experience might be that (i) congestion charging is implementable in a highly polyarchical society and (ii) congestion charging is an answer to severe congestion on the road.

Although many world cities are looking eagerly at the London experience, the selection of degree of intervention and the model of restraint must ultimately be carefully tailored to local circumstances. It may well be that the differences between London and US cities are sufficiently great to lead to a rejection both of the policy of heavy restraint and the model of implementation.

REFERENCES

Banister, D. (2003), 'Critical pragmatism and congestion charging in London', *International Social Science Journal*, **55**(176): 249–64.

Bowman, B. (2006), 'San Francisco – Move to charge toll for driving in core of downtown area – Country transit panel to receive $1 million from US for study', *San Francisco Chronicle*, 28 March, http://www.sfgate.com/cgi-bin/article.cgi?f=/c/a/2006/03/28/BAG5IHV4C21.DTL.

Chan, S. (2005), 'Driving around Manhattan, you pay, under one traffic idea', *New York Times*, 11 November, http://www.nytimes.com/2005/11/11/nyregion/11traffic.html.

Coombs, C.H. (1987), 'The structure of conflict', *American Psychologist*, **42**(4): 355–63.

Department for Transport (DfT) (2006), *Transport Trends, 2005 Edition*, Department for Transport, Transport Statistics, London: DfT.

Dolowitz, D. (2000), *Policy Transfer and British Social Policy: Learning from the USA?*, Buckingham, UK and Philadelphia, PA: Open University Press.

Dolowitz, D. and D. Marsh (1996), 'Who learns what from whom: a review of the policy transfer literature', *Political Studies*, **44**: 343–57.

Giuliano, G. (1998), 'Information technology, work patterns and intrametropolitan location: a case study', *Urban Studies*, **35**: 1077–95.

Gordon, P., A. Kumar and H. Richardson (1989), 'The influence of metropolitan spatial structure on commuting time', *Journal of Urban Economics*, **26**: 138–51.

Gordon, P., H. Richardson and M. Jun (1991), 'Commuting paradox: evidence from the top twenty', *Journal of the American Planning Association*, **57**: 416–20.

Lee, B. and P. Gordon (2006), 'The US context for highway congestion pricing', working paper, CA90089-0626, University of Southern California, Los Angeles.

Lee, S., J.-G. Seo and C. Webster (2007), 'Changing metropolis: economic diversity and commuting in the US suburbs', *Urban Studies*, **43**: 2525–49.

Local Transport Today (2004), 'Study says London congestion charge is harming retail sales', 22 April, p. 1.

Mossberger, K. and H. Wolman (2003), 'Policy transfer as a form of prospective policy evaluation: challenges and recommendations', *Public Administration Review*, **63**(4): 428–40.

Prud'homme, R. and J.P. Bocarejo (2005), 'The London congestion charge: a tentative approach', *Transport Policy*, **12**: 279–87.

Rubin, T., J. Moore II and S. Lee (1999a), 'Ten myths about urban rail systems', *Transport Policy*, **6**: 57–73.

Rubin, T., J. Moore II and S. Lee (1999b), 'A post-mortem analysis of the Los Angeles County Metropolitan Transportation Authority's 20-year Long-range Plan', *Public Works Management and Policy*, **3**(3): 187–206.

Simon, H. (1957), *Administrative Behaviour*, New York: Free Press.

Transport for London (TfL) (2005), 'Central London Congestion Charging Impacts Monitoring: Third Annual Report', Transport for London, http://www.ftl.gov.uk/tfl/cclondon/pdfs/ThirdAnnualReportFinal.pdf.

Transport for London (TfL) (2006), 'Central London Congestion Charging Impacts Monitoring: Fourth Annual Report', Transport for London, http://www.tfl.gov.uk/tfl/cclondon/pdfs/FourthAnnualReportFinal.pdf.

US Bureau of Economic Analysis (2000), '1969–2003 Regional Economic Information System', Bureau of Economic Analysis, Census Transportation Planning Package (CTPP 2000).

US Bureau of Transportation Statistics (2003), 'Traffic Congestion Rated a Problem By Two of Five U.S. Adults, BTS Survey Shows', http://www.bts.gov/press_releases/2003/bts018_03/html/bts018_03.html.

US Bureau of Transportation Statistics (2005), *Annual Report*, Washington, DC.

US Department of Transportation (2004), '2001 National Household Travel Survey Data', US DOT Bureau of Transportation Statistics and Federal Highway Administration.

US Department of Transportation (2006), 'Value Pricing Pilot Program', http://www.ops.fhwa.dot.gov/tolling_pricing/value_pricing/index.htm.

PART III

International examples

12. Inter-urban road goods vehicle pricing in Europe

Chris Nash, Batool Menaz and Bryan Matthews*

1 INTRODUCTION

The European Commission has long been concerned that distortions in charges for transport infrastructure use could distort competition between road hauliers based in different countries and more broadly between the economies of countries as a whole. Following on from the Green Paper of 1995 (CEC, 1995), the European Commission has sought to achieve a closer relationship between transport prices and the marginal social cost of transport. Because of its importance in European economics, heavy goods vehicle (HGV) charges have formed a central part of this policy (CEC, 2001) while charging for the private car is left as a matter for the member states. It is widely recognised that existing charges do not adequately reflect these costs. Annual licence fees may vary with the characteristics of the vehicle but not with where and when it is used, while the relationship between fuel tax and vehicle use is driven by technological rather than economic factors. The Eurovignette, which came into force in some European countries in the 1990s, is a time-based user charge, while the tolls which exist on motorways in some countries are usually more concerned with raising finance than with reflecting costs. A distance-based user charge is generally seen as a better solution to the problem of reflecting the costs, and a number of countries have now introduced such a system.

Section 2 looks at the new Eurovignette directive which was agreed late in 2005. Section 3 then examines the issue of measurement of costs, and Section 4 discusses the results of some modelling of the impacts of pricing reform for HGVs. Section 5 examines the reforms that have actually taken place in three European countries: Switzerland (which was the first country to introduce such a charge), Germany and Austria. Section 6 discusses the impacts of the new road infrastructure charges. Finally Section 7 concludes.

2 THE EUROVIGNETTE DIRECTIVE

Traditionally there have been two approaches to charging for the use of roads in Europe. The first has been through annual licence duty and fuel tax. The second has been to add to these specific tolls on certain roads. Needless to say, countries which adopted the latter generally had lower levels of annual licence duty and fuel tax than the former. The result is that when vehicles from such countries operate in countries with higher levels of taxes, they are at a considerable advantage, especially if they are able to buy much of their fuel before they come. Conversely, vehicles from high-tax countries operating overseas are at a disadvantage as they have to pay tolls on top of the high taxes in their own country.

The origin of the Eurovignette system was in an attempt to overcome this problem. Initially Germany, but following that also the Benelux countries, Sweden and Denmark, introduced a system whereby all vehicles using the motorway system of the country concerned had to buy a supplementary licence known as a vignette. This was valid for a given period of time – a day, month or year. The original Eurovignette Directive was designed to regulate this charge to ensure that it was not used to discriminate between vehicles of different countries or to exploit monopoly power.

After a long debate between themselves and the Council of Ministers, in December 2005, the European Parliament voted in favour of a new framework for road infrastructure charging based on that proposed by the European Commission in 2003. The new European road charging regime was then finalised in March 2006. The scope of this new Eurovignette Directive is broader and it was stated that it would 'encourage member states to introduce and develop tolls and charges which will make it possible to improve the management of commercial freight traffic, reduce pollution and generate funds for investment in new infrastructure' (European Commission, 2006, IP/06/383).

The new directive will allow greater variation in tolls to reflect congestion, and will also make mandatory from 2010, toll variations to reflect pollution caused by vehicles. As in the earlier directive, on average, user charges will be tied to the costs of construction, operation, maintenance and development of the network. Originally it was proposed also to include the uncovered costs of accidents, but this provision was dropped. The overall average charge is to be equal to average infrastructure costs, where infrastructure costs must be allocated to vehicle types on the basis of equivalence factors based on objective evidence.

The new directive allows the toll to be applied to HGVs weighing over 3.5 tonnes as from 2012, replacing the previous 12 or more tonnes. It will be applied to the trans-European network (TEN) and to other roads to which traffic might divert, but permitting application of pricing to other roads as well. Therefore the charges are potentially applicable to all roads, thereby differing greatly from the previous directive which only covered motorways.

The new directive allows member states the ability to increase tolls with a 'mark-up' (they can charge up to 15 per cent more, or 25 per cent on cross-border routes) on roads in particularly sensitive mountainous areas. The income from the mark-ups must then be used to fund alternative transport infrastructure. This introduces a new principle whereby tolls on HGVs can be used to finance infrastructure on alternative modes, such as rail.

The new directive aims to provide a more differentiated charging system. Tolls can vary according to a number of factors such as:

- the distance travelled;
- the location, as the probability of accidents differs between different areas (urban and rural), population density and weather conditions;
- infrastructure type and speed as expenditure on maintenance varies from that on a motorway to a trunk road and infrastructure type determines speed, which also affects the accident rate;
- the vehicle type, including characteristics such as axle weight and suspension type which influence infrastructure repairs and maintenance. Engine type, energy source and emission standards influence air pollution levels and vehicle size as larger vehicles make a bigger contribution to congestion;
- the time of day also affects congestion levels as it varies from peak and off-peak times; and
- tolls may also be differentiated according to specific routes. However, no charge may be more than 100 per cent higher than the minimum, so the degree of differentiation is heavily constrained.

Originally it was also proposed that there should be strict enforcement that revenue from the tolls is used for expenditure on roads, other transport networks, transport substitutes or the transport sector as a whole, but not general state expenditure such as spending on health or education. Each state was to create an independent transport infrastructure supervisory authority to guarantee that charges are being set and revenue is being used in the required way. However, this provision was also dropped due to opposition from the member states. The new directive states that 'revenues from tolls or user charges should be used for the maintenance of the infrastructure concerned and for the transport sector as a whole, in the interest of the balanced and sustainable development of transport networks' (European Commission, 2006). However this is a recommendation; it does not have the force of law.

The proposed new directive is not consistent with the policy of short-run marginal cost pricing adopted by the European Commission in the White Paper on Fair Payment for Infrastructure Use (CEC, 1998). The reason is that the decision to tie average user charges to the cost of 'constructing, operating, maintaining and developing the network' limits the extent to which the overall level of tolls can reflect environmental costs, external accident costs and marginal costs of congestion. There would obviously be a degree of double counting if both additional capacity and congestion costs were charged for. The exclusion of environmental costs from the total costs to be covered is explained by the Commission on the grounds that these are more uncertain than infrastructure and external accident costs, despite the enormous amount of work the Commission has funded on their measurement and valuation in recent years. However, additional regulatory charges to deal with congestion and environmental problems are permitted in specific circumstances.

In summary then, this new directive significantly improves on the existing situation. It represents a clear advance on the existing directive in a number of respects. It makes it clear that kilometre-based charges are a permitted form of user charge, and that they need not be confined to motorways; they may be levied on other competing roads, and indeed all roads in a particular country. It permits user charges on lighter goods vehicles and increases the degree of differentiation allowed, indeed making this compulsory from 2010. In terms of many of the decisions open to freight vehicle operators (type of vehicle, route, time of day), it is toll differentiation rather than the average level of toll that is the crucial factor. It also permits a surcharge in environmentally sensitive areas, which may be used for rail infrastructure enhancement rather than road. It appears that the limits on permitted levels of charges may lead to significant degrees of distortion. However, part of the compromise is that the European Commission is required to re-examine the issue and produce new proposals within 2 years.

3 MEASUREMENT OF MARGINAL SOCIAL COST

Short-run marginal cost (SMC) pricing ensures that prices are set to reflect the additional costs to society associated with an additional kilometre travelled or an additional trip made, given that the capacity of the transport network is held constant. In an ideal world, capacity would then be adjusted until SMC also equalled long-run marginal cost (the cost of carrying extra traffic given that capacity is appropriately adjusted). However, when time-lags or constraints prevent that from being achieved, SMC pricing makes the best use of the available capacity. ECMT (2003) emphasises that it is vital that pricing reforms charge close to the point of use of the infrastructure. They insist that such a system enables rational decisions to be undertaken by individuals and firms, informed by price signals of the full costs of their travel demands, to determine traffic levels and trends in transport demand. Traditional systems of charging for the use of roads through annual vehicle charges and fuel tax do not achieve this. They state that the goal is best achieved by kilometre-based charges, as implemented for HGVs in various European countries, but ideally differentiated by vehicle type, location and time.

The strong direction of EU policy towards marginal social cost pricing has played a major role in orientating research on external costs of transport (Nash and Matthews, 2005). Previous estimates were often based on a top-down approach, which resulted in average costs being used when allocating total external costs to the different modes and vehicle categories. In order to estimate marginal costs, research has turned to test and implement bottom-up methods, in particular econometric and engineering methods.

Econometric methods derive cost functions and then marginal cost estimates on the basis of detailed cost databases. As an alternative, engineering methods can be implemented as the starting-point for economic valuation: 'physical relationships' are derived from detailed databases of physical data, and monetary values are then attached to these. A third approach, the accounting approach, is to take total cost data and allocate each category of cost according to assumed cost drivers.

Infrastructure Costs

The traditional approach to analysing road infrastructure costs has been a cost allocation approach. Across Europe, there have been numerous cost allocation studies that have allocated expenditure on road to different vehicle types. Formulae used in cost allocation vary considerably between European studies. Link et al. (1999) applied the formulae of Austria, Denmark, Switzerland, Germany, the UK, Finland and the Netherlands to data from Austria, Germany and Switzerland. The methods used usually rely on the '4th power rule' (which takes wear and tear to be proportional to the 4th power of the axle weight) for the allocation of pavement maintenance costs but vary greatly in allocation of other costs according to parameters such as vehicle-km, passenger car units-km and gross vehicle-km. As an example, Sansom et al. (2001) derive marginal infrastructure costs (maintenance and renewals) for heavy goods vehicles, based on the British formula, of 3.8–7.6p per vehicle-km. The allocation procedure involved is illustrated in Table 12.1.

Except where there is evidence from engineering studies, as with the 4th power rule, cost allocation methods usually proceed on the basis of

Table 12.1 Road wear and tear costs (%)

Description	PCE km	Av gwt-km	Max gwt-km	Sa-km	Include in MC?
Long-life pavements				100	•
Resurfacing				100	•
Overlay				100	•
Surface dressing	20	80			•
Patching and minor repairs		20		80	•
Drainage	100				•
Bridges and remedial earthworks		100			
Footways, cycle tracks & kerbs		100			
Fences and barriers	33	67			
Verges, traffic signs and crossings	100				
Sweeping and cleaning	100				
Road markings	10	90			•
Winter maintenance & misc.	100				
Street lighting	100				
Policing and traffic wardens	100				

Note: PCE = passenger car equivalents; av.gwt – average gross vehicle weight; max gwt – maximum gross vehicle weight; sa – standard axles (a measure of the relative damage due to axle weights). The costs attributed to pedestrians for roads other than motorways (50% of the categories from fences and barriers through to street lighting) are removed prior to allocation to motorised vehicles.

Source: Sansom et al. (2001).

Table 12.2 Results of marginal infrastructure wear and tear cost estimation for roads

Study type	Country	Study	Costs considered	Elasticities	Costs per vehicle-km (eurocents)		
					Mean	Trucks	Passenger cars
Econometric	Germany	Link 2006	Renewal	0.05–1.17 (mean 0.87)		0.08–1.87	
	Austria	Herry and Sedlacek 2002	Maintenance and renewals	1.046	0.16	2.17	0.07
	Switzerland	Schreyer et al. 2002	Maintenance, renewals and upgrades	0.8	0.67–1.15	3.62–5.17	0.42–0.50
Bottom up	Sweden	Lindberg 2003	Renewals	0.1–0.8		0.77–1.86	

Sources: Link et al. (2002). Updated using Link (2006). Elasticity for Austria taken from Herry and Sedlacek (2002).

judgement. Thus it is desirable to have the firmer evidence of robust econometric studies.

Table 12.2 shows the implied marginal infrastructure wear and tear cost from econometric studies in Europe for roads. They show considerable variation in estimates of marginal cost. For example, the cost elasticity for maintenance and renewals/upgrades for Switzerland is approximately 0.8 (and even higher for Austria), for Sweden the renewals cost elasticity is on average 0.4 (ranging from 0.1 to 0.8 depending on traffic density) and for Germany, again for renewals only, the average is 0.87 (ranging from 0.05 to 1.17). Given the wide range of estimates of marginal costs and elasticities, generalisation of these results to other countries is difficult, although it would be expected that the elasticity would be higher where traffic levels are higher and pavements narrower. There are particular difficulties in the case of renewals, as these are related to traffic levels over the life of the road rather than current traffic levels. The Lindberg study (2003) overcomes this by regressing pavement life on cumulative traffic levels. These studies represent the first steps in this field (Link et al., 2002) and more research is clearly required before reliable generalisations can be made.

Environmental Costs

The most common approach to the calculation of environmental costs is the impact pathway approach (IPA), originally developed in the ExternE project series (see, for example, Friedrich and Bickel, 2001). The IPA adopts the principle of estimating costs, starting from the chain of events (physical changes) induced by the transport activity: the emission of a pollutant, its diffusion and chemical conversion in the environment, its impacts on the various receptors (humans, crops and so on). The impact is then 'monetised', that is, valued in monetary terms. In other words, information is generated on three levels: (i) the increase in pollution, (ii) the associated impact and (iii) the monetary valuation of this impact.

The dominant cost of air pollution is its impact on human health, which is typically valued in terms of the number of years of life lost, based on stated preference studies. Noise valuation may be based on stated preference analysis or on revealed preference in the form of house price studies. Global warming is valued based on studies of either damage costs, which are very uncertain, or the abatement costs of reaching specific targets (for example, Kyoto).

Table 12.3 shows environmental costs for the inter-urban case studies in UNITE. The UNITE project generally used conservative estimates of costs to obtain lower-bound figures; some other studies (for example, INFRAS/IWW, 2004) produce much higher values.

Table 12.3 Range of environmental costs of Euro 2 diesel HGV on inter-urban roads (eurocents per vehicle-km)

Study	Cost
Air pollution	2.0–7.5
Greenhouse gases	2.2–2.7
Noise	0.1–5.0

Source: Bickel et al. (2005).

Congestion Costs

There is general consensus about the methodology to be used to estimate marginal congestion costs or benefits. Three basic steps are usually undertaken, each of which is associated with specific data requirements:

- Determination of a volume–time relationship. This relationship is given by engineering speed–flow curves in inter-urban road transport.
- Derivation of values of time and other operating costs.
- Estimation of the price elasticity of demand: this is needed to determine the optimal level of congestion charge, as the imposition of a charge itself induces changes on the volume of traffic and on the level of congestion.

For road transport, extensive research on the required relationships exists; the main outstanding problem is that different models use different speed–flow relationships and therefore yield different results. Latest research increasingly uses detailed transportation modelling able to capture in a more refined way the possible directions of users' reactions (Lindberg, 2003). Among the more recent studies on this issue are the UNITE project (Doll, 2002), RECORDIT (2003) and the INFRAS/IWW study (INFRAS/IWW, 2004).

Table 12.4 shows estimates of marginal external costs of congestion for different types of road in Great Britain at the 1998 level of traffic, showing the enormous variation; similarly large differences are observed by time of day.

Accident Costs

The methodology for the estimation of traffic accident costs is well established and includes the estimation of the risk of accidents and of both the

Table 12.4 Estimates of the marginal external costs of congestion by road types (pence per vehicle-km, 1998 prices and values)

Category	Value	Category	Value
Central London		Urban 15–25 km^2	
Motorway	53.75	Trunk & principal	7.01
Trunk & principal	71.09	Other	0.00
Other	187.79		
Inner London		Urban 10–15 km^2	
Motorway	20.10	Trunk & principal	0.00
Trunk & principal	54.13	Other	0.00
Other	94.48		
Outer London		Urban 5–10 km^2	
Motorway	31.09	Trunk & principal	2.94
Trunk & principal	28.03	Other	0.00
Other	39.66		
Inner Conurbation		Urban 0.01–5 km^2	
Motorway	53.90	Trunk & principal	1.37
Trunk & principal	33.97	Other	0.00
Other	60.25		
Outer Conurbation		Rural	
Motorway	35.23	Motorway	4.01
Trunk & principal	12.28	Trunk & principal	8.48
Other	0.00	Other	1.28
Urban >25 km^2			
Trunk & principal	10.13		
Other	0.72		

Source: Sansom et al. (2001).

material (including property damage, administrative costs, medical and hospital costs, net lost production and congestion caused) and non-material (emotional and social costs of casualties resulting from transport accidents) costs (HLG, 1999).

Non-market estimation techniques such as the contingent valuation method (CVM) are the most commonly used methods to assess the risk value: they allow researchers to elicit the willingness to pay (WTP) of the users for a small reduction of risk (and hence the value of a statistical life).

The external element of accident costs varies according to the extent to which average accident costs are paid by users (either directly or through insurance) or by third parties (including taxpayers by funding police and hospitals) – this varies between countries – and the degree to which additional traffic changes the accident risk for existing users.

The biggest uncertainty in estimating external accident costs is in the relationship between traffic volume and risk. Most studies carried out in this area suggest a decreasing trend of accident risk with respect to traffic volume (Lindberg, 2003). Lindberg argues that possible reasons for the declining of risk with the increase in traffic volume are: (i) that an increased traffic volume implies reduced speeds, and as a consequence increased safety and (ii) increased traffic volumes can induce risk-reducing behaviours of unprotected users such as using other routes or using cars instead of non-motorised modes (for example, bicycles). On the other hand, there is clearly an effect of vehicle type: HGVs cause more serious accidents when involved in collisions with light vehicles or pedestrians.

Thus it is seen that while uncertainties still exist, considerable progress has been made on valuation of externalities. It seems that it should be possible at least to agree on lower-bound valuations.

4 MODELLING IMPACTS

Another major concern leading to opposition to pricing reform has been worry that such reform will have damaging effects on the economy, particularly in peripheral regions. Many projects have carried out work relevant to answering this question; two will be referred to here – IASON and TIPMAC.

The IASON project, which undertook an impact assessment of SMC in the road freight market throughout Europe, was carried out using the SCENES model (Tavasszy et al., 2004) and the valuation of externalities from UNITE. A computable general equilibrium model was used to assess regional impacts.

Major impacts of marginal social cost pricing were in terms of route choice (less traffic in urban areas) and type of vehicle, with a higher proportion of the larger trucks. There was some mode shifting – around 2 per cent of road freight tonnes and 6 per cent of tonne-km transferred to rail or water. But about half the reduction in total truck traffic stemmed from changes in patterns of trade, with inputs to production and consumer goods being sourced more locally. The reduction in truck traffic was greatest in the core and less at the periphery. On the other hand there was a clear tendency, ignoring use of revenue, for output and employment in peripheral countries to be reduced more than at the core. Given that most revenue would accrue to the core countries, this is clearly a worry. More efficient pricing without compensating mechanisms does appear to benefit core countries at the expense of the periphery. However, TIPMAC again used SCENES but this time with an input–output model to examine economic impacts and computed the effects if revenues were recycled in the form of reduced income tax. In this case all countries benefited, though some much more than others.

5 NEW CHARGING SYSTEMS

Although Norway and Sweden used to have kilometre-based HGV charges, these were abandoned when Sweden joined the European Union, and by the 1990s no such charges existed in Europe. Since then, three European countries have introduced kilometre-based charges: Switzerland, Austria and Germany. Britain also proposed a km charge for HGVs but has postponed it to coincide with the introduction of inter-urban road pricing for all vehicles, while other countries – such as the Netherlands and Sweden – are examining the issue.

Switzerland

Although not a member of the European Union, Switzerland was the first country to introduce a kilometre-based charge for HGVs. Of course this was not constrained to follow the Eurovignette Directive. The Swiss Heavy Vehicle Fee (HVF) came into operation on 1 January 2001. The charge was levied on the entire Swiss public road network, applying to both Swiss and foreign vehicles alike, weighing over 3.5 tonnes. It coincided with Switzerland giving way to pressure from the EU to permit heavier goods vehicles, with the weight limit rising from 28 tonnes, first to 34 and then to 40.

The charge level of the fee was calculated as the average uncovered cost per tonne-km. The first step was to calculate the uncovered costs of heavy traffic. This included uncovered road infrastructure costs and external costs caused by heavy vehicles. Damages caused by congestion or the greenhouse effect were not considered. The external costs were found from studies and were divided into three areas that could be given monetary values: air pollution, noise and accidents (Balmer, 2003).

This was then divided by tonne-km to obtain the level of charge. The fee varies according to three factors: distance (kilometres travelled on Swiss territory), weight (admissible weight of vehicle and trailer) and the emissions of the vehicle. Therefore the HVF is calculated by:

$$\text{Rate} \times \text{Distance travelled in Switzerland} \times \text{Weight of vehicle and trailer} \times \text{Emissions factor.}$$

Two systems were developed: one for domestic and one for foreign vehicles, in order to gather the relevant data. Each domestic vehicle has to be fitted with an onboard unit (OBU) which is connected to a tachograph, enabling the OBU to register the kilometres driven. The installation of an OBU is not mandatory for foreign vehicles, but is available on request. For an unequipped vehicle, the fee is registered by using an identification card at the special terminals for HVF clearance. Thus the technology is simple and relatively inexpensive, but can handle only a single charge per km for each vehicle type – no differentiation in time and space is possible.

Balmer states that there are three decisive reasons for the political implementation of the HVF. First, before the final implementation project started, it was criticised heavily, but the political deal of introducing the HVF to outbalance the negative effects of the higher weight limit ensured that the project was on safe political grounds again. Second, the way the revenue of a pricing project is used is important, as learned from the EU project PRIMA (Pricing Measures Acceptance). The project showed that acceptance was good when the revenue was reinvested in transport infrastructure in road and public transport. A large majority of people agreed that up to two-thirds of the revenue from the HVF should be used for projects in public transport. This decision fits well with the strategy of shifting goods from road to rail and helps finance the new railway lines. The remaining third goes to the cantons where it is used mainly for road purposes. And finally, one of the strongest arguments in favour of the HVF was its link to the polluter-pays principle.

Austria

Motorway tolling had existed in Austria since 1968 as the first toll motorway – A13 Brenner motorway – connected Austria and Italy. The Austrian HGV charge came into force on 1 January 2004. It applied to all vehicles exceeding 3.5 tonnes, using the Austrian motorways and expressways. It is based on the distance travelled and the number of axles. The main motivations for the charge were to finance the motorway network and to attribute costs more fairly according to use.

The Austrian system uses a dedicated short range communication (DSRC) system, which is the main type of microwave system used for road tolling. DSRC is based on 400 roadside beacons distributed across Austria's 2,000 km autobahn network. Onboard devices (Go-Boxes) are used to communicate with these beacons and track truck movements across the network and calculate the toll level. Tolls are either paid via a centrally registered account or prepaid by topping toll credit in advance through the internet or sale points. Go-Boxes must be fitted on all vehicles with a gross weight of over 3.5 tonnes travelling on the Austrian motorways (McKinnon, 2005).

Thus Austria represents more sophisticated technology where some differentiation in time and space is possible, but where it would be impractical to charge all roads.

Germany

The German HGV charge was introduced on 1 January 2005, applying to all lorries exceeding 12 tonnes gross weight. The tax is calculated based on the vehicle's environmental status (engine emission levels) and the number of axles.

Rothengatter (2002) explains that the objectives of the study into the HGV charge were to derive fair and efficient user charges for the different vehicle categories using the federal roads and to ensure that charges for infrastructure costs recovered all costs, including capital costs, and took into account future re-investment cycles, new investment and current expenditures. It was necessary that all users should bear exactly the costs that they were responsible for.

European law (European Parliament and Council, 1999: Directive 1999/62/EC of 17 June 1999) required that the toll rate had to be based on actual infrastructure costs: 'The weighted average tolls shall be related to the costs of constructing, operating and developing the infrastructure network concerned'. External costs were not included. The vehicle category charge had to be based on the category's average infrastructure cost. It was possible to differentiate the charge by the time of day (peak/off-peak) and by environmental performance (emission category). The German government decided initially to differentiate only according to environmental performance.

By introducing the HGV toll system, the German government believed that there would be more rigorous application of the user-pays principle to domestic and foreign users. HGVs are responsible for much of the cost of construction, maintenance and operation of motorways, and a distance-based toll will allow HGVs to make a contribution towards infrastructure costs. It was suggested that more efficient use would be made of transport infrastructure capacity due to the tolls (Hahn, 2002).

The German government decided to invite bids for a private sector operator to run the system of upgrading, maintenance, operation and financing. The idea was to have a combination of tolling and public–private partnership models and the operator has to pre-finance the system. This allows the private operator to receive a share of the tolls collected on a stretch of motorway. There was additional relief for public budgets by switching from tax- to user-funded infrastructure.

The German system mainly relies on satellite tracking to determine the distance trucks travel on the autobahn network. In mid-2005, around 70 per cent of the trucks on the network were fitted with OBUs which use GPS satellite signals and other positioning sensors to track vehicle movement, calculate the toll charge and communicate information to the agency responsible for collecting the toll. Toll revenue is then collected at the end of each month by direct debit from registered accounts, credit cards or fuel cards. For vehicles without OBUs, payments can be made for particular trips in advance either online or at any of the 3,500 toll-station terminals.

Thus Germany has the most sophisticated pricing system of the three countries, which in principle could be extended to cover all roads, and to differentiate in space and time as well as by vehicle type.

6 IMPACTS OF THE NEW ROAD INFRASTRUCTURE CHARGES

Switzerland

Switzerland has the highest charges of the three countries, averaging 1.6 eurocents per tonne-km (or for a lorry with a payload of 20 tonnes, 32 cents per vehicle-km). Balmer (2003) explains that the combination of the introduction of the HVF in Switzerland with the allowance of heavier vehicles led to remarkable changes within road transport. There was a change in fleet composition because in the year before the introduction of the HVF, sales of HGVs increased by 45 per cent. Truck owners saved money as new vehicles belong to the lowest and therefore cheapest emission class and the admissible weight of the trucks in the fleet could be better matched to the actual needs of the market. The HVF system led to a concentration in the haulier industry, either through mergers or closure of smaller firms. Larger firms were able to manage their vehicles more efficiently and avoid empty runs, as empty vehicles cost as much as fully loaded vehicles. In terms of road performance, nationally there was a change to the growth trend as annual increases of vehicles on motorways were replaced by a fall after the change from a flat fee to a distance-related fee. In transit traffic across the Alps, the higher weight limit led to an increase in articulated lorries, which was almost outbalanced by a decrease in lighter lorries. This meant that the total number of lorries crossing the Swiss Alps in 2001 was stable and is currently about equal to the level before the HVF.

A four-year study found that there were no significant changes in the modal split, rail retaining its unusually high market share. The study states:

> The new traffic regime has led to a sustained change in the road haulage sector. The trend towards an ever growing number of lorries on the roads has been

> broken and the negative effect on the environment shows a significant decrease. The rail sector's share of freight remained steady. (Swiss Federal Office for Spatial Development, 2004).

McKinnon (2005) stated that once the new trans-Alpine rail tunnels which are largely funded by HVF revenue are opened in 2007 and 2014/15, rail would capture a much larger share of the Swiss freight market.

Balmer (2003) concludes that the system works well overall. Truck traffic has been reduced and there is an incentive to buy cleaner vehicles. The rail market share has been protected despite the advent of heavier goods vehicles.

Austria

After two years of operation, over 590,000 Go-Boxes had been distributed and there are 3,000 user contracts with Swiss TRIPON-Box. The performance rate is believed to be high at over 99.9 per cent. On an average workday, 1.8 million toll transactions are estimated. There are estimated to be a daily average of 800 toll violators which amounts a small rate of 1.4 per cent of toll dodgers.

In Austria, charges are also relatively high, averaging 22 eurocents per vehicle-km. The Austrian toll generated €600 million in 2004. Revenue for 2005 was expected to rise further to €780 million. The costs of the system were estimated to amount to approximately 12 per cent of the revenue.

User acceptance is believed to be high due to a user-friendly system, although there are some problems of local traffic diversion to untolled roads. There does not yet appear to be evidence of other effects.

Germany

The German charges are the lowest of the three, averaging 12.4 eurocents per vehicle-km. The scheme was expected to raise around €3 billion a year, which is proposed to be spent on road and rail infrastructure. One year on since it was introduced, Kossak (2006) states that revenue of €2.86 billion had been generated, which is in line with expectations. Some 23 billion vehicle-km had been travelled in the first 11.5 months, where 35 per cent was travelled by foreign trucks. Approximately 1 million tolling transactions were estimated to have taken place each day, and the toll violator rate was 2 per cent (nearly 300,000 penalty proceedings).

According to Kossak, it was found that there has been no apparent increase in transport charges and no significant impact on the structure of the trucking industry. Also there seems to be no apparent impact on consumer prices; however, according to model calculations, the increase would be 0.15 per cent on average if the toll were fully compensated. There has been a significant tendency towards a higher average load factor.

So far, there has not been any significant shift from road to rail or inland waterways. However in some areas, many trucks use alternative toll-free routes which causes environmental and safety problems. These routes are expected to be tolled in 2006. It has been estimated in Germany that 'up to 5% of truck traffic has diverted away from motorways to minor roads to avoid the charges' (*International Freighting Weekly*, 2005). There has been an increase in sales of trucks with higher environmental standards.

7 CONCLUSION

In this chapter we have critically examined the proposed revisions to the Eurovignette Directive, and the new systems of HGV charging in Switzerland, Austria and Germany. The first conclusion to be drawn is that none of these proposals really implements the EC's stated policy of marginal social cost pricing. In each case they rely on some sort of allocation of total cost to determine the average level of toll. For EU members, the new directive, like the old, ties charges to average infrastructure costs. In Switzerland, charges include environmental costs but not congestion. In Austria and Germany (where the charges were introduced under the old legislation) neither environmental nor congestion costs are taken into account in the overall level of charges, but charges are differentiated according to the pollution category of the vehicle. In all cases the use of revenue is tied to the transport sector.

The principal reasons given for not adopting marginal social cost pricing are the difficulty of valuing marginal social cost and fears as to the impact of such pricing, particularly in peripheral countries. There has been extensive research on these issues which suggests that neither argument is valid. Methods exist to get estimates at least of the lower bound of marginal social cost, while the evidence is that, provided that revenue is used efficiently, all countries in Europe would benefit.

The emerging systems offer the potential for charging which reflects the costs of road use much more accurately, by permitting a charge directly related to kilometres travelled, and which may be differentiated by vehicle type and, depending on the technology, in time and space. On balance, it appears likely that all these developments will significantly improve the efficiency with which HGVs are charged for their costs, and thus give better incentives in terms of the types of vehicles used, the times and locations of their use and the competitive conditions between vehicles registered in different countries. But fully efficient charging would require extension of the charging system to all types of vehicles (currently seen as an issue for member states under the subsidiarity principle) and achievement of a level playing field with rail – where Directive 2001/14 already requires charges based on direct cost, with provision for charging for all external costs when this is achieved on other modes, and mark-ups where needed for financial reasons.

NOTE

* This chapter draws heavily on work undertaken for the IMPRINT-Europe and IMPRINT-net projects and we wish to acknowledge the support of the European Commission and the amount we have learned from participants in those meetings, especially Uli Balmer, Wolfgang Hahn, Andreas Kossak, Friedrich Schwarz-Herda and Christophe Deblanc. Responsibility for its content is, however, solely our own.

REFERENCES

Balmer, U. (2003), *Practice and Experience with Implementing Transport Pricing Reform in Heavy Goods Transport in Switzerland*, IMPRINT.

Bickel, P., S. Schmid and R. Friedrick (2005), 'Environmental costs', in Nash and Matthews, (eds).

Commission of the European Communities (CEC) (1995), Green Paper: *Towards Fair and Efficient Pricing in Transport Policy: Options for Internalizing the External Cost of Transport in the European Union*, Luxembourg: Office for Official Publications of the ECU.

Commission of the European Communities (CEC) (1998), White Paper: *Fair Payment for Infrastructure Use: A Phased Approach to a Common Transport Infrastructure Charging Framework in the European Union*, Luxembourg: Office for Official Publications of the ECU.

Commission of the European Communities (CEC) (2001), White Paper: *European Transport Policy for 2010: Time to Decide*, Luxembourg: Office for Official Publications of the ECU.

Doll, C. (2002), *Transport User Cost and Benefit Case Studies*, Deliverable 7, UNITE (UNIfication of accounts and marginal costs for Transport Efficiency), Funded by EU 5th Framework RTD Programme, Institute of Transport Studies, University of Leeds, and Karlsruhe.

ECMT (2003), 'Reforming transport taxes', European Conference of Ministers of Transport, Paris.

European Commission (2006), *IP/06/383: Sustainable Transport: Towards Fair and Efficient Infrastructure Charging*, Brussels.

European Parliament and Council (1999), *Directive 1999/62/EC of the European Parliament and of the Council of 17 June 1999 on the charging of heavy goods vehicles for the use of certain infrastructures*, Strasbourg.

Friedrich, R. and P. Bickel (eds) (2001), *Environmental External Costs of Transport*, Heidelberg: Springer Verlag.

Hahn, W. (2002), *Implementing Transport Pricing Reform in Germany*, Federal Ministry of Transport Building and Housing, IMPRINT-Europe.

Herry, M. and N. Sedlacek (2002), *Road Econometrics: Case Study Austria*, Deliverable 10 Annex A1c, UNITE (UNIfication of accounts and marginal cost for Transport Efficiency), Funded by EU 5th Framework RTD Programme, Institute of Transport Studies, University of Leeds, Leeds.
HLG (1999), *Calculating Transport Accident Costs*, Final Report of the Expert Advisors to the High Level Group on Infrastructure Charging, Brussels: European Commission, 27 April.
INFRAS/IWW (2004), *External Costs of Transport: Accident, Environmental and Congestion costs of Transport In Western Europe*, Zurich/Karlsruhe.
International Freighting Weekly, 4 July 2005.
Kossak, I.A. (2006), 'The German experience of HGV tolling', Paper presented at IMPRINT-Net Interurban Road Expert Group Meeting 1, 25 April.
Lindberg, G. (2003), 'Estimating external cost', Paper presented at IMPRINT-Europe seminar, Brussels, 1 October, http://www.imprint-eu.org/public/Papers/IMPRINTHGV_lindberg.pdf.
Link, H. (2006), 'An econometric analysis of motorway renewal costs in Germany', *Transportation Research Part 1*, **40**: 19–34.
Link, H., J. Dodgson, M. Maibach and M. Herry (1999), *The Costs of Road Infrastructure and Congestion in Europe*, Heidelberg: Physica/Springer.
Link, H., J. Dodgson, M. Maibach and M. Herry (2002), *Case Studies in Marginal Infrastructure Costs*, Deliverable 10, UNITE (UNIfication of accounts and marginal costs for Transport Efficiency), Funded by EU 5th Framework RTD Programme, Institute of Transport Studies, University of Leeds.
McKinnon, A.C. (2005), *Application of road-user charging to trucking operations in Europe: A review of the tolling schemes and assessment of their possible impact on logistics systems*, Conference Proceedings Logistic Research Network 2005, Plymouth.
Nash, C. and B. Matthews (eds) (2005), *Measuring the Marginal Social Cost of Transport*, Research in Transportation Economics, Vol. 14, Elsevier.
RECORDIT (2003), *Final report. RECORDIT: Real Cost Reduction of Door-to-Door Intermodal Transport*, Funded by EU 5th Framework RTD Programme, Institute of Studies for the Integration of Systems (ISIS) Rome.
Rothengatter, W. (2002), *Charging systems for the use of transportation infrastructure*, Institute for Economic Policy Research (IWW), Project for European Commission, Karlsruhe, Germany.
Sansom, T., C. Nash, P. Mackie, J.D. Shires and P. Watkiss (2001), *Surface Transport Costs and Charges: Great Britain 1998*, A Report for the Department of the Environment Transport and the Regions (DETR), London.
Swiss Federal Office for Spatial Development (ARE) (2004), *Fair and Efficient: The Distance-related Heavy Vehicle Fee (HVF) in Switzerland*, Berne: ARE.
Tavasszy, L., G. Renes and A. Burgess (2004), *Final report for publication: Conclusions and recommendations for the assessment of economic impacts of transport projects and policies*, Deliverable 10, IASON (Integrated Appraisal of Spatial economic and Network effects of transport investments and policies), Funded by EU 5th Framework RTD Programme, Delft, Netherlands.

13. Worse than a congestion charge: Paris traffic restraint policy

Rémy Prud'homme and Pierre Kopp

1 INTRODUCTION

Congestion charges are good in theory, but can be bad in practice. Studying the case of London, Prud'homme and Bocarejo (2005) concluded that implementation costs were higher than the time gains for remaining car users (net of the welfare loss of evicted car users) and the time gains for bus users and environmental gains. Studying the case of Stockholm, Prud'homme and Kopp (2006) reached a similar conclusion. However, the shrinking of road space policy followed in Paris is much worse: it is bad in theory and bad in practice.

Let us begin by clarifying the meaning of 'Paris'. The expression is used, and can be used, to designate two different realities: the 'Paris agglomeration' and the 'Paris municipality'. The Paris agglomeration is an economic and social entity with a population of 11 million people, which functions as a largely integrated labour market, in part thanks to good highway and public transport systems. It comprises over 1,000 municipalities, the basic French politico-administrative unit. The Paris municipality, with about 2 million people, is one of them. It is obviously the most important one, and the heart of the agglomeration, but it represents only a fifth of the agglomeration in terms of population, and much less in area. The many studies that compare Paris defined as a municipality (2 million people) with London (7 million people) or New York (9 million people) in terms of age distribution or employment structure or productivity are meaningless. In this chapter, we shall use the word 'Paris' to mean the Paris municipality.

Why focus on the Paris municipality, when the significant socio-economic – and transport – realities relate to the Paris agglomeration? The answer is because a number of policy decisions are taken at the municipality level. In 2001, a new team was elected in the Paris municipality, consisting of Socialists, Communists and Greens, and headed by Bertrand Delanoe, a Socialist. Although the Socialists are the dominant force, they need the support of the Greens. Green politicians were put in charge of transport policy (and of environmental policy), and the transport policy followed reflects their priorities rather than those of the Socialists (although the latter did endorse it). It is this policy which is examined in this chapter, by comparing data for 2000 with data for 2004. Obviously, outcomes are not entirely the result of the policies followed. They are also influenced by all sorts of factors or trends that we shall try to identify and take into consideration. Figure 13.1 presents the analytical framework used.

The impact of the transport policy on street supply (a reduction) led to changes in car speeds (a reduction) which had consequences on car trip costs (an increase). In principle, transport policy could also impact on public transport supply, which would lead to changes in public transport

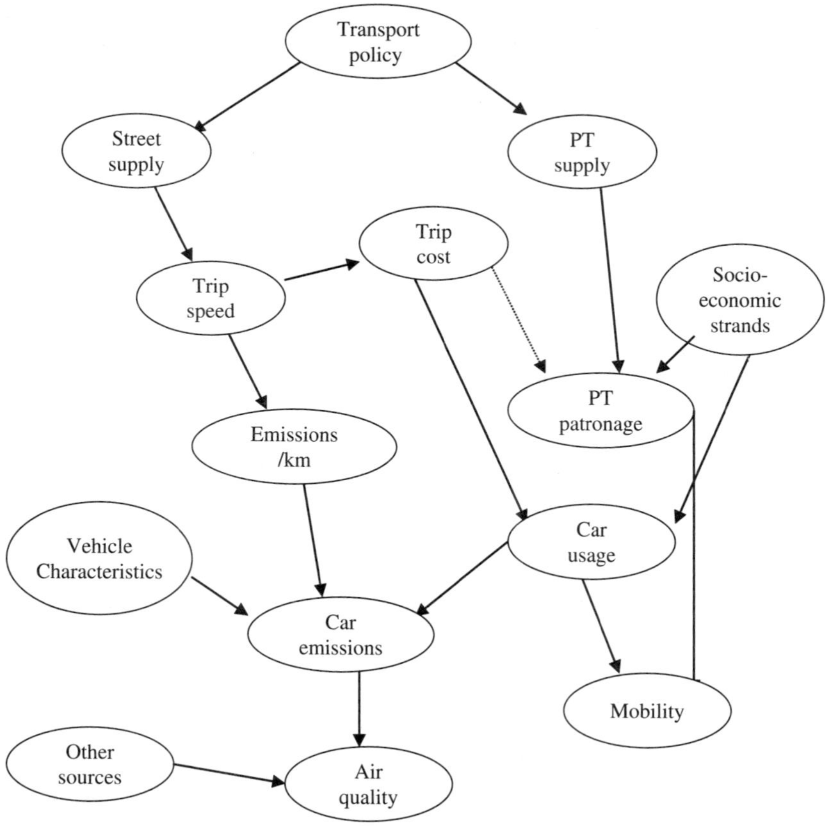

Note: PT = Public transport.

Figure 13.1 Analytic framework

patronage, which would combine with changes in car trip costs to produce changes in car usage. As we shall see, this did not happen in Paris. It follows that changes in car usage (a reduction) can be attributed to the changes in car trip costs (and probably also to changes in exogenous socio-economic factors) that also affected public transport patronage. Changes in car usage and in public transport patronage let to changes in mobility (a decrease).

Changes in speed also had an influence on unit pollutant emissions, that is emissions per kilometre (an increase). Combined with changes in car usage, and with the exogenous changes in vehicle characteristics, this led to changes in pollutant emissions (a decrease). Combined with changes in other pollution sources, this led to changes in air quality.

Section 2 offers a brief description of the transportation situation in 2000 as a benchmark, before the new policy was introduced. Section 3 continues with a presentation of the policies undertaken. The 2000–04 changes are then presented and analysed, and related to the policy undertaken. This makes it possible to estimate the costs and benefits of the policy (Section 4). Section 5 concludes.

2 PARIS TRANSPORT IN 2000

It is useful to try to characterise the transport situation before the new policy was introduced. To this end, five points can be emphasised.

Rich Public Transport Supply

A first point relates to the abundance and quality of public transport supply. With 16 lines and 380 metro stations (of which 327 are in the Paris municipality), and a high frequency, the Metro offers 68 million seat-km per day. With 60 lines and 1,270 bus stops, much lower frequency and capacity, Paris municipality buses offer 8 million seat-km per day (suburban bus lines offer more than twice that amount). The picture is completed by trains, in particular the RER (high-speed regional trains), linking suburban areas to the Paris municipality.

Important Street Network

Thanks in particular to the arterial roads created in the late nineteenth century by Baron Haussmann, Paris is also well endowed with boulevards and avenues, more so than most similarly dense cities. The road network comprises about 1,500 km, including about 400 km of 30-metre wide arterials. There are about 750,000 parking spaces, 75 per cent of which are private (underground private garages or parking lots). Free street parking accounts for about 7 per cent of the total. To this road network should be added the 35 km of ring road that circles the Paris municipality (and legally belongs to it). This ring road is the most travelled road in Europe. It is estimated that about 65 per cent of its traffic is Paris-municipality related (the rest being non-Paris to non-Paris traffic).

Decrease in Car Traffic

Contrary to what most people believed and to what was repeatedly stated by politicians, car use in the Paris municipality did not increase in the 1990s, but decreased. Two different sources give different estimates of this decrease: 0.5 per cent per year according to the 1991 and 2001 transport surveys, 2 per cent per year according to the Paris municipality Observatoire des Transport (Transport Observatory). This decrease is not surprising: during the period, the Paris municipality population stagnated and employment declined. The number of trips with Paris as the origin and/or destination could hardly not diminish.

Balanced Modal Distribution

Surprisingly, it is not easy to allocate Paris municipality trips to the various modes. The transport survey gives useful orders of magnitude, but it excludes trips by non-Paris region residents (all tourist trips, for instance, are ignored) and most goods vehicle trips; also it does not discriminate between the various public transport modes. RATP (Régie autonome des transports Parisians), the Paris transport authority, publishes detailed statistics on public transport, but little is known about car trips. The Transport Observatory monitors car movements only on 190 km of the 1,500 km road network. Table 13.1 presents the modal distribution of trips in the Paris municipality. It is in passenger-km per day, not in number of trips, and it ignores walking, because this is more meaningful for an analysis of transport policy. It also ignores goods transport, a more serious limitation.

It appears that slightly more than half of motorised passenger transport in the Paris municipality is taking place in dedicated lanes, mostly underground. The other half takes place in the streets of Paris, which, in addition, carry practically all goods vehicles. For the most part, public transport and cars travel in different lanes, and are therefore not in conflict. The exception is buses, which in 2000 accounted for 4 per cent of passenger transport, or 9 per cent of transport in the streets of Paris – less if we consider total transport including goods vehicles.

Table 13.1 Modal distribution of motorised transport in Paris municipality, 2000

Transport mode	Pass-km/day (m)	%
Below ground		
Metro	17.7	29.0
Train (SNCF + RER)	14.2	23.3
Total	31.9	52.3
Above ground		
Bus	2.7	4.4
Motor vehicles	24.7	40.6
Taxis	0.6	0.9
Motorbikes	1.1	1.8
Total	29.1	47.7
Total (% rounded)	61.0	100.0

Sources: Authors' estimates, based on data from the Transport Observatory and from RATP.

Table 13.1 describes only motorised transport. It does not include bicycles, whose importance is small (estimated to be 0.2 per cent of passenger-km on the streets and 0.1 per cent of all passenger-km). The share of Paris municipality residents is not known precisely, but it is likely to be less than 50 per cent. It is higher for buses, and probably the Metro, but lower for cars (particularly on the ring road) and for the RER and trains.

Declining Pollution

According to repeated claims by the media and politicians, air quality was deteriorating rapidly. According to a December 2000 survey, 94 per cent of Paris agglomeration residents were convinced that air quality had deteriorated during the previous decade. However, the reality was the exact opposite. Air pollution declined markedly in Paris in the 1990s, a fact that could easily be verified on the website of Airparif, the agency in charge of monitoring air quality in the Paris region (www.airparif.asso.fr). Table 13.2 compares the rates for six pollutants in 1991 and 2000. These figures relate to the Paris agglomeration but are representative of the Paris municipality, particularly for the assessment of change. The table shows that in the nine years preceding the traffic restraint policy, the decline was significant for all primary pollutants; indeed, lead disappeared altogether from Paris air. Only ozone, a secondary pollutant, increased. CO_2 emissions also increased, but CO_2, a greenhouse gas with worldwide effects, is not monitored at the local level.

Table 13.2 Pollution in Paris, 1991 and 2000

	1991 (mgr/m^3)	2000 (mgr/m^3)	Change (%)
Sulphur dioxide (SO_2)	27	9	−67%
Nitrogen oxides (NOx)	105	70	−33%
Benzene	4.9	1.9	−61%
Ozone (O_3)	19	35	+84%
Particulates	34	15	−56%
Lead	0.16	0	−100

Source: See airparif.asso.fr.

3 POLICIES UNDERTAKEN

What Policy Changes Were Introduced by the Municipal Team Elected in 2001?

Public transport supply

The first point worth noting is that policy changes had little effect on the supply of public transport. As shown in Table 13.3, public transport increased only marginally between 2000 and 2004. The number, frequency and comfort of the Metro, buses and trains did not increase significantly. The Paris municipality is not responsible for this, because it does not control public transport supply on its territory. This supply is determined by STIF (Syndicat des Transports de l'Île de France) which was then controlled by the government (it is now controlled by the Paris Île de France region), with the SNCF (Société nationale des chemins de fer français), the national rail company, and RATP, the national bus and Metro company, as executing agencies. The Paris municipality, with the help of STIF, did invest in a tramway line, but this line was only recently opened and is obviously not considered here.

Anti-car policies

The policy followed was not pro-public transport, it was anti-car. It was clearly articulated in 1999 by Ms Chantal Duchêne, who was in charge of transport at DREIF, the Paris region directorate for equipment of the central government (and simultaneously elected as a Green): 'It will be necessary to reduce the space available for automobiles. Thanks to bus lanes, cycle tracks, and wider pavements, car speeds will decrease, making other modes more attractive' (*Journal du Dimanche*, 9 September 1999).

Table 13.3 Public transport in Paris, 2000 and 2004

	2000	2004	Change (%)
Metro			
Length of lines (km)	167.7	168.4	+0.4
Stations	326	327	+0.2
Seat-km per year (billion)	24.8	26.2	+5.6
Buses			
Lines	59	59	0.0
Length of lines	565.5	567.8	+0.4
Seat-km per year (billion)	3.0	3.1	+3.3

Source: RATP.

The deputy mayor in charge of the environment in the new municipal team said the same thing in more forceful terms: 'It is only by making it hell for car drivers that we will force them to give up their damned cars'. It is this traffic restraint policy, announced in no uncertain terms before the elections, that has been implemented since 2001.

The main thrust of the policy was the reduction in road space for private cars and goods vehicles. Dedicated bus lanes were increased and expanded. In many cases, the widening of these bus lanes, and the construction of 60 cm high walls to separate them from the rest of the road, resulted in the elimination of one, and in some cases two, car lanes. Four-lane avenues were transformed into three- or even two-lane avenues. Rue Beaubourg was downgraded from three lanes to one lane (on one particular stretch, but it is that stretch that regulates speed and flow). Road space was also reduced in streets not served by any bus, by means of cycle tracks, or by the widening of pavements. Some streets were also closed to car traffic. The relative importance of this road space policy – proudly put forward by the municipal team – is difficult to assess. Fifteen per cent (one-third of space on 40 per cent of streets in terms of traffic) might be a likely estimate.

Another element of the policy relates to parking, which was made more difficult. Street parking fees for 'residents' (of the Paris municipality or of the area) were greatly reduced (by 80 per cent), encouraging car-owning residents to leave their cars parked in the street. Simultaneously, parking fees for non-residents, including the many suburban dwellers working or shopping in the Paris municipality, were increased (by 30 per cent). Private construction of underground parking space was also discouraged (but this measure will take time to be effective). As a result, the vacancy rates of parking spaces, which were low, decreased by about 50 per cent.

Table 13.4 Paris bus speed and patronage, 2000 and 2004

	2000	2004	Change (%)
Speed (km/h)			
On 16 best protected lines	12.4	12.2	−2
On rest of network	12.8	12.7	−1
On entire network	12.7	12.5	−2
Traffic (m trips/year)			
On 16 best protected lines	135	134	−1
On rest of network	219	214	−2
On entire network	355	348	−2

Source: Calculated from RATP data.

Pro-cycle policies

Additional cycle lanes were created, and bicycles were allowed to use dedicated bus lanes. Since this could have reduced the speed of buses, bus lanes were widened from 3 to 4.5 metres, in order to enable buses to overtake bicycles more safely.

Anti-two-wheeler policies

Unlike bicycles, two-wheelers (motorbikes, scooters) were not allowed to use bus lanes, and parking was made more difficult. Fines for illegal parking increased by 180 per cent between 2003 and 2004.

Policy Outcomes

Decline in bus speed and patronage

One could have expected increased and widened dedicated bus lanes to produce an increase in the bus speeds, in the quality of the bus service, and consequently in bus patronage. However, as Table 13.4 shows, this did not happen. The table is based on bus line by bus line RATP data.[1] Even on the 16 better protected lines, speeds actually worsened.

This might sound surprising. One is used to seeing buses in their dedicated lanes, often empty, and travelling faster than the cars in the congested adjacent lanes. Four tentative explanations can be offered. First, the speed at which buses travel when they are moving is much higher than the average speed at which people are driven in buses, because buses stop at traffic lights and above all because they stop to let passengers get on and off; a 20 per cent increase in peak speed would result in only an 8–9 per cent increase in average speed (and a much lower increase in origin to destination speed for users, because it would not affect access and waiting time). Second, one bus in a bus lane can delay the following buses; at a bus stop, the time lost by one bus cannot be less than the time lost by the preceding bus. Third, the management of certain intersections has been made more complicated by dedicated bus lanes; in some places, traffic lights go green successively for cars and then buses, automatically increasing waiting time for both types of vehicles. Finally, bus lanes, unlike the Metro, are not entirely dedicated, and many buses lose in 'ordinary' traffic jams the time they might gain in their dedicated lanes.

What is certain is that bus patronage did not increase. Since bus supply did not increase either in quantity (no additional lines) or in quality (no increased speed or frequency), this stagnation is hardly surprising. Bus demand could have been sensitive to increases in the cost of car travel. But it seems that the cross-elasticity of bus demand to the price of alternative modes is very low. A large share of car traffic comes from non-Paris municipality residents, for whom buses are not an attractive alternative – you never see a plumber from a suburban municipality boarding a bus with a sink under his arm. The bus, with its frequent stops, is well adapted to very short trips (the average length of a bus trip is 2.2 km). Metro and car trips are on average much longer.

Decrease in car speed and traffic

The shrinking of road space and other road traffic impediments resulted in lower speeds and therefore less traffic. The causal link should be emphasised. Why would traffic decline, if not because of lower speed and more generally higher costs of car transport? It did not decline because of improvements in the quality of public transport: there were no such improvements. It did not decline because of increases in fuel prices: between 2000 and 2004, fuel prices in France actually decreased (by about 8 per cent, largely because the euro appreciated relative to the dollar). It must have declined because of lower speeds, and also because of a trend in the decline in the demand for road transport in Paris.

Traffic decline between 2000 and 2004 is known, although imperfectly. The Transport Observatory, which monitors traffic on about 200 km of streets (50 per cent of main streets and 15 per cent of all streets), puts it at 13.3 per cent. Let us accept this number. It must be deconstruct into two parts: the first reflects the general change in population and activity, and the second is the consequence of the policy under analysis. We have two numbers reflecting past trends: one, from previous transport surveys, would put the decline at 2 per cent (over the four-year period), and another one, from the Transport Observatory, would put the decline at 8 per cent. We shall assume a 'natural' 4.3 per cent decline. The impact of the transport policy can therefore be estimated to have produced a 9 per cent traffic decline.

Speed decline between 2000 and 2004 is unfortunately not well documented. The Transport Observatory estimates a 6.3 per cent decline, or 1.6 per cent per year. This is very hard to believe. First, it does not square at all with the daily experience of Parisians, particularly of workers and enterprises involved in goods transport, who report a marked deterioration in driving conditions and time required for delivery. Second, it is hardly compatible with the established stagnation of speeds on bus lines. Above all, it cannot explain the measured traffic decline.[2] This is why we propose our own estimates of changes in car speeds, based on price elasticities of the demand for car travel. Our starting-point is the price elasticity calculated in the case of London: –0.83 (Prud'homme and Bocarejo, 2005), which is already rather high relative to accepted urban car transport elasticities. The price considered is the generalised cost, consisting of the money cost plus the time cost. The money cost is estimated to be 0.15 €/km. The time cost is a function of speed s, the number of passengers per car n, and the value of time t. Taking $n = 1.3$ persons per car, and $t = 9$ €/h (the French official value of time), we have for the generalised cost c:

$$c = 0.15 + 9*1.3/s,$$

which yields:

$$s = 11.7/(c - 0.15).$$

Combined with given values of elasticities of traffic to cost, this makes it possible to estimate the decline in speed associated with a given decline in traffic. This is done in Table 13.5, for two estimates of elasticities (−0.8 and −0.5) and two estimates of traffic reduction (−13 per cent and −9 per cent).

Table 13.5 is built upon the hypothesis – reasonable in the absence of an improvement in public transport supply – that it is the reduction in speed, and the cost increase that comes with it, that explain traffic reduction. It appears that speed reductions (as a percentage) are always higher than traffic reductions. This is not surprising, since time is the main component of generalised cost, and elasticities of speed relative to cost are greater than –1 (more precisely: between –1 and 0). The speed change associated with the recorded traffic decline of 13 per cent is –17 per cent with a –0.8 elasticity. The decline is greater (–25 per cent) with a not implausible elasticity of –0.5. If one considers the traffic reduction induced by the policies (–9 per cent), then speed reduction must be –12 per cent, with a –0.8 elasticity, and –18 per cent with a –0.5 elasticity. We shall retain this –12 per cent estimate of speed change. Note that it is a very conservative estimate, the lowest of our four estimates, based on a high elasticity (–0.8) and a low traffic decline (–9 per cent).

Table 13.5 Speed reduction associated with various traffic reductions, 2000 and 2004

	$e=-0.8$	$e=-0.5$
Speed in 2000 (km/h)	17.4	17.4
Generalised cost in 2000 (€/km)	0.822	0.822
With a 13% traffic decline (2004)		
Change in generalised cost	+16.2%	+26%
Generalised cost 2004 (€/km)	0.955	1.036
Speed in 2004 (km/h)	14.5	13.1
Change in speed 2000–04	−17%	−25%
With a 9% traffic decline (2004)		
Change in generalised cost	+11.25%	+18%
Generalised cost 2004 (€/km)	0.914	0.970
Speed in 2004 (km/h)	15.3	14.3
Change in speed 2000–04	−12%	−18%

Note: Generalised costs are costs per vehicle-km. To obtain cost per passenger-km, divide by 1.3.

Source: Authors' calculations.

Table 13.6 Rail patronage, 1996–2004 (%)

	1996–2000	2000–2004
Metro	+15.9	+5.5
RER (high-speed subways)	+7.3	+17.9
Suburban trains	+15.0	+8.4

Increase in rail usage

Metro patronage, and even more so RER and suburban train patronage, increased in the 2000–04 period, as shown in Table 13.6.

RER and suburban train numbers must be treated with caution, because they include changes in suburb-to-suburb trips that have nothing to do with the Paris municipality. The increase in Metro travel, more than 5 per cent, is unambiguous. The question is whether and to what extent it is the result of the Paris municipality traffic restraint policy. Some car drivers discouraged by increased traffic jams generated by this policy must have abandoned their car for the Metro (since they did not do so for the bus). It is nevertheless noteworthy that the increase in Metro patronage was more important in the preceding period (1966–2000) than in the period studied. This suggests that the hoped-for modal shift, if it existed, was probably very limited.

Increase in two-wheeler usage

A marked increase in the usage of motorised two-wheelers (motorbikes, scooters and so on) took place in the 2000–04 period. However, available data on this issue leave much to be desired. Where two-wheeler usage is measured by the Transport Observatory, a 10 per cent increase took place. For people who travel in the Paris municipality, this number seems an underestimate.

Increase in bicycle usage

According to Transport Observatory records, bicycle usage in the Paris municipality increased by 41 per cent between 2000 and 2004. It would have increased from 60,000 passenger-km to 85,000 passenger-km, that is from 0.10 per cent to 0.14 per cent of total passenger transport.

Negative impact on pollution

In 2004, relative to 2000, and as a consequence of the policies followed, there are fewer vehicles in Paris streets driving at lower speeds. Other things being equal, this has a negative impact on pollution in Paris. For most pollutants (and for fuel consumption as well) emissions per km are a function of speed. The general shape of this relationship is well known: it is U-shaped. Emissions per km are high at very low speed, then they decrease, reaching a low plateau in the 40–80 km/h range, then increase significantly. The precise equation of this relationship is not well known. Part of the difficulty is related to the notion of 'speed'. A vehicle driving 'at 20 km/h' in a city is not a vehicle driving at a constant 20 km/h. It is a vehicle driving at 35 km/h, then slowing down, then accelerating, then stopping, then accelerating. From an emissions viewpoint, it is the profile of the speed curve that counts, not its average. This is why emission norms are defined for standard profiles, although these profiles are too simplistic to be realistic. Estimating the elasticity of emissions to 'speed' is therefore a difficult exercise.

A study undertaken by L'Union Technique de l'Automobile, du Motorcycle et du Cycle (UTAC), the agency that controls the conformity of new vehicles to EU norms, on behalf of Lucas Diésel, a diesel engine manufacturer, compared emissions for similar vehicles driven according to (i) the European norm speed profile, resulting in a 19 km/h average speed, and (ii) an actual speed profile, resulting in an 11.5 km/h average speed. We used these measurements to produce the elasticities presented in the first two rows of Table 13.7. These high elasticities are questionable. The purpose of the study was to compare diesel and gasoline engines, and it was designed to that effect, not to compare emissions at different speeds: therefore the recorded differences could reflect differences in speed profiles as

Table 13.7 Elasticities of emissions to speed, various pollutants

	Gasoline (calculated)	Diesel (calculated)	Average (retained)
CO	−15.9	−2.3	−4.6
Particulates	–	−5.4	−2.7
HC	−8.7	−3.1	−3.0
NOx	−4.3	−1.5	−1.5

Note: Actual emisions were measured for same vehicles driven at 19 km/h according to the EU speed profile, and at 11.5 km/h according to an effective speed profile. Retained values are 50% of the average diesel–gasoline numbers.

Source: Calculated from measurements made by UTAC for Lucas Diésel.

Table 13.8 Impact of policies on motor-vehicle pollutant emissions, 2000–2004 (%)

Traffic change	Effective	Policy induced
Traffic change	−13	−9
Speed	−17	−12
Emissions		
CO	+68	+50
Particulates	+38	+29
Hydrocarbons	+44	+33
NOx	+22	+16
Average	+43	+32

Note: The notion of 'average' has no rigorous meaning and is calculated only to give a broad order of magnitude.

Source: Calculated on the basis of Table 13.7.

much as or more than differences in average speeds. To be on the safe side, for the rest of our evaluation we retained values equal to half the calculated elasticities (we also gave an equal weight to gasoline and diesel cars, a reasonable assumption in the case of the Paris municipality).

This made it possible to produce Table 13.8, which shows the policy-induced changes in motor-vehicle pollutant emissions resulting from the combination of decreased traffic and speeds. Note that these numbers do not take into account the changes in the pollution efficiency of vehicles, and the fact that vehicles were cleaner in 2004 than in 2000, the third determinant of emissions as discussed below. Table 13.8 does not signify that vehicle emissions

Table 13.9 Changes in pollution levels in Paris, 2000/1996 and 2004/2000 (%)

	2000/1996	2004/2000
NOx	−25	−17
SO_2	−40	−22
Particulates	−25	−13
Benzene	−62	−15

Note: Data refer to the Paris agglomeration. The pollutants selected are those for which comparable data are available for 1996, 2000 and 2004.

Source: See airparif.org.

actually increased between 2000 and 2004, rather that the impact of Paris municipality policies on emissions was negative. This was compounded by another factor: the replacement of cars by more polluting two-wheelers.

This policy-induced increase in motor-vehicle pollutant emissions did not result in an increase in registered pollution levels in the Paris municipality, for three reasons. First, motor vehicles are not the only source of pollution emissions: they account for only 17 per cent of particulate emissions, 19 per cent of HC and 49 per cent of NOx. Second, pollution levels recorded in the Paris municipality are a function of pollution emissions in the entire Paris agglomeration, including many municipalities that did not institute traffic restraint policies. The third reason, and the most important, is that the marked decline in pollution levels registered in the 1990s (and shown in Table 13.2, above) resulted from the spread of cleaner cars. Vehicles put on the market in 2000 were 10 to 20 times less pollutant than the vehicles put on the market in 1990. A stock effect slows down the decline in emissions, or rather spreads it over time. The considerable progress recorded in the preceding period continued in the 2000–04 period, as shown in Table 13.9.

It appears that air pollution levels in Paris continued to decrease in the 2000–04 period, although they did not decrease as fast as in the preceding four-year period. However, these numbers do not prove much. It is the argument developed that makes it possible to conclude that the traffic restraint policy had a negative impact. The contrast with London is striking. In London, as a result of the policy followed, there were fewer cars driving faster: two reasons for a decline in emissions. In Paris there were fewer cars driving more slowly: one reason for a decline and one – more powerful – for an increase.

We know that fuel consumption, and associated CO_2 emissions, increase with decreased speeds, but we have no estimate of the elasticity of CO_2 emissions to speed. Let us assume an elasticity of −1. This would imply a 12–17 per cent emission/km increase, to be combined with a 9–13 per cent decrease in traffic. This would produce a 3–4 per cent increase in CO_2 emissions. In view of the uncertainties surrounding this estimate, we shall assume that the policy has been roughly neutral relative to CO_2 emissions.

Mediocre performance in road safety

The decrease in speed, other things being equal, should have contributed to a decline in road accidents and in their seriousness. The increase in two-wheeler traffic certainly contributed to an increase in road accidents. Road safety must also be seen in a more global context. Table 13.10 shows road casualty figures in Paris and France in 1996, 2000 and 2004.

The table shows that in the 2000–04 period, casualties decreased *more slowly* in Paris than in France as a whole, whereas in the preceding period it decreased *faster* in Paris than in France. One must also relate road accidents to road traffic, which declined in Paris wheras it increased in France. It appears that the number of casualties per vehicle-km decreased much less in Paris (12 per cent) than in France (44 per cent).

4 COSTS AND BENEFITS

We can now try to estimate the costs and the benefits of the traffic restraint policy introduced by the Paris municipality.

Table 13.10 Road casualties, Paris and France, 1996, 2000 and 2004

	Paris	France
Effective number		
1996	82	8,080
2000	67	7,643
2004	50	5,232
2000/1996 (%)	−18	−5
2004/2000 (%)	−25	−32
Number relative to traffic		
2004/2000 (%)	−12	−44

Note: Numbers relative to traffic are obtained by relating casualty to traffic changes: −13% in Paris, +6% in France.

Sources: Transport Observatory for Paris municipality; Union Routiere de France (URF) (www.urf.asso.fr) for France.

Financial Cost of Public Works Undertaken

Public expenditures on dedicated bus lanes are reported to amount to €65 million for the 2001–04 period. This is not an annual cost – which is the opportunity cost of capital (estimated to be 10 per cent) and the depreciation cost (also estimated to be 10 per cent), amounting to €13 million.

Loss for Car Users

Figure 13.2 shows the losses inflicted on car users. *D* (*q*) is the demand for road usage as a function of the unit costs of road usage. $I_1(q)$ indicates the unit cost of road usage as a function of road usage (road usage determines speed, which determines time spent and therefore time cost) in 2000, before the traffic restraint policy was instituted. Point A is an equilibrium point (with daily traffic of 15.5 million vehicles and a speed of 17.4 km/h, implying a unit cost of €0.822).

The traffic restraint policy shifts curve $I_1(q)$ to $I_2(q)$. A new equilibrium is found in B (with 13.5 million passenger-km and a speed of 14.5 km/h, implying a unit cost of 0.955 €/km). The welfare loss of car users – which is a welfare loss for society at large – is the area EBAD. It is equal to €1.92 million/day, or (by multiplying by 321 days) €618 million per year.

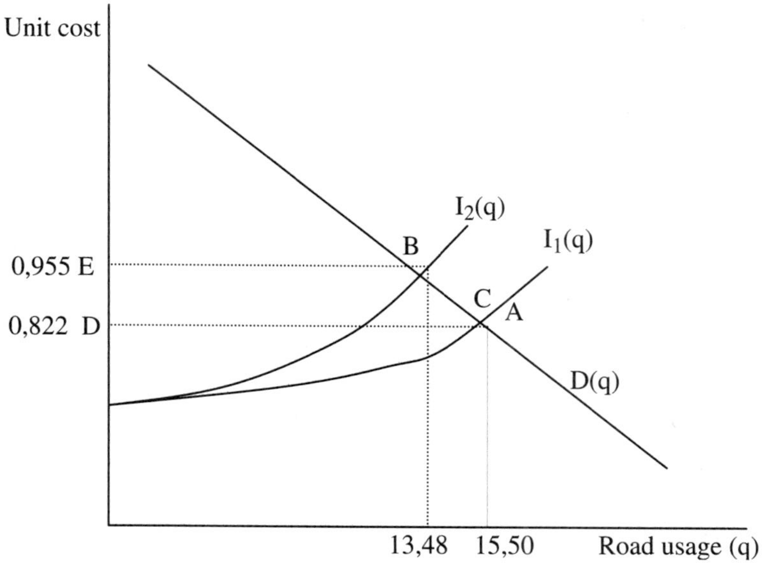

Figure 13.2 Losses associated with traffic restraint policy

Loss for Goods Delivery

Trucks and goods delivery vehicles in the Paris municipality in 2000 were estimated at 1.6 million vehicle-km per day. Let us ignore the fact that some of this has been eliminated (at a cost) and assume that this traffic is still present in 2004. Like cars, delivery vehicles take 0.69 minutes more to drive one km, which is a loss of 5.9 million hours per year. The value of time of goods delivery vehicles in Paris is not known, but cannot be less than €30 per hour, the same as for trucks in France at large. This represents a loss of €177 million per year. Note that this is a conservative estimate, which ignores completely the time lost because of increased parking difficulties.

Environmental Losses

We have seen that the traffic restraint policy led, other things being equal, to a 32–43 per cent increase in emissions (note that pollution levels decreased in 2000–04, and that this increase is an estimate, admittedly a crude one, of the impact of the policy). We can get an idea of the cost of this increase: the official Boiteux report values the cost of pollution in urban areas at €29 per 1,000 vehicle-km. This puts the cost of automotive pollution in 2000 in the Paris municipality at €144 million per year. A 32 per cent increase (to retain the lowest value) implies a cost – a surcharge – of €46 million per year.

Impacts for Bus and Bicycle Users

We have seen that there were no changes in bus speed or in bus frequency between 2000 and 2004 (this is consistent with the lack of change in bus usage). The policy has therefore been neutral for bus users.

If cycle usage increased, as it did (by 40 per cent), this means that the unit cost of usage declined, and that cyclists benefited from the policy. To get an idea of that welfare gain, let us assume that the per km cost in 2000 was equal to the per km cost of bus travel, that is €0.4 per km. Let us further assume a −1 price elasticity of demand for bicycle travel. It follows that the unit user cost changed from €0.4 to €0.24, and that the daily welfare gain is €11,600 ((0.4–0.24)*(60,000 + 85,000/2)). Multiplying by 300 days, this amounts to €3.5 million per year.

Table 13.11 sums up these gains and costs. It is worth emphasising that these estimates have been obtained on the basis of a somewhat cautious hypothesis, and that the numbers presented are in the low range of estimates. Speed decline was estimated with a generous −0.8 elasticity. With a lower elasticity, it is likely that speed decline – and the associated costs – would be

Table 13.11 Costs and benefits of traffic restraint policy

	€m per year
Annual cost of investments	−13
Time loss for car users	−618
Time loss for goods delivery	−117
Environmental costs	−46
Cost/benefit for bus users	0
Benefit for bicycle users	+3
Total	−791

much larger. The values of time used (€9 per hour for passengers and €30 per hour for trucks) are low for the Paris municipality. In the study on the London charge, we utilised a €15.6 per hour value, which was criticised by many (see, for instance, Raux, 2005) as being too low. Obviously, higher values of time would produce higher car-user costs. The combination of a lower elasticity (−0.5) and a higher value of time (€12 per hr) would lead to costs well above €1.5 billion per year.

5 CONCLUSION

The purpose of this study was to analyse the effects of the traffic restraint policy introduced in 2001 in and by the Paris municipality, and to sketch a comparison with the policy introduced in London at about the same time – with the proviso that the geographic, economic, demographic, institutional, historic and political contexts are very different. Table 13.12 presents the results obtained.

The results do not speak highly in favour of the French policy. In both cases, the stated objective of reducing traffic has been achieved. But success in itself is not enough. One must also see how and at what cost it has been obtained. In London, it has been achieved by means of a toll, and accompanied by an increase in motor vehicle speed, generating a time gain for car users, and also for bus users, and a small environmental gain. In Paris, the main policy instrument has been the reduction in road space for motor vehicles, which led to a decrease in motor vehicle speed, but failed to increase bus speed and patronage. This policy generated a considerable time loss for car users and for goods delivery vehicles, and even environmental losses, without gains for public transport users. The only benefit seems to have been a small one for bicycle users: 0.4 per cent of all losses.

Table 13.12 Impacts of transport policies in London and Paris

	London	Paris municipality
Policy instruments	Toll More buses	Less road space Less parking
Physical impacts (%)		
Traffic	−15	−9
Car speed	+17	−12
Bus supply	+250 buses	0
Bus speed	+7	0
Bus patronage	+38	0
Car pollution	−34	+32
Economic impacts (€m/year)		
Implementation costs	−177	−13
Gains/losses for car users	+69	−618
Gains/losses for goods delivery	na	−117
Gains/losses for bus users	+31	0
Cost of bus subsidies	−5	0
Environmental gains/losses	+5	−46
Gain/losses for bicycle users	0	+3
Total	−73	−791

Note: The policy conducted in London concerns only a small part of London: the numbers given here relate registered changes to the magnitudes for the charge zone only. The changes in Paris are for 2000–04; in London they are for 2000–03, but the changes brought by the toll were mostly instantaneous. Time gains in London are calculated using a value of time of €16 per hr; and in Paris, €9 per hr.

Source: London: Prud'homme and Bocarejo (2005).

The downside of the London policy is the implementation cost, which is high, but which can be expected to decline over time.

Figures for Paris cannot readily be compared with those for London because the areas covered by the two policies differ in size. It is nevertheless striking that the Paris policy largely generates only costs, whereas the London policy generates both costs and benefits. To reduce car traffic is not an objective *per se*, but a means to an end. What matters is to improve mobility and to reduce pollution. These objectives, which have not been reached in Paris, have been achieved in London, albeit at a high cost.

One should emphasise the limits of this study. It focused only on the short-term changes introduced by the traffic restraint policy. It therefore ignores long-term modifications, particularly on the location decisions of households and enterprises – which are probably important and negative. The study also focused only on changes introduced within the Paris municipality, ignoring the consequences of the policy on traffic in adjacent municipalities. The study nevertheless takes into account the costs and benefits (mostly costs) incurred in the Paris municipality by residents of other municipalities. These costs are important, since more than half are borne by non-residents. Our analysis also ignored gains and losses in reliability of transport time. On a given route, transport time is not constant: it is best represented by a distribution, with an average and a standard error. Users are sensitive to changes in the average (valued by time gained or lost). They are equally sensitive to changes in the standard error, which also varies. Taking into account these changes in reliability would increase car user gains in London, and also car user losses in Paris. In a recent paper, Santos and Bharkarb (2006) introduce a concern that we have also ignored. They say that the psychological cost of being caught in a traffic jam is so high that the value of time spent in traffic jams is much higher than the value of time spent in transport in general. The (difficult) introduction of this interesting consideration would increase gains in London and losses in Paris.

This study could therefore be significantly improved. But it is unlikely that these improvements would change the main lesson that can be derived from it. This lesson, familiar to any first-year economics student, is that price regulation is more efficient than quantity regulation. The London toll may not be the success often claimed, but the Paris traffic restraint policy is certainly a failure.

NOTES

1. We had some difficulty in obtaining these figures. A RATP official first denied their existence, so we obtained them 'laterally'. We published them in French papers, mentioning the initial refusal. RATP then explained that there had been a misunderstanding or a mistake, and pledged cooperation in the future.
2. Why accept the Transport Observatory estimates of traffic, and reject their estimates of speed? Both estimates come from point measurements. To measure the flow on a given route, the exact location of the measurement instrument does not matter much. To measure speed, however it does matter. Rue Beaubourg, for instance, is one lane for one stretch – which is nearly always jammed – then two lanes where traffic is more fluid. Measuring speed on the second stretch does not give a meaningful idea of the overall speed on the street.

REFERENCES

Prud'homme, Rémy and Juan Pablo Bocarejo (2005), 'The London congestion charge: a tentative appraisal', *Transport Policy*, **12**, 279–87.

Prud'homme, Rémy and Pierre Kopp (2007), 'Le péage de Stockholm: Evaluation et enseignements', *Transports*, **443**, 175–189..
Prud'homme, Rémy, Pierre Kopp and Juan Pablo Bocarejo (2005), 'Evaluation économique de la politique parisìenne des transports', *Transports*, **434**, 346–359.
Raux, Charles (2005), 'A propos de l'article "l'expérience du péage de Londres" ', *Transport*, **431**, 174–78.
Santos, Georgina and Jasvinder Bharkarb (2006), 'The impact of the London Congestion Charging Scheme on the generalised cost of car commuters to the City of London from a value of travel time savings perspective', *Transport Policy*, **13**, 22–33.

14. The European and Asian experience of implementing congestion charging: its applicability to the United States

Tom Rye and Stephen Ison*

1 PURPOSE AND STRUCTURE OF CHAPTER

The main purpose of this chapter is to consider the lessons that the experiences of implementing congestion charging in Europe and Asia can tell us about its 'implementability' in US cities. This means the relative ease with which it can be implemented, which is a product of several factors, including:

- the political and legislative context in a given urban area;
- local factors, such as the degree of congestion experienced in the area;
- scheme design; and
- the nature of the people and organisations that are responsible for funding and implementing a scheme.

It is useful to consider such factors that combine to make implementation more or less likely in the form of a conceptual framework for policy implementation. Several authors in various fields have previously developed such frameworks. Section 2 begins by reviewing this literature and, from that review, developing a conceptual framework more specific to transport policy implementation.

Section 3 then looks at the experience of certain European and Asian cities in trying to implement congestion charging. Use of the conceptual framework for analysis will allow us to draw conclusions about the way in which factors must combine in order for implementation to have the greatest chance of success.

Section 4 then turns to the USA and considers its context for charging and the experience there of charging to date. In the light of the conceptual framework, in Section 5, conclusions are then drawn about the transferability to the US context of European and Asian experience.

One final point to make in this introductory section is that in this chapter we define congestion charging as the area-wide charging of existing roads that were previously free to use, rather than the addition of extra capacity, the use of which is charged.

2 CONCEPTUAL FRAMEWORKS FOR POLICY IMPLEMENTATION: FINDINGS FROM THE LITERATURE

The study of implementation is a study of change, how that change occurs and possibly how it is induced (Parsons, 1995). There are a number of approaches which can be taken with respect to the study of implementation – the *analysis of failure*, with the work of Pressman and Wildavsky (1973); *rational models* (or the *top-down approach*) which identify factors which influence successful implementation, with work such as Gunn (1978) and Sabatier and Mazmanian (1979); and the *bottom-up approach* which stresses the importance of other actors and the interactions within organisations (see, for example, Elmore, 1978).

An example of the top-down approach is the work of Gunn (1978), who sets out 10 conditions in the form of a framework of questions which might be asked about the implementation of a particular policy. While Gunn's top-down approach can be criticised, not least for perhaps placing too much emphasis on the definition of goals by 'the top', the 10 conditions still act as one useful structure through which to analyse transport policy implementation. They are as follows:

1. *The circumstances external to the implementing agency do not impose crippling constraints* For example, the policy may be unacceptable to various interests, which have the power to veto them.
2. *That adequate time and sufficient resources are made available to the programme* This could also be viewed as an external constraint. Politicians *may* also will the policy 'end' but not the 'means' and as such restrictions in terms of expenditure may starve the project of adequate resources.
3. *That the required combination of resources is actually available* This follows on from the previous point in that at each stage in the implementation process the appropriate combination of resources is actually available.
4. *That the policy to be implemented is based upon a valid theory of cause and effect* This means that policies are sometimes ineffective not

because they are badly implemented but because they are bad policies. The failure of implementation could be a failure of policy making. The 'problem of implementation which can only be tackled by better analysis at the issue definition and options analysis stages of the policy-making process' (Hogwood and Gunn, 1984, p. 201), and there are limits to policy makers' ability to understand and deal with complex economic problems (Bardach, 1977).

5. *That the relationship between cause and effect is direct and that there are few, if any, intervening links* In other words, the more links in the chain, the greater the risk that some of them will prove to be poorly conceived or badly executed.
6. *That there is a single implementing agency which need not depend upon other agencies for success or, if other agencies must be involved, that the dependency relationships are minimal in number and importance* Thus 'where implementation requires not only a complex series of events and linkages but also agreement at each event among a large number of participants, then the probability of a successful or even a predictable outcome must be further reduced' (Hogwood and Gunn, 1984, p. 202).
7. *That there is complete understanding of, and agreement upon, the objectives to be achieved; and that these conditions persist throughout the implementation process* Hogwood and Gunn (1984, p. 204) states that, in reality, 'even *official* objectives, where they exist, may not be compatible with one another, and the possibility of conflict or confusion is increased when professional or other groups proliferate their own *unofficial* goals within a programme'.
8. *That tasks are fully specified in correct sequence* As such, when moving towards agreed objectives it is possible to specify, precisely, the tasks to be performed by each participant.
9. *That there is perfect communication and coordination between the various elements or agencies involved in the programme* Overall, communication has an important contribution to make to coordination and implementation.
10. *That those in authority can demand and obtain perfect compliance* In other words, those in authority should also be those in power, with the ability to secure total and immediate compliance from others whose consent and cooperation is required.

The list above is presented in the same order as set out by Hogwood and Gunn (1984). Gunn provided no hierarchy or order of conditions, thus implying that they are all equally important. Clearly, though, one could in fact argue that some are relatively more important than others. For example, the need for clearly stated objectives is, from the authors' perspective, a central condition of implementation. It can also be argued that the preconditions break down into two groups: the first, related to gaining agreement on policy direction (for example, condition 5); and the second, then ensuring that the practical means of implementation are in place (for example, condition 8).

In contrast to and indeed in advance of Gunn, Pressman and Wildavsky (1973) studied the issue of implementation and introduced the idea of 'decision points' and 'clearances' in which an act of agreement has to be reached before a programme can continue. They suggested that the greater the number of decision points and clearances in terms of a particular policy, the greater the chances of failure in terms of meeting the policy's objectives. Thus in order to reach agreement it is argued that policies need to be relatively simple, designed with problems of implementation in mind and if possible involving relatively few decision points and clearances. This point is similar to that made by Gunn in condition 6.

Weimer and Vining (1992) detailed the implementation phase of the policy process and the factors affecting success and failure. They consider three general factors, namely: the logic of the policy, the nature of the cooperation it requires and the availability of 'fixers' or those who will manage the process. While analysing policy implementation less from the 'top-down' viewpoint of Gunn, they none the less reiterate some of the points that he made.

For example, they were concerned that the logic of the policy should be 'reasonable' for the policy to be more likely to be implemented. They also stated that it was important that objectives should be clear and well understood. They also pointed out that the nature of cooperation required affects a policy's chances of success. The more numerous and varied are the components to the policy being considered and the number of parties involved then, they argued, the greater the implementation problems. Implementers should be prepared to use compromise in order to keep the various parties on board and this may involve offering concessions. They also suggested consideration of the resources that the implementer has available – including legislation. As they stated (Weimer and Vining, 1992, p. 327): 'Clear legal authority is almost always a valuable resource for implementation. It may not be sufficient by itself to guarantee cooperation, however'. As for the fixers, these authors argue that fixers are individuals 'who favour the policy and [are] willing to expend time, energy, and resources to see it put into effect'. This latter is a point raised by neither Gunn (1978), nor Pressman and Wildavsky (1973), but it is an important one to which we shall return later.

While the policy implementation frameworks reviewed above may be categorised in terms of analysis of failure, top down or bottom up, there are clearly many elements of commonality between them. We now move on to attempt to synthesise these elements into a modified conceptual framework suitable for the analysis of congestion charging implementation.

Moving towards a Modified Framework for Policy Implementation

Taking into account the frameworks reviewed above, we now move on to develop a new conceptual framework – one that is, essentially, a hybrid of the others – with which we can then analyse European and Asian congestion charging experience. This 'new' framework recognises the bottom-up and analysis of failure frameworks, but it is most closely related to Gunn's (1978) framework. However, it attempts to build on his work by adding new preconditions and ordering and prioritising them.

This 'new' framework begins by splitting the implementation process into two – first, reaching agreement on policy direction and, second, organising the means to implement the policy. Clearly, the former is more important than the latter because, if there is no agreement on a policy among (at least) those with the power to implement it, then having the means to implement it is largely irrelevant.

The key elements of the first stage in the process, that of reaching agreement on policy direction, are as follows, listed in order of priority.

First, we argue that the existence of strong political leadership (often referred to as a 'champion' – although experience shows that the 'champion' may in fact be more than one person, as in some Norwegian examples of the implementation of congestion pricing), and the perception of the problem to be addressed by the policy, are of equal greatest importance. Many cities have a problem with congestion, but this does not translate into a congestion charging scheme without political leadership that is willing to take the risk of implementing what is, initially at least, often an unpopular policy. Without a problem to address, however, even strong leadership may be unable to implement a policy for which a justification is hard to find.

The logic and internal consistency of the policy is the next key factor in reaching agreement on policy direction. It is obviously easier to reach agreement on a policy that is logical and where it is not difficult to see that its implementation has a high probability of achieving the desired objectives. In the case of congestion pricing, there will of course always be those who argue that it will *not* have the desired impact (normally that of reducing congestion). However, it is a relatively obvious counter argument that, if the price of something increases, demand for it is likely to fall – so the policy can be seen to have an internal logic and this makes it less difficult to defend. In contrast, for example, it is more difficult to show that the policy of levying a local tax on off-street parking spaces (see Enoch, 2001 for a full description) will reduce congestion, since it really depends on whether the charge is passed on to users, and also because parking charging does not affect through-trips. This issue is related to that of clarity of objectives, since a policy that is not logically consistent is unlikely to have very clear objectives. In a survey of transport decision makers in the UK, 92 per cent responded that they were very or fairly concerned with respect to the need for clearly stated objectives for congestion charging schemes (see Ison, 2000).

Even with a recognised problem, a champion and a logical and internally consistent policy with clear objectives, if the reason for the policy is not well communicated both to the general public and to 'key stakeholders', then it is much more likely to fail. Therefore we would highlight the importance of communication in helping to achieve and to maintain agreement on policy direction.

The number of times agreement must be reached and with whom, also has an effect on how easily policy direction can be agreed, although we argue that it is subsidiary in importance to the factors explained above. This issue is analogous to the 'decision points' and 'clearances' of Pressman and Wildavsky's (1973) work, and navigating a policy through these difficulties will depend on 'fixers' and 'champions' to smooth negotiations along the way and, at times, in the case of 'champions', to impose a course of action where there is no consensus.

Subsidiary to reaching agreement on policy, the second category of conditions that must be met are as follows, all related to the means to actually implement policy once its direction has been agreed.

1. Enabling legislation is a prerequisite. If there is no enabling legislation, a policy is effectively illegal and cannot be implemented. It is most unlikely that a policy that includes the levying of a charge on people for public services for which they have not previously paid – that is, congestion charging – could be pursued without legislation.
2. Once enabling legislation is in place, resources are required. Congestion charging is cheap neither in its planning stages, nor to put into actual operation, as the European and Asian case studies will demonstrate.

The complexity of the actual implementation chain is the next most important factor. It is a basic axiom of project planning that the more inputs that are required, and the more people there are who are responsible for these, then the greater the risk that implementation will be delayed or stopped completely. One way to reduce this risk is to have a single implementing agency that is not dependent on other parties for any aspect of implementation. It is none the less likely that careful negotiation plus, at times, the intervention of a champion to move the policy through a particular 'break point' in the implementation chain, will be required.

Equal in importance to the complexity of the implementation change is the way in which implementation is communicated. This is obviously related to the way in which policy direction is communicated. However, once policy direction has been agreed, it is then key to ensure that the steps in the implementation process are communicated simply, clearly and effectively, in order to reduce the risk of negative publicity caused by misunderstandings of the planned congestion charging system.

Having now outlined this empirical framework we go on to examine actual cases of congestion charging projects in European and Asian cities, in order to validate the framework.

3 EMPIRICAL VALIDATION OF THE CONCEPTUAL FRAMEWORK

This section considers the experience of implementing various charging schemes in Asia and Europe, and uses this experience to validate the conceptual framework developed above. The schemes considered include those that have been implemented successfully, and others that were planned but, for various reasons, not implemented. The consideration of this latter group of schemes is still useful as a means of validating the framework, since reasons for non-implementation can be identified. The schemes on the experience of which we draw here are summarised in Table 14.1. The important thing to note is that they are all urban congestion charging (or tolling) schemes rather than charges to use specific road links. Therefore, any person wishing to travel into and/or within a charged area would have no choice other than to pay to drive, or to use a different, uncharged, mode.

Cambridge, UK

Cambridgeshire County Council proposed the idea of a congestion metering scheme in the early 1990s. It was intended that all vehicles within a 12–15 mile radius of the city should be fitted with an electronic meter connected to the odometer, with the vehicle owner being issued with a smartcard. The smartcard would be inserted in the meter which would be activated by roadside beacons when the vehicle entered the city centre. Charges would take place (by monetary units deducted from the smartcard) when it was deemed – by means of average speed – that the vehicle was in a congested situation. The director of transportation within the County Council was responsible for initiating the scheme and it was intended to charge for congestion when and where it occurred. The scheme was not advanced beyond the trial and the reasons suggested for this were:

Table 14.1 Summary of European and Asian schemes

	Year of introduction or proposal	Area covered	Charge type	City population
Cambridge, UK	1990	Central city and inner suburbs	Distance/speed based	105,000
Edinburgh, UK	2006	Whole city	Cordon charge	500,000
Hong Kong	1983, 1991	Central area	Distance based	4,000,000
Bergen, Norway	1986	City centre, broadly defined	Cordon charge	237,000
Stockholm, Sweden	2006	City centre	Cordon charge	1,500,000
Central London, UK (including proposed western extension)	2003, 2007	City centre (and inner western suburbs)	Area licence	7,500,000 (of which 250,000 in scheme area)

- concern as to how the technology would operate;
- the cost of installation;
- concern as to the use of the revenue raised;
- change in the political complexion of the authority, and with it the political push for charging;
- retirement of the chief protagonist (the director of transportation);
- a view that all the alternative measures had not been exhausted;
- lack of clearly stated objectives; and
- widespread opinion that the level of congestion was simply not severe enough to merit a congestion charge.

See Ison (1996 and 1998) for more detail.

While congestion metering was not implemented at that time, Cambridgeshire is now investigating the idea of congestion charging again as a possible measure to reduce traffic in the city.[1]

The reasons for reconsidering congestion charging are the growing levels of congestion, set to increase with substantial development planned in the area, and as such the unreliability of public transport. Consultation is seen as being central to the potential implementation of congestion charging.

A view is still being put forward, however, that existing schemes to improve public transport and discourage car use should be pursued before considering a 'draconian' congestion charge (*Cambridge Evening News*, 2005). Such a scheme is not imminent, for if the investigation leads to an actual project it would be 2011/12 before it starts.

Edinburgh, UK

Here, the City Council proposed a twin cordon-crossing scheme which would have levied a maximum £2 charge for crossing one or both of the cordons, in-bound only, during charging hours (07:30–18:30 for the city centre cordon and 07:00–10:00 for the cordon on the edge of the city). The scheme should reduce the forecast growth in congestion by some 25 per cent compared to a no-charging scenario, and to raise a net £968 million over 20 years. The scheme was in detailed design stage when it was abandoned after a referendum of the city's population found only 25 per cent of people in favour, on a 65 per cent turnout (see Gaunt et al., 2007 for a more detailed analysis of the scheme and why voters opposed it).

In relation to the conceptual framework, it is clear that Edinburgh faced many problems. In terms of political leadership, in the later years of the scheme's development, it was clear that it lacked a clear champion. On the other hand, opinion surveys showed consistently that the people of Edinburgh did consider congestion to be a serious problem (Saunders and Lewin, 2005). The policy did, however, fall down somewhat on its logical consistency and the clarity of its objectives. While, legally, it was necessary for the Council to demonstrate that the scheme would reduce congestion, it was also clear that a key reason for the scheme was its ability to generate revenues to fund additional investment and subsidy for the city's transport system. Certain interest groups disagreed that the scheme *would* reduce congestion, arguing that it would instead lead to traffic diversion. In addition, councils outside Edinburgh but bordering on the city area criticised the scheme for charging their residents to drive into the city, while leaving many city residents free to drive at no charge in the large zone between the two cordons.

In relation to the implementation process itself, congestion charging legislation exists in Scotland under the Transport (Scotland) Act (2005). However, Edinburgh was the first authority to attempt to use it and this absorbed a great deal of time and effort. Resources were a rather different matter: for a scheme of this size, development funds of around £8 million over the three years leading up to the referendum were, in retrospect, insufficient to fund all the development and (especially) communications work required.

There was not a single implementing agency in Edinburgh. Instead, the City Council set up a further agency for 'day to day' implementation work, but it used many consultants. In addition, the Scottish Executive (government) played a very active part. The City Council is not able to directly manage the bus system in Edinburgh and so could only hope that deregulated bus operators would choose to offer additional services at the introduction of congestion charging. Finally, the reasons for the scheme (gaining support for the policy), or how the scheme was supposed to work, were not communicated very effectively.

Hong Kong

An electronic road pricing (ERP) trial was undertaken in Hong Kong between 1983 and 1985 involving automatic vehicle identification (AVI), with vehicles having electronic number plates mounted under the vehicle. On crossing a boundary into the central area the AVI would be energised with a message sent to a roadside recorder (Hau, 1990). The trial, although not comprising an actual charge, involved a bill being sent monthly to the motorists taking part in the trial.

The following reasons help to explain the non-continuation of the scheme:

- the Automobile Association was of the view that the congestion problem had been exaggerated;
- there was a decline in economic growth at the time of the trial;
- there was failure to 'sell' the ERP scheme to the general public (Borins, 1988; Hau, 1990);
- there was a perception that ERP would be used as a means of taxing the motorist excessively (Borins, 1988);
- the cost, as viewed by the general public, particularly given that spending on other transport items (for example, footbridges) was being cut or deferred while the ERP pilot was running;
- the invasion of road users' privacy, with each vehicle being identified in order for a charge to be applied; and
- the lack of a strong political leader. In Hong Kong, the scheme was headed by the transport secretary, who had an image of promoting unpopular policies (Borins, 1988).

Hong Kong has subsequently revisited congestion charging with the conclusion that 'drastic restraint measures (such as ERP) are not warranted on traffic management grounds before 2006 for Hong Kong Island and 2011 for Kowloon at the earliest if the growth of the private vehicle fleet is no more than 3% per year' (Legislative Council Brief, 2001, p. 2).

Stockholm, Sweden

The Swedish capital implemented an experimental cordon charge in the form of a single ring around its centre in January 2006, and switched the cordon off in July 2006. It is likely to be made permanent after a referendum of city residents on 17 September 2006 approved it by a margin of 5 to 4.

In the case of this scheme, there has been a political champion – although not one particular person – in the form of the Swedish Green (environmental) party in the national parliament. Their commitment to the scheme came from a widespread belief that congestion was a major problem, particularly at peak times, in the central part of Stockholm. The policy has a clear objective, that of reducing congestion, while raising revenue has always been a secondary consideration, not least because it was realised that the revenues would take a considerable period even to pay back the implementation costs of the scheme, estimated at about €350 million. The logical consistency of the policy as a measure to tackle congestion was reinforced by the decision to have a higher charge at peak times, and to charge each time a cordon is crossed (rather than a flat daily charge – although Stockholm does have a capped maximum daily charge in place).

Initially the legislation to permit the scheme did not exist, but the Green Party saw a new law through the national parliament which enabled the scheme's implementation. Clearly, resources have been made available at the national level to fund the scheme, and revenues revert to the national government. The implementation agency was the national road administration and the implementation process, although not straightforward, was clearly mapped out and did not involve a multitude of actors. Communication of the Stockholm scheme has been excellent at every stage.

Bergen, Norway

This subsection is based on Kocak (2005). Bergen is a small coastal city of 237,000 people. Bergen's pricing scheme was the first Norwegian toll ring to be introduced in 1986 to match funds for a comprehensive road widening and tunnelling programme. It is a simple inbound cordon crossing scheme that covered an area slightly larger than Bergen's central business district with seven (initially six) toll gates. Because the major objective was to raise revenue, not to deter traffic, there were no effective by-pass routes. It operates between 06:00 and 22:00 hours, Monday to Friday. The toll is NOK 10 (about £0.80), payable in cash, by ticket, or by period pass to use non-stop lanes.

Although the Bergen scheme permitted no alternative routes due to the city's unique topology and the scheme's design, it has had little impact on traffic. There was some evidence of travel time shift (Lewis, 1993), and car-pooling increased a little, but there was no evidence of impacts on public transport use. In the first year (1986–87) city centre bound traffic reduced by 6–7 per cent, but has since increased (ILS, 2000).

The scheme was effective in raising revenue. The initial investment to establish the ring was NOK 15 million (approx £1.27 million). The yearly income has proved to be far higher than expected, and now amounts to NOK 70 million (£6 million) per annum, of this NOK 50 million (£4.23 million) is allocated to road construction, NOK 14 million (£1.2 million) for operating costs and NOK 6 million (£0.5 million) is put aside in a fund for other improvements (Skulstad, 2005).

Bergen satisfied the majority of the elements of the conceptual framework for charging implementation. The scheme was a response to a recognised problem – that of insufficient funds for road building – and, as such, had clear objectives and the internal logic that the money raised would be used to pay for new infrastructure. Legislation was in place, and the County Council had the funds to pay for the (in this case) modest set-up costs. The County Council was also the sole agency involved, and has control of local public transport. There was not a great deal of public support for the scheme and it remains relatively unpopular even though new roads have been built with the money generated; but it did have a clear political champion to see its implementation through.

London, UK

The London scheme has been extremely well documented elsewhere, and continues to be so – including in other chapters of this book. In relation to the conceptual framework, the original scheme area satisfies all of the conditions to a greater or lesser extent. The western extension has endured considerable criticism on the grounds that the problem is not as severe as in central London, and that there is a logical inconsistency in the scheme (in that it will allow residents of the new area to drive at a 90 per cent discount within the existing scheme area, while they currently (2006) have to pay full price). The scheme is to go ahead largely because of the political will, and power, of the mayor showing, once again, the key importance of a powerful champion. See Ison and Rye (2005) for more detail.

Conclusion to Review of European and Asian Schemes

Table 14.2 compares those schemes that were implemented and those that were not with the conceptual framework. While there is a degree of interpretation in deciding whether or not a condition is satisfied in each

Table 14.2 Comparison of schemes with conceptual framework for implementation

	Champion	Severity of the problem	Logic of policy	Clear objectives	Legislation	Resources	Single agency	Steps in process	Communication
Cambridge	✓	×	×	×	×	×	×	×	×
Edinburgh	×	✓	×	×	✓	×	×	×	×
Hong Kong	×	×	×	×	✓	✓	✓	×	×
Bergen	✓	✓	✓	✓	✓	✓	✓	✓	Somewhat
Stockholm	✓	✓	✓	✓	Later	✓	✓	✓	✓
London	✓	✓	✓	✓	✓	✓	✓	×	✓

location, it is clear that those schemes that have been implemented satisfied many more of the conditions in the framework than those that did not. Even among the group of non-implemented schemes, Hong Kong and Edinburgh came closer to implementation than Cambridge – and we see that the last satisfied fewer conditions of the framework. Therefore, the framework does provide a useful means for analysing these implementation processes.

Following this analysis of the European and Asian experience, we now go on to consider experience of charging in the USA and how the conceptual framework can help us to assess the likelihood of congestion charging schemes on the European/Asian model being implemented in the USA.

4 US CONTEXT FOR CHARGING

History of Charging in the USA

The first steps towards charging in the USA were taken in the 1970s in a number of studies of the feasibility of charging in US cities that were funded by the Federal Highway Administration (FHWA). Higgins (1986) notes how, in 1976, the then secretary of state for transportation wrote to a number of cities' mayors to encourage them to consider the idea of congestion charging for their downtown areas, on the model of Singapore, whose area licence scheme had started one year previously. The cities included Baltimore, Atlanta, Rochester (NY), Berkeley, Ann Arbor, Honolulu, Seattle and Madison. The majority did not reply or replied in the negative, suggesting that their cities were congestion free and/or in such competition with out-of-town developments that a congestion charging scheme could not be contemplated because of its likely effects on the city's economy. However, Berkeley, Madison and Honolulu accepted the offer of money to carry out studies, and these took place in 1976 and 1977. They all concluded that there was insufficient support to take forward any kind of scheme; key arguments against were the regressive effects of a scheme, its interference with fundamental rights to mobility, and its economic impacts.

After this brief consideration of urban congestion charging, the next stage in the development of charging in the USA originated at state, not federal level, and moved its focus to new road construction. A shortage of public sector funds for road building led to the 1989 California Bill 680 – permitting private finance for new roads – which led directly to the construction of tolled express lanes on SR91 and of the SR125 (variable) toll road, both in Orange County to the south of Los Angeles. These gave new credibility and impetus to the idea of charging users to avoid congestion.

The federal government renewed its interest in what is now termed 'value pricing' with the 1991 Intermodal Surface Transportation Efficiency Act (ISTEA). This was continued in later legislation such as the 1998 Transportation Efficiency Act (TEA), the 2001 TEA-21, and the 2005 SAFETEA-LU. All this legislation makes available matching funds for value (road) pricing pilot projects through the US Congestion (now Value) Pricing programme. Federal funding for 2004 to 2008 is about US$11 million per year.

According to *Local Transport Today* (2006) in March 2006 the San Francisco County Transportation Authority (SFCTA) was the recipient of a US$1.04 million grant from the FHWA to look into how to implement congestion charging. The study will report back in 2008. In Seattle, a trial is currently under way assessing the effectiveness of a congestion charging scheme. This involves 500 vehicles fitted with an on-board unit. With traffic expected to grow substantially over the next decade or so there is a growing realisation that innovative incentives and disincentives will have to be considered with the dual objectives of raising revenue in order to finance infrastructure investment and improving the flow of traffic.

From this extremely brief review, we can see that much of the impetus for congestion charging in the USA has come from the federal level, although it does not have competence to actually implement charging at the state or lower levels of jurisdiction. Only in California is there evidence of 'home-grown' congestion charging, but mainly as a response to a lack of funding for road construction, not as a means to manage congestion.

Sullivan (2003) lists more than 20 projects in place or being implemented. However, as DeCorla-Souza (2003) points out, all of the existing and planned projects fall into two main categories: the pricing of additional freeway (motorway) capacity; or variable bridge tolls. There are none that can be seen to be similar to the European and Asian experience of congestion pricing of existing roads; the closest are variable tolls, where peak tolls have been raised and off-peak tolls lowered (for example, on the Hudson River crossings into Manhattan and the New Jersey Turnpike) – but this is a change of price for an already-priced facility, rather than a completely new charge. It is worth noting that the effect of the price change on traffic levels on the Hudson River crossings was similar to that seen in European charging schemes (Litman, 2006).

Success Factors in the Implementation of Pricing Schemes in the USA

It is not the intention here to provide lengthy descriptions of the schemes that have been implemented and that are planned in the USA – for detailed information, the reader is referred to the FHWA value pricing website (http://www.ops.fhwa.dot.gov/tolling_pricing/value_pricing/index.htm) and to Ward (2001) and Litman (2006). However, it is important here to consider the factors that have contributed to the successful implementation of the value pricing schemes that we see today in the USA. This analysis is based on Ward (2001), Sullivan (2003) and DeCorla-Souza (2003).

It is clear from the literature that US pricing schemes have faced opposition, often on the grounds of inequity (the impact on poorer drivers) and a belief among opponents that money should be used to provide general road capacity, not capacity targeted at those who can afford to pay. However, the fact that in almost all extant US schemes, those who choose not to pay can still drive along the same route – albeit more slowly – has been a key factor in aiding public acceptability, and a benefit that European and Asian congestion charging schemes have not enjoyed.

Ward (2001) highlights that those projects that have been implemented had an obvious problem with which they had to deal. For example, on SR91 in California and the Katy Freeway in Houston, there was obvious congestion and limited money with which to add new capacity. She contrasts this situation with other studies of area-wide pricing in Portland (Oregon) and Minnesota, where market research found a much lesser public perception of a problem serious enough to merit congestion/value pricing.

The literature also suggests that strong implementation agencies, headed by political champions, have been very important in those schemes that have been implemented in the USA. This contrast becomes clear when comparing the schemes that have been implemented with others that have not. Sullivan (2003) also notes that certain agencies, such as turnpike authorities, are not directly elected and, therefore, are somewhat insulated from day to day politics, allowing them, perhaps, to make more controversial decisions slightly more easily than their directly-elected counterparts.

Excellent marketing has, according to DeCorla-Souza (2003), been a crucial part of the schemes that have been implemented. Other factors have included the availability of federal funds, obvious benefits once schemes have been implemented and, finally, the hypothecation of revenues to public transport.

Where schemes have not gone ahead, this is due to a number of factors that have echoes in the conceptual framework. There has been disagreement about the objectives of proposed schemes, and about the credibility of predictions of outcomes made by scheme proponents. Fear about economic development impacts – compounded by the economic weakness of US city centres relative to suburban areas to an extent far greater than in Europe – and possible consequential litigation from businesses and parking operators have also been factors that have deterred scheme implementation. Finally, institutional fragmentation has been a problem: US regions are often characterised by a very large number of municipal governments, and the regional governments that have tended to have a greater interest in pricing initiatives have little formal power in most states.

5 CONCLUSIONS: COMPARING THE US CONTEXT WITH THE CONCEPTUAL FRAMEWORK

This section takes the conceptual framework that was developed earlier in the chapter in relation to European and Asian schemes, and applies it to the US situation. The purpose of this exercise is twofold: both to analyse the elements of the implementation framework that are perhaps most lacking in the US context; and to identify any deficiencies in the framework itself.

Reaching Agreement on Policy Direction

For route-based or bridge toll pricing schemes, there has been a clarity of objectives and agreement that there is a problem, at least in those schemes that have gone ahead. The objectives have been to raise money for new construction; and to offer users a less-congested alternative to the existing route.

The presence of strong leadership or a champion to assist in reaching agreement on policy direction has, according to other commentators, varied across the schemes that have been proposed in the US, but they also note that those schemes that have gone ahead often have strong political leadership. For these schemes, such leadership has helped to keep the scheme on track when the logic and internal consistency of the project has been questioned, particularly on grounds of equity, which appears to be a greater concern in congestion charging in the US than it is in Europe and Asia. The logic and internal consistency of the project has been questioned to a greater extent in schemes that have proposed cordon charges or area licences for particular areas of cities, as opposed to route length or bridge charging schemes. In most cases, communication of scheme proposals and schemes once implemented has been excellent. Finally, US schemes do suffer from the number of times that agreement must be reached and with such a large range of bodies, due to the institutional fragmentation that tends to exist in local and regional government.

Means to Implement Policy

Legislation for congestion charging in the US is required at the state rather than the federal level and it has been forthcoming in those areas that have implemented route-based charging schemes, such as California and Texas.

However, it is not automatic and it is not clear that the same legislation, even in those states that have it, would be sufficient for cordon charge/area licence schemes on the European or Asian model.

The level of resources devoted to pricing in the US is very small at the federal level – as noted above, around $11 million per year for starting pilot schemes. Given the start-up costs of recent European schemes, such funding is clearly inadequate. Also, in view of the large number of implementing agencies in the US, the implementation chain for a pricing scheme is likely to be complex and difficult to manage.

In most ways, therefore, the conditions in our conceptual framework for pricing implementation are far from ideal in the USA, and certainly less conducive than in European and Asian cities where charging has been implemented. The only exception to this is the way in which policy change is communicated, which is generally excellent according to commentators, and an area from which Europe and Asia could learn.

The Applicability of the Conceptual Framework to the US Situation

The conceptual framework is useful in that it has helped to demonstrate that the US context is perhaps (even) less suited to the implementation of cordon/area licence-based charging than the European and Asian contexts. While it is possible to reach agreement on policy direction with regard to pricing single routes and bridges (especially where there is an unpriced alternative), such agreement has, to date, been impossible to reach for area-based schemes. There appear to be few champions for such schemes and the institutional context is fragmented and therefore problematic in terms of reaching agreement on policy direction. In terms of the means to implement charging, few of the conceptual framework's preconditions are satisfied in the US.

The authors would contend that one of the main barriers to the implementation of area-based congestion charging in the USA is the difficulty of reaching agreement on the policy direction in a situation where it is difficult to identify the area that should be charged. City centres still tend to be highly congested but they are not necessarily the most congested parts of an urban region, and they are not as dominant economically as they are in Europe and Asia. Therefore, reaching agreement that such areas should be charged may be impossible in all but a handful of US cities, if any.

NOTES

* Thanks are due to David Banister and Chris Nash for earlier comments on aspects of the ideas in this chapter.

1. The UK government made £385,000 available to Cambridgeshire County Council as pump prime funding over the 2005/06–2007/08 period to support planning for local demand management schemes where pricing is a major element.

REFERENCES

Bardach, E. (1977), *The Implementation Game*, Cambridge, MA: MIT Press.

Borins, S. (1988), Electronic road pricing: an idea whose time may never come', *Transportation Research A*, **22**, 37–44.

Cambridge Evening News (2005), 'Council chiefs eye "pay to drive" scheme', http://www.cambridge-news.co.uk/news/city/2005/09/08/4fdece22-36c0-4cfebfd1-cc7f39b1b6ea.lpf, 5 October 2006.

DeCorla-Souza, P. (2003), 'The value of pricing the use of roads', *Public Works Management & Policy*, **7**(4), 267–76.

Elmore, R. (1978), 'Organizational models of social program implementation', *Public Policy*, **26**, 185–228.

Enoch, M. (2001), 'Workplace parking charges down under: schemes in Perth and Sydney', *Traffic Engineering and Control*, **42**(10), 357–60.

Gaunt, M., T. Rye and S. Allen (2007), 'Public acceptability of road user charging: the case of Edinburgh and the 2005 referendum', *Transport Reviews*, **27**(1), 85–102.

Gunn, L.A. (1978), 'Why is implementation so difficult?', Management Services in Government, **33**, 169–76.

Hau, T. (1990), Electronic road pricing: developments in Hong Kong 1983–1989', *Journal of Transport Economics and Policy*, **24**, 203–14.

Higgins, T.J. (1986), 'Road pricing attempts in the United States', *Transportation Research A*, **20** (20), 145–50.

Hogwood, B.W. and L.A. Gunn (1984), *Policy Analysis for the Real World*, Oxford: Oxford University Press.

ILS (2000), 'Legal and regulatory measures for sustainable transport in cities', Final Report, Institute for Landesentwicklung Nordrhein Westphalen, Dortmund, http://www.leda.ils.nrw.de/.

Ison, S.G. (1996), 'Pricing road space: back to the future? The Cambridge experience', *Transport Reviews*, **16**(1), 109–26.

Ison, S.G. (1998), 'A concept in the right place at the wrong time: congestion metering in the city of Cambridge', *Transport Policy*, **5**(3), July, 139–46.

Ison, S. (2000), 'Local authority and academic attitudes to urban road pricing: a UK perspective', *Transport Policy*, **7**, 269–77.

Ison, S.G. and T. Rye (2005), 'Implementing road user charging: the lessons learnt from Hong Kong, Cambridge and Central London', *Transport Reviews*, **25**(4), 451–65.

Kocak, N.A. (2005), 'Road user charging: tools for scheme option generation', Unpublished PhD thesis, University of Westminster, London.

Legislative Council Brief (2001), Electronic Road Pricing, TBCR 2/1/2061/89 Pt 16, http://www.legco.gov.hk/yr 00-01/english/panels/tp/papers/legco-erp.pdf, accessed 21 December 2007.

Lewis, N.C. (1993), *Road Pricing: Theory and Practice*, London: Thomas Telford.

Litman, T. (2006), 'Victoria Transport Policy Institute', www.vtpi.org, 15 August 2006.

Local Transport Today (2006), 'Road user charging', Autumn, 17–18.
Parsons, W. (1995), *Public Policy: An Introduction to the Theory and Practice of Policy Analysis*, Aldershot, UK and Brookfield, US: Edward Elgar.
Pressman, J. and A. Wildavsky (1973), *Implementation*, Berkeley, CA: University of California Press.
Sabatier, P.A. and D. Mazmanian (1979), 'The conditions of effective implementation: a guide to accomplishing policy objectives', *Policy Analysis*, **5**, 481–504.
Saunders, J. and K. Lewin (2005), 'Congestion charging in Edinburgh: AS genstation with complications', Paper presented at conference on Road Pricing with Emphasis on Financing, Regulation and Equity, Cancún, Mexico, 11–13 April.
Skulstad, Tore (2005), 'Urban road charging in Norway', Presentation to International Transportation Finance Summit, Nice, France, April.
Sullivan, E.C. (2003), 'Implementing value pricing for U.S. roadways', *European Journal of Transport Infrastructure Research*, **3**(4), 401–13.
Ward, J.L. (2001), 'Value pricing: a synthesis of lessons learned', Transportation Research Board, 80th Annual Meeting, Washington DC, January, 1–18.
Weimer, D.L. and A.R. Vining (1992), *Policy Analysis: Concepts and Practice*, Englewood Cliffs, NJ: Prentice-Hall.

15. The Stockholm congestion charging system: a summary of the effects

Jonas Eliasson, Karin Brundell-Freij and Muriel Beser Hugosson

1 INTRODUCTION

The Stockholm Trial consisted of two parts: a congestion charging scheme, which operated between 3 January and 31 July 2006, and an enhanced public transport scheme, which ran between 31 August 2005 and 31 December 2006. Initially, the trial was meant to consist only of a congestion charging scheme. However, it was later decided that this should be complemented with public transport improvements – several new bus lines, additional capacity on commuter trains and subways, and more park-and-ride facilities.

The congestion charging scheme was originally meant to be a 'full-scale trial for several years', and was part of an agreement between the Social Democrats and the leftist and Green parties on the national level following the election in Autumn 2002. For various reasons, particularly legal objections regarding the technology procurement process, the congestion charge period was ultimately considerably shorter than was initally planned.

The Stockholm Trial was followed by referendums in the City of Stockholm and in about half of the neighbouring municipalities. In the City of Stockholm referendum, a majority favoured keeping the charges, but a majority of the voters in the county were against them. However, the results are somewhat misleading, since most of the municipalities where polls showed greater support for the charges did not hold a referendum at all.

After considering for a few weeks how to interpret the outcome of the referendums, the new national government decided that congestion charges should be reintroduced during 2007 (provisionally in July). At the time of writing, a negotiator appointed by the national government is trying to strike a deal among the Social Democratic, Left and Green Parties, the municipalities and the county of Stockholm. The aim is to produce a 'package' whereby the charge revenues will be used to finance a number of road investments – possibly also including additional government funding.

2 FACTS ABOUT THE STOCKHOLM TRIAL

The stated goals of the congestion charge trial were to 'test whether the efficiency of the traffic system could be enhanced by congestion charges'; to 'reduce congestion, increase accessibility and improve the environment' (both the perceived living environment and the measurable emissions from car traffic); and to reduce traffic across the cordon by 10–15 per cent. These goals were (loosely) based on previous studies on how congestion charging studies should be designed.

The Charging System

The charging system related to an inner-city zone with time-differentiated charges, covering about 30 km^2 (Figure 15.1).

There were 18 control points located at Stockholm city entrances and exits. Vehicles were registered automatically by cameras that photograph

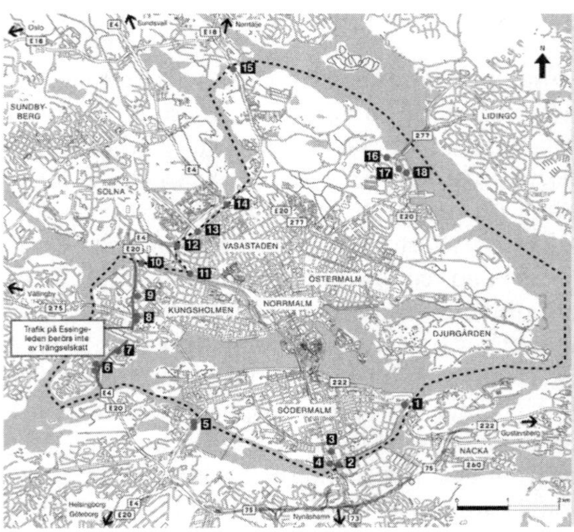

Time	(SEK) Amount
06.30–06.59	10 kr
07.00–07.29	15 kr
07.30–08.29	20 kr
08.30–08.59	15 kr
09.00–15.29	10 kr
15.30–15.59	15 kr
16.00–17.29	20 kr
17.30–17.59	15 kr
18.00–18.29	10 kr
18.30–06.29	0 kr

Source: Tollroadsnews (2006), 'Central Stockholm toll reducing traffic 20% +, doubling peakhour speeds', 17 March, http://tollroadsnews.info/artman/publish/article_1298.shtml.

Figure 15.1 The inner-city zone and time-of-day toll rates

the number plates. Those vehicles equipped with an electronic on-board unit (transponder) for direct debit payment were also identified through this means.

There was no opportunity to pay at the control points. More than 60 per cent of the payments were made automatically through transponder/direct debit. The rest were paid either at local shops (7-eleven, for example) or through bank transfers – by direct transfer or by credit card using a 'virtual shop' on the internet.

The cost for passing a control point was 10, 15 or 20 Swedish kronor (SEK) depending on the time of day (see Figure 15.1). The cost was the same in both directions. The maximum amount payable per vehicle per day is SEK 60. No congestion charge was levied in the evenings or at night, or on Saturdays, Sundays, public holidays or the day before a public holiday. Various exemptions, for example, for taxis, buses, alternative-fuel cars and for traffic between the island of Lidingö and the rest of the county, meant that about 30 per cent of the passages were free of charge.

There was no congestion tax levied on vehicles driving on the E4/E20 (the Essinge bypass) past Stockholm. This was the only free-of-charge passage between the northern and southern parts of the county. The Essinge bypass was heavily congested even before the charges, so from a pure traffic perspective, there was a strong argument for also charging vehicles on the bypass. The opposition from the surrounding municipalities was so strong, however, that the City of Stockholm politicians decided that the Essinge bypass had to be free of charge.

Extended Public Transport

Public transport facilities were extended with 197 new buses and 16 new bus lines. This provided an effective and fast alternative for travelling at peak hours from the municipalities surrounding Stockholm into the inner city. Where possible, existing bus, underground and commuter train lines were reinforced with additional departures. New park-and-ride facilities were built in the region, increasing the park-and-ride capacity by about 25 per cent. Existing park-and-ride facilities were also made more attractive.

Costs of the Trial

All costs were paid by the national government, and the total budget for the trial was SEK 3.8 billion (about $420 million).

The total cost for the charging system was approximately SEK 1.9 billion. SEK 1.05 billion was incurred prior to the start of the operation. A significant part of these start-up costs was for extensive testing. The system would be operational for only 7 months, making it absolutely necessary that everything worked right from the start. The start-up costs also included, in addition to purely technical investments, system development in a wide sense, educating and training staff, testing, public information and so on, and certain other additional minor costs, such as those for traffic signals, and the services of the Swedish Enforcement Agency and the Tax Agency.

The remaining costs (SEK 850 million) were operating costs and additional development costs during 2006. Extra costs were incurred during 2006 beyond operating costs: the system was improved in several ways during the spring of 2006. Also included are the National Road Administration's costs for closing down the system and evaluating the results during the second half of 2006.

Operating costs decreased significantly month by month, when it quickly became obvious that things in fact were going better than planned: the number of complaints and legal actions, for example, was considerably lower than had been anticipated, reducing costs for legal and tax administration. Furthermore, the number of calls to the call centre (the largest single item in operating costs) was 5 per cent of what had been anticipated, about 1,500 calls per day instead of 30,000. In the spring the call centre was downsized, with a considerable reduction in operating costs. Therefore investment costs could probably have been reduced quite substantially if the conditions (and not least the time constraints) had been different. This point may be especially important to other cities considering similar schemes. The National Road Administration estimates future operating costs to be around SEK 220 million per year.

The cost for the extended public transport services consisted of purchases of new buses (SEK 580 million), new and enhanced bus services (SEK 460 million), investments in new bus stops and garages (SEK 140 million), investments in park-and-ride facilities (SEK 70 million) and additional train/subway capacity (SEK 80 million). Hence, the new bus services accounted for most of the costs. More than half of the cost, however, is for investment in new buses, which of course has value beyond the period of the trial.

3 EFFECTS ON TRAFFIC AND TRAVEL PATTERNS

Traffic Volumes

The goal was that the number of vehicles crossing the cordon should decrease by 10–15 per cent. Forecasts for the impact on traffic made by a

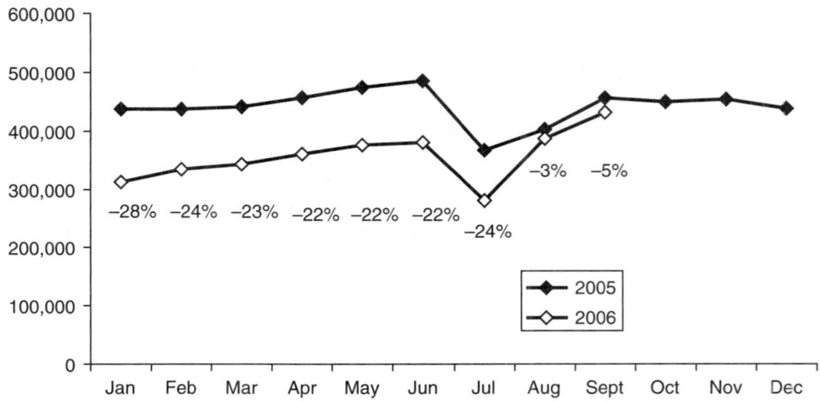

Figure 15.2 Vehicles across the cordon per day (6:00–18:00 weekdays)

team led by one of the authors (Eliasson) pointed to a significantly larger decrease, about 20–25 per cent. This was such a large decrease that it seemed unreasonable at the time. Further, it was unclear what effects on accessibility the charges would really have. While forecasts for traffic volumes are in general fairly credible, forecasts for travel times on heavily congested roads are not, since the only available forecasting tools were static network equilibrium models. Hence, it was decided that the system was probably appropriately designed.

In the event, traffic flows across the cordon decreased almost exactly as predicted by the model. The decrease compared to Spring 2005 remained stable at around 22 per cent each month after the first month, taking account of normal seasonal variations during spring (see Figure 15.2). Once the charges were abolished in August, traffic returned to previous volumes. It is possible that there is a residual effect of the charges after they were abolished (traffic during Autumn 2006 was a few percentage points lower than in Autumn 2005), but it is too early to tell, especially as there are alternative explanations.

The biggest traffic decline was during the afternoon peak, probably because of the larger share of discretionary trips compared to the morning peak, and perhaps because homeward journeys have a more flexible departure time than journeys to work. Traffic also declined during the evenings after the charge period, a natural phenomenon since fewer trips into the city during the charge period results in fewer return trips during the evening.

In addition, effects of the trial were felt further away from the charge zone than expected, with traffic volumes declining at locations far from the zone. Consequently, many of the feared side-effects, on link roads on the city's outskirts for example, were unfounded.

Traffic flows on major inner-city streets declined during the charge period, but not as much as over the charge cordon. This is natural because the traffic flow in the inner city also includes residents' vehicles, many of which do not leave the charge zone but are used for travel within the zone. There are also indications from studies other than traffic monitoring that motorists who do not need to pass through the cordon benefit from the decline in congestion and, in consequence, now use their cars more often. This could partly explain why the traffic flow decline in the inner city is lower.

Because the Essinge bypass was heavily congested even before the charges, there were fears that congestion would increase dramatically. The same fear also applied to the southern link – a ring road tunnel outside the cordon connecting the south-west suburbs with the south-east and the Essinge bypass. In fact, the traffic increase on the Essinge bypass was limited, varying across months between 0 and 4 per cent. Average travel times increased a little, but well within normal variation.

On the southern link, both traffic volumes and travel times increased significantly compared to Spring 2005. However, it is unclear how much of the increase was because of the charges and how much was because of an autonomous increase in traffic. The link opened in October 2004, and traffic increased steadily during the first year, until a boat hit a vital bridge on the bypass, reducing its capacity and hence its traffic volumes. The bridge was restored to full capacity at the same time the charges were introduced, and the flows on both the Essinge bypass and the southern link increased overnight. Because of this, it is impossible to determine how much of the traffic increase on the southern link is because of the charges. Based on the short time series that we have, the increase is between 7 and 10 per cent (Figure 15.3).

Travel Times

A consequence of vehicle traffic declining is that accessibility improved and travel times fell. This had a large, positive influence on the reliability of travel times, that is, travellers were now more certain that a journey could be made within a given period. Travel times for vehicle traffic declined significantly in and near the inner city. Particularly large declines were seen on approach roads, on which queue times fell by one-third during the morning peak period and by one-half during the afternoon/evening peak period.

Not only were average travel times shorter, but reliability also improved. The 10 per cent worst travel times were reduced even more than average

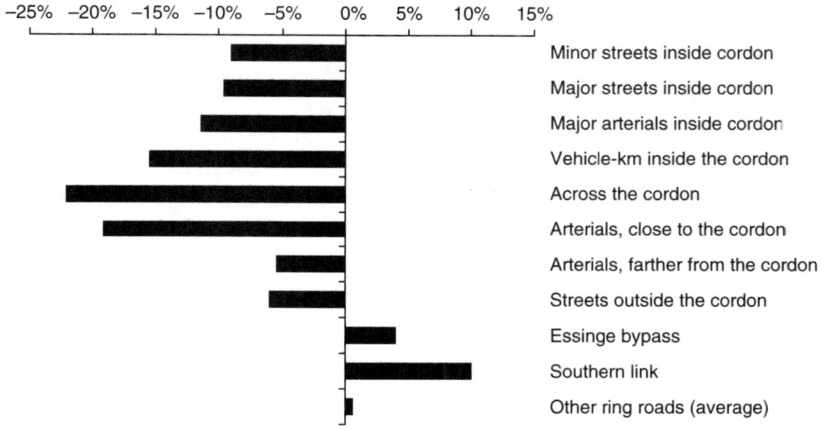

Figure 15.3 Changes in traffic volume during the Stockholm Trial

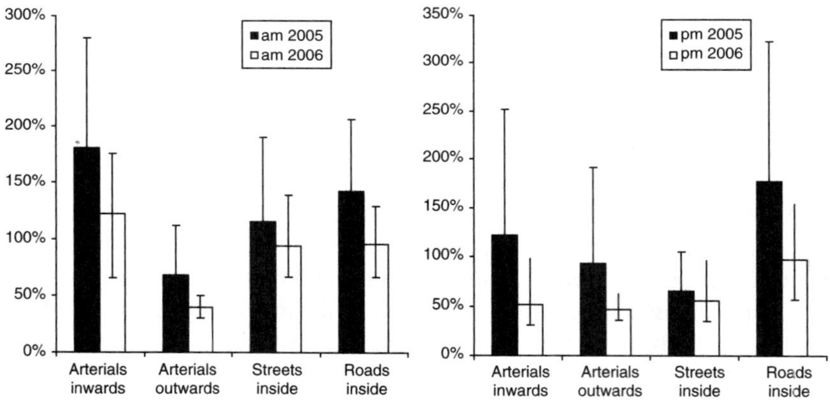

Note: 0% indicates free-flow travel.

Figure 15.4 Relative increases in travel times for various links

travel times for some categories of roads (such as arterials during the evening peak), by a factor of 3 or more.

The shaded bars in Figure 15.4 show average travel times while the 'margin of error bars' indicate the worst 10 per cent and the best 10 per cent travel times. Measurements were taken from all weekdays for six weeks in April and May 2006. The 'morning peak' refers to 7:30–9:00, while the 'afternoon peak' refers to 16:00–18:00.

Travel times on the Essinge bypass have not increased significantly. On the southern link, travel times were much higher in Spring 2006 than in Spring 2005, but as explained above, it is unclear how much of this increase is because of the charges.

Several studies show that the decline in traffic volumes and improved accessibility were very visible. For example, the perceived work environment for commercial drivers (lorry and bus drivers) improved, and the number of people viewing congestion as a severe problem decreased significantly.

Where Did the Car Drivers Go?

Two travel surveys were carried out, one before the charges were imposed (Autumn 2004) and one during the period of the charges (Spring 2006). The surveys involved the completion of 40,000 one-day travel diaries. Initially, the charges were planned to start in August 2005, and the second travel survey was planned to be carried out one year after the first. When the charges were postponed (because of legal objections), this meant that the two travel surveys took place during different seasons and quite a long time apart (18 months). This resulted in problems for the analysis, for example, many respondents had moved or changed jobs, trip frequencies and modal splits are different during autumn and spring and so on.

About half of the 'lost' car drivers consisted of those driving to work and school. Almost all transferred to public transport (which meant that public transport trips increased by around 6 per cent). There was no increased car-pooling or working at home. The other half consisted mainly of those making discretionary trips. Almost none of these transferred to public transport, and it is very hard to say precisely what happened to them. Some diverted to other destinations (not crossing the cordon), but many trips were either cancelled or bundled into trip chains.

However, the 'amount of travel' is not a static fixed number to be replaced by one alternative, but instead there is a large adjustment potential to reduce travel in a variety of ways. The travel surveys point to a reduction in the frequency of travel – people were not making as many trips as they did prior to the introduction of the congestion charge.

Adjustments in the form of taking advantage of reduced congestion on roads can also be observed. For example, commuters travelling only inside the cordon began to travel more during the peak period, and car share also increased. This is one example of how many people avoid having to make payments themselves, but are still able to take advantage of improved accessibility.

4 ENVIRONMENTAL EFFECTS

Emissions

The Stockholm Trial led to reduced emissions of both carbon dioxide and health-related pollutants (for example, particulates). The reduction in carbon dioxide is approximately proportionate to the decline in vehicle-kilometres travelled, which means that carbon dioxide emissions from traffic dropped by 2–3 per cent in the County of Stockholm. As a result of the single measure, this is a significant reduction, even if it is only an interim step towards meeting national climate goals. Emissions inside the cordon decreased by 10–14 per cent, depending upon the type of emissions. The emission reduction was calculated using traffic volume measurements, together with vehicle emission factors.

The Urban Environment

One of the goals for the trial was to 'improve the perceived urban environment'. This is a complex and diffuse concept. It is difficult both to find a common, clear-cut definition of what is meant by a 'good' or 'improved' urban environment, and also to measure these effects. Drawing conclusions from the study is made more complicated not only by the above-mentioned general problems but also by the completely different weather conditions during the two monitoring periods. Our conclusions are therefore very tentative.

The results point to perceived improvements of precisely those factors for which measured changes can be demonstrated, that is, those connected to traffic reduction. In the urban environment study, people noticed an improvement in traffic speed, air quality and vehicle accessibility. This was confirmed in interviews with cyclists in the inner city and children living there. Inner-city children's perception of the urban environment had very clearly improved and many cyclists commented that there were fewer cars in the inner city and that traffic flow had improved. Perceptions of deterioration mainly concerned accessibility, especially by foot, bicycle and public transport. The results do not support any clear-cut or unequivocal appraisal of whether the urban environment in general had improved. Perceptions of accessibility by foot or bicycle are strongly influenced by the weather and the season, and monitoring took place at different periods. However, the conclusion is that effects associated with traffic changes have influenced perceptions of the urban environment.

Road Safety

Since the congestion charges were imposed for such a short time, it is difficult to draw conclusions on the basis of follow-ups of actual and reported accidents during the trial. Evaluations of the road-safety effects of the trial are therefore based on estimated relationships between road safety and changes in traffic volumes, traffic flows and speed levels. A cautious estimate is that the effects on speed and traffic volume would, on average, imply a reduction in the number of traffic-related injuries inside the cordon by 5–10 per cent.

5 PUBLIC TRANSPORT

Public transport use was about 6 per cent higher in Spring 2006 than in Spring 2005. The congestion charge seems to have increased use by about 4.5 per cent, while higher petrol prices and other external factors are probably responsible for the remaining 1.5 per cent. Congestion on public transport (measured by the number of standing passengers) increased somewhat on the subway but decreased on commuter trains. Overall, congestion seems to be unchanged, probably because of the expansion in public transport.

Accessibility for bus traffic to/from and in the inner-city area increased. Because inner-city timetables were not adjusted during the trial period, improved accessibility did not significantly shorten travel times. However, there are signs that punctuality improved. Buses leaving the charge zone, which do not have fixed timetables once they have passed the cordon, have experienced considerably shorter travel times.

Efforts to improve public transport facilities (park-and-ride sites, expanded bus and light rapid-transit train services) did not, on the basis of current documentation, yield any visible effect on the total number of journeys during Autumn 2005 before the start of the Stockholm Trial. That is not to say that there was no effect, only that it may have been too small to register in the passenger statistics or in the travel-habit survey conducted in Autumn 2005. Indeed, it is unlikely that the expansion in services would not have had any effect on the total number of public transport journeys, but adequately detailed analyses and statistics enabling such an increase to be identified are not yet available. On-board surveys on the new buses indicated that motorists have been encouraged to switch to public transport but their number is still too small to make an impression on public transport generally. Overall, travel was about 2 per cent higher in Autumn 2005 compared to Autumn 2004, but that increase was probably because of higher gasoline prices.

Another question is whether the congestion charge would, in fact, have reduced vehicle traffic even if public transport facilities had not been expanded. Such extra facilities, as mentioned above, have not as yet provided much evidence of an increase in the number of public transport journeys, but it is quite conceivable that they boosted the effect of the congestion charge by making the switch from car to public transport easier. If that is the case, part of the effects of the congestion charge should instead be registered as an effect of enhanced public transport.

Nevertheless, the effect must be small. The on-board surveys on the new buses showed that between Autumn 2005 and Spring 2006 the number of new passengers who earlier used their cars for transport was tiny compared to the reduction in the number of journeys over the charge zone. Of the vehicle-traffic reduction of 22 per cent over the zone, at most 0.1 per cent can be ascribed to increased bus traffic.

6 TECHNICAL ASPECTS

The technical system worked very well, and full information was available. People knew what to do, how to pay and so on. Compliance with payments was very high, and the number of complaints was much lower than expected. On an average day in May 2006, 371,300 journeys took place over the charge zone, resulting in 115,100 tax decisions and income of more than SEK 3 million. Of these 115,100 tax decisions, 100 were investigated by the Tax Agency and five were appealed. The road administration customer-service unit received 2,200 calls on an average day in May, as opposed to an expected 30,000 calls. Based on this, our assessment is that the system worked well.

Studies showed that many companies were having problems with the administration of the charges. The system for book-keeping and keeping track of vehicles was not well designed for their business needs, especially at first.

7 EFFECTS ON THE ECONOMY

Effects on Location and the Regional Economy

The regional economy may be affected both in the short and the long terms. The effects on the economy depend to a large extent on whether, and in what way, the congestion charge is reinstated. The effects of the Stockholm Trial on the economy have been investigated in several studies. Most important, an economic analysis of the trade outlook and trade development has been carried out in Stockholm county. Moreover, studies of the retail market, visitor-intensive activities, handicraft companies, driving schools, refuse collection, delivery traffic, taxis, transport for the sick and disabled and courier services are also included. It is clear that the economy is dependent on a functioning road transport system.

Model calculations of the changed attractiveness of different areas are very sensitive to the value of time. The analysis shows many small changes that are uncertain because of their sensitivity to assumptions. The changes are also small in comparison with generally increased pressure from a growing number of citizens and workplaces in the region. Even the influence on house prices is not of great significance. The long-term effects according to the model are no greater than the normal price variations between two areas.

Effects on Retail Trade

There had been fears that retail trade in the zone would be adversely affected, but studies of the retail markets were not able to show any effects of the congestion charges. For example, the durable goods survey in shopping centres, malls and department stores during the Stockholm Trial shows that these have developed at the same rate as the rest of the country. The same holds for other retail sectors.

On an aggregate level, the effects are likely to be very small. For households, the congestion charge has an effect of about 0.1 per cent of total disposable income per year. This means that purchasing power in the county has not been significantly affected, even if for individual households the charge can have tangible consequences.

Cost–Benefit Analysis

A cost–benefit analysis is a means of systematically trying to compare the effects and costs of a particular measure. The analysis is carried out to establish whether a policy measure is justified, in other words whether the benefits it generates are greater than its costs.

Even if it is established that congestion pricing yields a social surplus, it is not evident that it will be enough to cover investment and operational costs, nor that a congestion pricing system is practical and that political limitations will be socially profitable.

The cost–benefit analysis shows that the Stockholm system yields a large social surplus, large enough to cover both investment and operational costs. A permanent congestion charge is calculated to yield an annual social surplus of about SEK 760 million (after deducting operating costs). It would take four years to pay back the system's investment costs in the form of socio-economic benefits. This is a very short payback time compared, for example, with highway or public transport investments which, even in favourable scenarios, have a payback time of 15–25 years. From a cost–benefit perspective, the most relevant basis for a decision is to ignore the investment cost; the Stockholm Trial cannot be undone and the investment undertaken cannot be recouped. But the congestion charge still generates a positive net present value, even when the cost of the investment is taken into account.

Increased bus traffic is unprofitable from a cost–benefit perspective, both during the Stockholm Trial and if it were made permanent. Benefits are calculated to reach SEK 180 million per year, compared to an operational cost of SEK 340 million per year. The result should be treated with caution, however, because it is not unusual for public transport to be considered unprofitable according to a strict cost–benefit analysis, while still deemed worthwhile for various other reasons.

Cost–benefit analysis looks at the average effects on all individuals in the community. For particular individuals, the consequences of the congestion charge could be both positive and negative. The net effect for different individuals depends to a large extent on how the revenues generated are used.

8 ATTITUDES AND OPINIONS

Attitudes Before and During the Trial

Before the start of the trial, public opinion was negative about the trial itself and about congestion charges in general. With regard to opinions about the charges 'in general', the phrasing of the question is crucial. Several studies have shown that if the purpose of the charge and the use of the revenues are specified, support for a congestion charge goes up. This was also confirmed in Stockholm, because several independent studies were made during the years before the trial. Perhaps not surprisingly, actors with different agendas chose different formulations when asking whether people supported a congestion charge.

Attitudes of the general public and of business (separate business attitude surveys were conducted) became increasingly positive once the charge was in place. The media picture also changed dramatically: from 'Prepare for hell!' headlines immediately before the start, to 'Stockholm says: "We love the charges"' a few weeks after it began. Several newspaper editorials reversed their attitude towards the charge, most prominently the *Expressen*, which publicly announced that it was now changing its mind: a congestion charge was a good idea after all. Judging from attitude surveys and from media coverage, we believe that the two most important reasons for this were the visible reduction in congestion and that the technical system worked well.

This development is also consistent with general findings in attitude research. Without the benefit of experience, people focus almost exclusively on obstacles and costs, but with personal experience they begin to appreciate the advantages and benefits gained for the costs incurred.

Not only attitudes towards charges in general, but also attitudes towards the Stockholm Trial itself became more positive during the trial. In Autumn 2005, about 55 per cent of all county residents believed that it was a 'rather/very bad decision' to conduct the trial. After the congestion charge was introduced in January 2006, this percentage fell steadily. In April and May 2006, 53 per cent believed that it was a 'rather/very good decision' while 41 per cent believed that it was a 'rather/very bad decision'. Significantly, even those travelling by car to/from the inner city during the charge period in the most recent two 24-hour periods became more positive by several percentage points.

As with the general population, companies have moved from being primarily negative to more positive, both about the trial and the congestion charge as a permanent measure, but especially about the trial.

The Referendum

The Stockholm Trial was followed by referendums in both the City of Stockholm and about half of the neighbouring municipalities. The initiative for the referendums came initially from the opponents of the congestion charge, but once the idea was proposed, it was generally accepted by all political parties. This was more or less inevitable, especially since the Green party advocates political referendums. Because opinions against the charge were so strong, suggesting a referendum seemed a guarantee of being able to block the charge for a long time.

Initially, only the City of Stockholm was planning a referendum. (The City of Stockholm is by far the largest municipality, accounting for almost half of the inhabitants of Stockholm County, with the remaining population divided among 25 other municipalities.) In Autumn 2005, opponents to the congestion charge suggested that the surrounding municipalities should also arrange referendums. The zone lies entirely within the City boundaries, and because municipalities have responsibility for 'local transport and roads' the City argued that it was entirely up to the City to make a decision about the charge. Several surrounding municipalities, especially those governed by liberal/conservative majorities, argued that the issue affected their inhabitants as much as it affected the inhabitants of the City (not true, according to traffic and travel survey data). The City responded that because it paid for the roads within its boundaries, why should it not be allowed to charge for their use? Moreover, the City argued that it was its inhabitants that were most adversely affected by traffic noise and emissions.

Ultimately, 14 of the surrounding municipalities arranged their own referendums, usually the liberal/conservative-run municipalities opposing the charges rather than those run by Social Democratic/Green majorities.

In the City of Stockholm referendum, a majority were in favour of keeping the charges (53 per cent for, 47 per cent against). However, in the referendums in the neighbouring municipalities (accounting for about a quarter of the county inhabitants) a majority were against keeping the charges (40 per cent for, 60 per cent against). About a quarter of the county inhabitants live in municipalities that did not have a referendum. In total, a majority of the voters were against the charges (48 per cent for, 52 per cent against), but the results are misleading, because many municipalities where polls indicated a majority in favour of the charges did not hold a referendum at all, but instead deferred to the City of Stockholm for a decision.

Hence, the results of the referendums are difficult to interpret. The legal power over the charge lies with the national government (since it is a tax from a legal point of view, and municipalities are only allowed to tax their own inhabitants). From the outset it was unclear how the national government would interpret the result. The political debate over the charge changed visibly during the trial. Before the trial, the opposition viewed the charge as a way of guaranteeing victory in the City, while the Social Democrats more or less tried to distance themselves from the charge. However, during the trial everything changed. The liberals/conservatives were no longer so keen on highlighting the charge issue (although this differed within the parties), while the Social Democrats were suddenly eager to capitalise on the 'success' of the charge during the election campaign.

The referendums coincided with general elections which resulted in new majorities at the national, county and city levels, liberal/conservative instead of the previous Social Democratic/Green majorities. After lengthy consideration on how to interpret the outcome of the referendums, the national government decided on the reintroduction of congestion charges, but the revenues will be applied for road investments. The logic was that this should relieve the negative impacts on the municipalities surounding the City of Stockholm. At the time of writing, a negotiator appointed by the national government is trying to strike a deal between the municipalities and the county of Stockholm. The aim is to produce a 'package' whereby the charge revenues will be used for road investments, with the possibility of additional government funding.

9 CONCLUSIONS

Comparison with Other Measures and Investments

That vehicle traffic declines as driving becomes more expensive is hardly surprising. An interesting question, however, is how great the effect of the Stockholm Trial is compared to other types of measures. The answer is that the reduction in traffic congestion and travel times is very large compared to other traffic measures which have been implemented or considered in Stockholm, for example:

- a new eastern bypass is estimated to reduce the number of vehicles passing over inner-city bridges by approximately 14 per cent;
- a new western bypass is estimated to reduce traffic across inner-city bridges by 11 per cent; and
- if public transport were made free of charge in the county, this is estimated to reduce vehicle-kilometres travelled in the county by 3 per cent.

Of course, road investments are expensive and roads take a long time to build. Many desirable investments in Stockholm fall into the several billion kronor category. For example, it has been estimated that the western bypass would cost SEK 25 billion, the eastern bypass SEK 15–20 billion and free public transport around SEK 5 billion per year. Because the congestion charge would result in a financial surplus of SEK 500–600 million each year (after the deduction of operating costs), it is unreasonable to compare these investments with the congestion charge. Both financially and from a traffic perspective, they are more complements than substitutes.

At the same time, it should be pointed out that the congestion charge, even if the net effect for society is positive, would entail sacrifices for many, which should be set against the positive accessibility and environmental effects of the charge.

The Significance of the Stockholm Trial

The Stockholm Trial resulted in a unique collection of data about traffic and its effects in Stockholm. Knowledge and competence in this area therefore increased. We briefly present some of the lessons learned.

For example, improvements in travel times were so tangible that they were noticed by the general public, with many improvements occurring far from the inner city.

One surprise was that no more than about half of the motorists who 'disappeared' reappeared in the public transport system. Judging from the travel surveys, the general impression is that the number of trips is not a fixed number that can be divided into different destinations, modes of transport or times. Travel patterns may adjust in very complicated or subtle ways. Problems with 'moving congestion' to other roads or to public transport were less than anticipated.

Adjustments to the congestion charge were made and took place quickly. Before the Stockholm Trial, especially when it became clear that the trial period would be reduced to six months, there was some doubt as to whether the traffic reductions would take place. Would the trial be considered to be so brief and transient that it was not worthwhile changing behaviour, with people deciding instead to 'sit out' the trial period without adjusting their travel habits? We now know that the trial had an immediate effect.

The Stockholm Trial provides interesting insights into what a road charging system should look like, an experience that may be helpful to other cities. Traffic economists have long discussed the extent to which a congestion charge zone of the kind used in Stockholm is sufficient for controlling traffic in an entire city. Traffic changes from street to street and from minute to minute. When the zone is as large as it is in Stockholm, there was concern that even if it had a big effect on travel beyond the cordon, streets inside the zone would soon be full of motorists already in the zone, increasing travel on the less-congested roads. Alternative solutions were discussed for several years prior to the Stockholm Trial, involving several sub-zones with varying charge rates. None of the existing road-charging systems sheds much light on this question. In London, the zone involves a small area in the city centre, in Singapore access to cars is strictly regulated, while in Oslo and Bergen the system is designed to affect traffic as little as possible. The Stockholm Trial confirms that a simple charge zone creates significant effects over a large area.

It is also clear that increased investment in public transport cannot alone be used as a means of reducing congestion, especially when the public transport system is already relatively well developed as in Stockholm. Improved and new bus lines did not result in any measurable reduction in vehicle traffic. However, a well-functioning public transport system is a prerequisite for being able to manage the increasing number of passengers. It also increases the effect of a given charging system, because it makes the threshold over to a 'second-best' alternative smaller for many travellers.

PART IV

The United States

16. The Puget Sound (Seattle) congestion pricing pilot experiment

Chang-Hee Christine Bae and Alon Bassok

1 INTRODUCTION

This chapter reports on a federally-sponsored pilot project on road pricing in the Seattle metropolitan area. This is one of several such experiments in the United States (for example, in Georgia, Iowa, Minnesota and Oregon), and it takes place against a backcloth of much more attention being paid to road pricing than some years ago. While it is true that the primary driver is the transportation funding problem, road pricing in urban areas would have substantial congestion-reduction effects. The most interesting aspect of the Seattle experiment is its use of GPS (global positioning system) technology rather than the more standard transponder plus road sensors. Although the experiment is small scale, it offers opportunities to judge the feasibility of the GPS approach. If it works effectively, it is more suitable for a system-wide approach (that is, freeways plus arterials) than the alternatives.

2 RECENT TRAFFIC TRENDS

The recent traffic experiences of the Seattle metropolitan area were summarized in a Washington State Department of Transportation report (WSDOT, 2006). The WSDOT bases its analysis on the empirically supported assumptions that a speed of 51 miles per hour (mph) maximizes traffic throughput, while congestion increases below 40 mph and becomes severe under 35 mph. In the two years (2003–05) of the WSDOT study, travel times increased on 34 out of major 35 commute routes in the peaks (defined somewhat traditionally as 6–9 am and 3–7 pm), with a tendency to larger increases in the afternoon peaks. On the most congested routes, the increases were startling for such a short period: 28 percent on the Redmond–Seattle and Bellevue–Tukwila routes in the afternoon and

25 percent on the Seattle–Bellevue route in the morning (for those unfamiliar with Seattle, these routes link downtown Seattle and the SEA-TAC airport area with the very prosperous Eastside; also the dominant traffic flows are increasingly reverse flow). Using another congestion measure (the ratio of peak travel time to maximum throughput travel time), the am ratio increased system-wide from 1:25 to 1:44 while the pm ratio rose from 1:39 to 1:58, once again by significant amounts.

Increasing congestion has affected travel behavior. Unlike elsewhere in the United States, high occupancy vehicle (HOV) use has increased (on all but one route). About 200 miles of HOV lanes have been constructed since 1970, but many of them are now congested in peak hours. Transit use has increased on some key routes, for example, by 20 percent over the past decade between Bellevue and Seattle. Returning to the 2003–05 period, peak vehicle volumes declined (slightly) for the first time, although *total* daily vehicle volumes have increased, again slightly. Also, for the first time, an increase was associated with a moderate decline in vehicle-miles traveled. Furthermore, there was a decline in discretionary non-peak trips, although more likely because of rising gasoline prices rather than as a consequence of congestion.

3 THE POLITICAL BACKGROUND TO TRANSPORTATION IN WASHINGTON STATE

Transportation is an important political issue in Washington State. One problem, not without a solution by the legislature, is that current Washington State legislation requires that tolls be removed once the bonds for road and bridge construction are paid off. Furthermore, a series of ballot initiatives, most of them started by an activist named Tim Eynman, illustrate strong resistance to taxes for transportation purposes. Initiatives have an important role in the political process in Washington because their results can override legislation. Also, the very balanced state legislature (approximately 50:50 Democratic and Republican) has found it very difficult to pass transportation bills, although they did pass a 9 cents per gallon increase in the gasoline tax. There have been other problems: a voter-approved Monorail, later abandoned when the cost estimates doubled; a light rail program drastically cut back because of cost escalations; and a political struggle about whether to replace the Alaskan Way viaduct damaged in the 1998 earthquake with another elevated highway or a tunnel.

4 RATIONALE

The surge in road pricing schemes has less to do with a desire to reduce congestion than with financing new roads. States have become concerned about whether the gasoline tax has enough long-term potential to pay for highway construction. So, a major reason for road pricing is to raise revenue. Nevertheless, even so, congestion reduction is a powerful secondary motive. Also, whereas years ago, the objections to road pricing were primarily based on political feasibility, these objections have ebbed as the road financing and congestion issues have become severe.

5 FISCAL BACKGROUND

As pointed out above, the road financing situation in Washington has been the prime motivation for interest in road pricing. The system is operating poorly for several reasons: the fiscal elasticity of the gas tax is very weak; there is a need for new highway capacity as the state continues to grow while the condition of existing roads deteriorates; efforts to improve public transit have been slow to implement with minimal impacts on ridership; and land-use regulations (for example, the growth management regime) may have aggravated rather than alleviated transportation problems. Although a phased 9 cents per gallon increase in gasoline taxes was passed by the legislature in 2002, there is little public support for further increases. For example, there is no correlation between who benefits from new roads and who pays for them, while the constant tax per gallon does not address the problem of congestion that may create the need for new capacity. Demand or value pricing, on the other hand, levies charges selectively on certain vehicle-miles traveled, helps to control peak-hour congestion, and forges a revenue link between the creation of congestion and its relief by capacity expansion.

The modal split shown in the pie chart of Figure 16.1, with the heavy reliance on solo driving (about three-quarters of all trips), is very similar to that of other US metropolitan areas. The implication is that the case for addressing this problem, if it is perceived to be a problem, is not unique to Seattle, so the lessons learned from its experimental scheme may be applicable elsewhere.

Freeway traffic volumes by time of day on different freeways and locations are shown in Figure 16.2. They illustrate the typical diurnal pattern of morning and even higher afternoon peaks. However, they also show a sizable quasi-peak in the mid-day hours, a phenomenon not fully explained but often reflecting lunchtime travel and an early end to the school day.

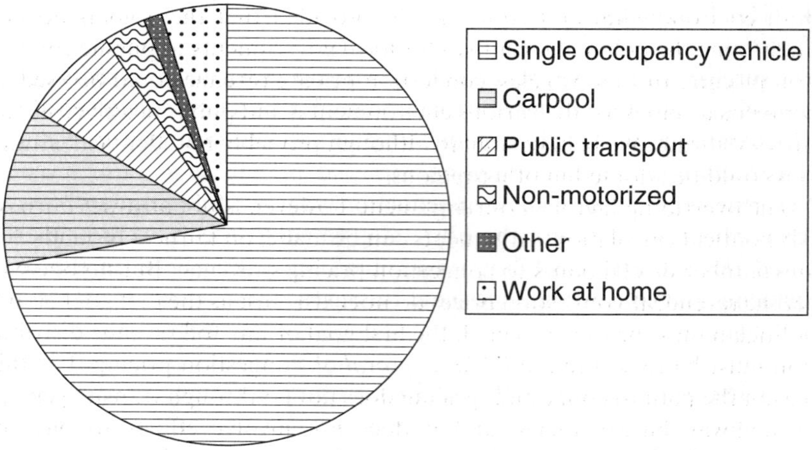

Figure 16.1 Modal split in the Seattle metropolitan region

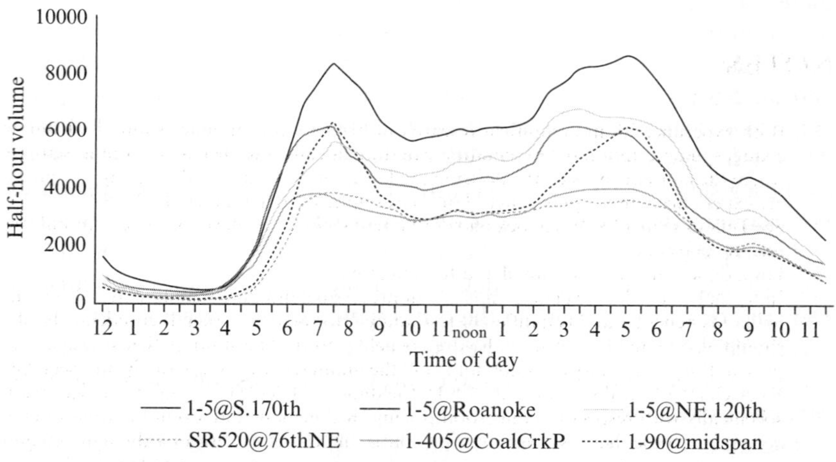

Figure 16.2 Freeway traffic volumes by time of day

Combined with the negligible travel overnight and modest volumes in the late evening, the data make a strong case for time-of-day differential pricing.

As illustrated in Figure 16.3, the growth in vehicle miles traveled (VMT) has consistently been much faster than the growth in population and employment over the past quarter of a century, and for most of the period jobs have expanded faster than population. Moreover, the rapid increase in

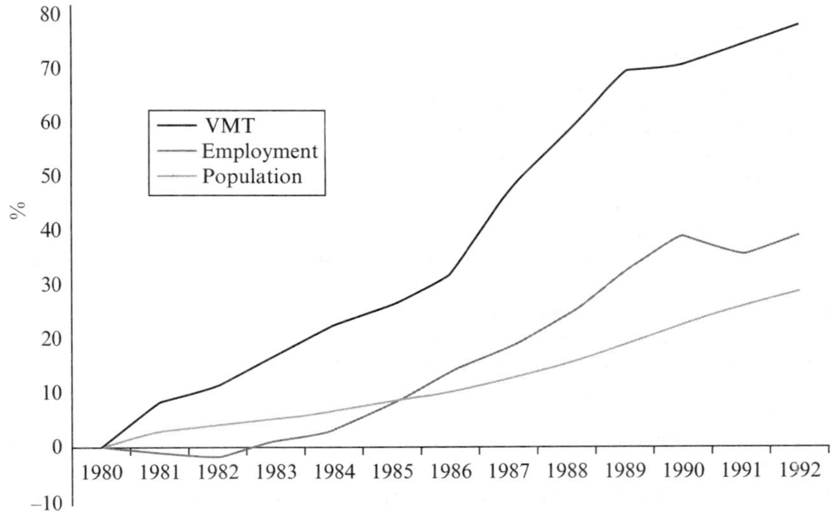

Figure 16.3 *VMT growth compared with population and employment growth*

VMT has been matched with a negligible growth in highway capacity. Even though there is some evidence that the VMT growth curve is flattening out, these relationships are unsustainable without intervention to control the congestion associated with VMT growth.

6 EXPERIMENTS

As shown in Table 16.1, there are about 25 experimental congestion pricing schemes either planned or operational throughout the country, and the number is increasing month by month. The first five in place were: SR91, Riverside County, California; I15, San Diego County; Katy Freeway (I10) and Northwest Freeway (US290), both in Harris County, Texas; and I394, Minneapolis. With the exception of the Southern Californian examples, these are very recent. Many of them are motivated by road construction financing considerations. Their scope is also enlarged in areas where there have been substantial investments in HOV lanes where expansion into HOT (high occupancy toll) lanes is relatively straightforward, permitting solo drivers to pay a toll for a faster commute.

The items in Table 16.1 are not a comprehensive list of all the toll roads in the United States because there are long-established toll roads, usually inter-urban freeways, in many states. In fact, 26 states have toll roads,

Table 16.1 Operational and planned toll roads

State	Location	Facility	Status
Arizona	Phoenix	All freeways	Study
California	Alameda Co.	I680, I880	Study
	Contra Costa	SR4W	Study
	Los Angeles	Various	Post-study
	Orange Co.	SR91 Express Lanes	Operational
	Orange Co.	SR57	Study
	Riverside Co.	SR91 extension	Study
	San Diego Co.	I15	Operational
	Santa Cruz Co.	SR1	Authorized
	Sonoma Co.	US101	Post-study
Colorado	Denver	I25	Study
Florida	Miami	I95, SR836	Study
	Orlando	I4	Study
Maryland	Baltimore suburbs	Various	Study
Minnesota	Minneapolis	All freeways	Study
Oregon	Portland	Various	Study
Pennsylvania	Philadelphia	US1	Study
Texas	Austin	I35	Study
	Dallas	I635	MIS
	Houston	I10	Operational
	Houston	I10 extension	MIS
Virginia	Hampton Roads	I64	Approved
Wisconsin	Milwaukee	I94	Proposed study

Source: Poole and Orski (2000), Regulation, http://www.cato.org/pubs/regulation/regv23n1/poole.pdf.

ranging from one mile in Utah to 597 miles in Oklahoma. Familiar examples include the New York Thruway and the New Jersey Turnpike. Eight states (Oklahoma, New York, Pennsylvania, Ohio, Florida, New Jersey, Illinois and Kansas) have more than 237 miles, the first five more than 360 miles (Federal Highway Administration (FHWA) data). California has 82 miles of toll roads with 590,000 weekday trips, all of them in Southern California and very recently built; Washington State has none. In terms of future developments, Texas has a plan to construct a 4,000-mile network, much of it aimed at increasing the competitiveness of the Port of Houston. The network would be built primarily by the private sector. Indeed, 21 states have passed legislation permitting, even promoting, public–private road-building partnerships. A few (for example, Indiana and Illinois) have turned over operations to international consortia.

7 CORDON, CORRIDOR AND SYSTEM-WIDE SCHEMES

The spatial structure of most US metropolitan areas makes cordon schemes on the London and Stockholm pattern an unlikely method of implementing congestion pricing schemes. New York City (or more precisely Manhattan) has some potential because of its very few entry routes. Yet a recent study recommending it was originally dismissed by the mayor on the ground that it was a 'commuting tax', with the prediction that the state government in Albany would reject the idea. However, in light of recent US Department of Transportation support, it is now back on track. It would probably need to diverge from London and Stockholm by charging for travel within the cordon. Because of its topography, San Francisco may be a candidate for a cordon scheme, and there has even been some discussion of a downtown cordon in Seattle, perhaps as one component in a hybrid cordon-corridor scheme.

Thus, segment corridor schemes are likely to be much more common, especially given the road financing angle. System-wide schemes may run into political opposition, both by politicians afraid of electoral retribution and the driving public. If adopted, such schemes may be more palatable if restricted to freeways than extended to most arterials. It is likely that the greatest promise lies in converting existing HOV lanes into HOT lanes that allow solo drivers to use carpool lanes at a price (http://www.cato.org; http://itsdoc.fhwa.gov).

8 THE SEATTLE PILOT STUDY

The US Department of Transportation financed several pilot schemes in the United States, of which Seattle is one. The FHWA awarded a $1.88 million grant in 2002, with supplemental funding of $600,000 in 2005. Its importance is primarily as a test of GPS technology. However, there is some experience with bridge tolls in Seattle. For example, there was a 35 cents toll on the SR520 floating bridge across Lake Washington between 1963 and 1979 before it was retired once the bridge was paid for. In the first year after the toll was dropped, bridge traffic increased by 19.3 percent. Now, the bridge is obsolete, in a poor state of repair and of limited capacity. A plan exists to build a new bridge at a cost of up to $3.4 billion; there have been some discussions about the possibility of tolls, but no decision has been made. Another important project is to add a new twin bridge to the existing Tacoma Narrows Bridge. Some $800 million of tax-exempt bond financing is expected soon to pay for it. In this case, however, an initial $3 toll is planned, generating revenue of $45 million a year. The technology will probably be electronic transponders offering access to express lanes.

The participants in the Pilot Study were randomly selected from a volunteer pool. The study design permitted the analysts to identify behavioral responses to time-of-day variations in network tolls. The tolls charged in the pilot study vary by day of the week, time of day and between freeways and arterials (Table 16.2). There were no tolls at the weekend, higher tolls in the morning and afternoon peaks, and double the toll on the freeway compared to the arterials, which is how it should be if congestion reduction is the goal. Approximating the charges to the difference between the private and the social marginal costs of congestion is the best way to achieve this. The tolled roads included the freeways and major state arterials (Figure 16.4).

One of the possible problems with the Seattle Pilot Study is that the 275-plus participants, with more than 400 vehicles, receive a modest endowment account (based on baseline travel behavior) from which they can draw to pay tolls as they require. At the end of the experiment, they retain any balance left. The problem with this approach is that they do not incur out-of-pocket costs so, in a sense, the decision making is artificial, somewhat similar to contingent valuation where individuals are asked a hypothetical question about willingness to pay rather than a revealed preference.

However, there has been some discussion about credit-based payment systems which are not too different from the Seattle Pilot Study practice. The concept is to address the equity concerns of the effects of congestion road pricing. By issuing credits to moderate-income households out of which congestion pricing tolls might be paid, an agency could deal with the equity problem. The credits would be financed out of toll revenues, a significant cross-subsidization solution.

There are at least four major ways of handling the technological aspects of a road pricing scheme: paper permits, cash or credit card tollgates, transponders and GPS on-board units. The use of paper is technically obsolete (it was a feature of the original Singapore District Licensing Scheme), and tollgates (while still in use) slow down traffic too much. Transponders are becoming quite popular, and they permit off-site billing. Their drawback is that, if used on more than a few corridor freeways, they require expensive road sensors. The Seattle experiment uses the GPS approach, certainly the most promising technological solution because of its flexibility. Although on-board GPS units are currently expensive, their

Table 16.2 *Tolls in the Seattle Pilot Study*

Time of day	Toll rates per mile*			
	Monday–Friday		Saturday & Sunday	
	Freeways	Non-freeways	Freeways	Non-freeways
Midnight				
1:00 am				
2:00 am		No charge		No charge
3:00 am				
4:00 am				
5:00 am				
6:00 am				
7:00 am	$0.40	$0.20	$0.10	$0.05
8:00 am				
9:00 am				
10:00 am				
11:00 am				
Noon	$0.15	$0.075		
1:00 pm				
2:00 pm			$0.20	$0.10
3:00 pm				
4:00 pm				
5:00 pm	$0.50	$0.25		
6:00 pm				
7:00 pm				
8:00 pm	$0.10	$0.05	$0.10	$0.05
9:00 pm				
10:00 pm	No charge		No charge	
11:00 pm				

Note: *Downtown Seattle zone is toll free.

cost could be drastically reduced if fitted on all new cars. Also, they have the advantage that with a remote monitoring system, they can be used on all roads, both freeways and arterials, without having to incur the cost of road sensors. Two possible downsides are the concerns about their privacy implications and doubts about their universal coverage because of topographical- or structure-induced fadeout (technological advancements should alleviate this problem relatively soon).

The on-board GPS equipment used in Seattle showing the route and the per mile charge was obtained from Siemens, with cellular communication to a back office which monitors the flows and bills for service. This latter

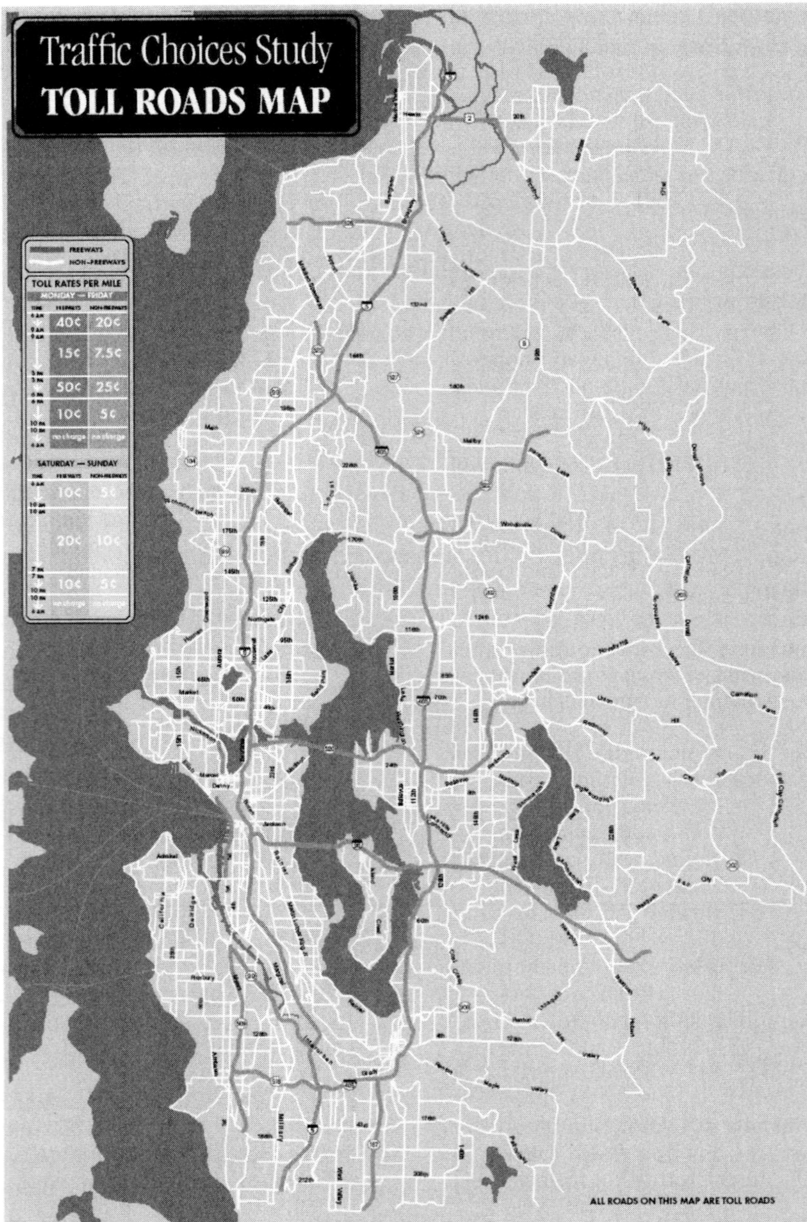

Source: PSRC.

Figure 16.4 Tolled roads in the Seattle region

component would be the most expensive cost item in a fully implemented metropolitan-wide system.

9 RESULTS

The experiment ran for 18 months (January 2005 to end of June 2006). With 450 on-board units, the experiment covered almost 247,000 individual trips with tolls paid amounting to $230,200 (about $837 per participant). More than two-thirds of VMT were on tolled roads. The experiment also resulted in hundreds of customer service calls, backed up with household surveys and focus groups. It also showed that tolls can affect travel behavior, although the results may have been distorted by the absence of out-of-pocket costs. The toll scheme influenced the morning and evening peaks differently. In the morning peak, it showed a potential reduction of household automobile trips by up to 10 percent and of VMT by up to 4 percent. However, in the evening peak, automobile trips fell by less (6 percent) while VMT fell by more (11 percent). In general, the roads and times with the highest tolls experienced the largest trip reductions. In addition, about three-quarters of participants responded to the tolls either by driving less or by reducing their tolling liability by changing routes and/or times. Of course, the toll environment was artificial because only the participants were paying for congestion; the congestion reduction would potentially be greater with universal participation, possibly weakening the incentive for any individual to pay the toll. Thus, the effects of changes in travel behavior under the experiment might be exaggerated. Nevertheless, 70 percent of survey respondents reported that they changed their behavior as a result of the tolls.

Table 16.3 presents some descriptive statistics by time of day during the experiment. They illustrate that the evening peak is the major problem (for example, much lower speeds), yet the tolls paid per trip were higher in the morning peak. The other interesting point (also observed in other studies) is the importance of the quasi-peak during the mid-day period, with slower average speeds than in the morning peak. One possibility is that the mid-day tolls are underpriced (15 cents per mile on freeways and 7.5 cents per mile on arterials). However, mid-day trips were more likely to be taken on arterials, which had lower tolls, rather than freeways.

Table 16.4 shows the trip elasticities by time of day. Peak trip elasticity is greatest during the morning peak (–16.10 percent), although VMT elasticity is much lower (–5.31 percent). This suggests that variable road pricing could reduce the number of discretionary morning peak hour trips but with a much lesser impact on VMT (that is, the trips eliminated tend to be short). Mid-day elasticities are high, but the sign on the VMT elasticity is positive

Table 16.3 Puget Sound Regional Council, GPS Study: descriptive statistics for the experimental period (weekdays only)

	Morning peak	Mid-day	Evening peak	Evening	Night and early am
Average number of trips per HH per period	1.02	2.73	1.77	0.81	0.24
VMT per HH per period	8.25	15.62	10.75	4.76	2.47
Vehicle-hours per HH per period	0.47	1.06	0.95	0.26	0.07
Avg Speed (miles/hour)	17.6	14.7	11.3	18.1	34.0
Total number of trips (experimental period)	38,278	102,449	66,423	30,434	9,006
Total VMT (experimental period)	309,598	586,172	403,415	178,629	92,692
Tolls paid (per trip) (experimental period) ($)	1.80	0.47	1.52	0.30	0.31
Tolls paid (total) (experimental period) ($)	69,050	48,410	100,948	9,006	2,777

Note: HH = households.

(18.06 percent). This suggests that the mid-day period attracts longer discretionary trips from the peaks, because the number of trips has the expected negative sign. Not surprisingly, the VMT elasticity is highest for the evening peak (–18.03 percent) with a trip elasticity of –9.96 percent. A more detailed examination of all the data shows greater trip reductions on the roads with the highest tolls.

Overall, the experiment worked, demonstrating that the core technology for a satellite-based toll system is effective. Translating it to a permanent

Table 16.4 Estimated weekday elasticities in response to automobile variable costs

	Am peak	Mid-day	Pm peak	Evening	Night and early am
Trips	−0.1610	−0.1560	−0.0996	−0.1290	0.0532
VMT	−0.0531	0.1806	−0.1803	0.0491	0.3326
Hours	−0.0019	0.2483	0.0795	0.0970	0.5346

metropolitan-wide system is merely a problem of scale (with a viable business model), legal enforceability and public acceptance (efforts by the WSDOT to 'sell' the concept to the general public have not been very successful).

10 OBSTACLES TO A PERMANENT SCHEME

Regardless of the relative success of the pilot schemes, there are major obstacles in the way of implementing a permanent scheme. A standard objection is on equity grounds (Richardson and Bae, 1998). One alleviation is the availability of unpriced arterial roads to which low-income drivers can move when tolls are onerous. However, because of chronic underinvestment in roads and topographical constraints, the Seattle area has very limited highway redundancy, so this type of relief is poor compared with most other US metropolitan areas. Also, even if the cost of on-board GPS monitors falls on the individual motorist, the system will be quite expensive, perhaps \$13 billion. Equity concerns can also be addressed by differential charging among vehicle classes and by location (for example, a statewide road pricing scheme might charge urban drivers more than rural drivers). Overall, however, the revenue generation associated with a road pricing scheme offers some prospects of using some resources to address the equity issue, although in the new environment much of the revenue would probably be eaten up by road construction and maintenance costs.

There is a long tradition of opposition to transportation funding in Washington State, and this could undercut any proposal, regardless of whether it was via ballot initiative or legislative action. Modal shift prospects are limited, certainly until the rail system is in place (2010) and even then its geographical coverage is small, and given the infrequent and erratic bus service in the suburbs except on a few key routes. Another obstacle is that many planners believe that they can deal with the congestion problem by land use and similar non-pricing prescriptions, for example, reducing automobile dependence via promoting density, and extreme traffic-calming measures as a disincentive to drive. Finally, legal and constitutional difficulties (for example, under current law highway financing instruments can be in place only until the costs of the roads have been paid off) may inhibit the implementation of a scheme.

11 CONCLUSIONS

Despite its limited scope and the artificial toll environment, the Seattle Pilot Study was an important step on the path towards widespread road congestion pricing. The main reason is its demonstration of GPS-based technology as an alternative to the transponder approach. Provided that scale economies can bring down the costs of on-board modules significantly, this approach gives much more flexibility in adjusting the geographical scope of a scheme without investments in road sensors, although the satellite and central system back-up are by no means cheap. However, given the transportation history of Washington State, it is very problematic whether the Seattle experiment will ever be converted into a permanent scheme in Seattle itself. Despite the severe traffic congestion there (perhaps as bad as in Los Angeles and Atlanta) and the underinvestment in highways, there is almost no popular pressure for a road pricing scheme. Much depends on the toll structure and on how toll revenues are used; for example, the peak charge on the SR91 freeway in Southern California is now $9.50 for a 10-mile stretch aimed at ensuring free flow on the toll lanes and repaying the bonds that financed the road construction. However, tolls at this level are hardly likely to garner popular support.

REFERENCES

Richardson, Harry W. and Chang-Hee Christine Bae (1998), 'Road pricing and income distribution', in K.J. Button and E.T. Verhoef (eds), *Road Pricing, Traffic Congestion and the Environment: Issues of Efficiency and Social Feasibility*, Heidelberg: Springer, pp 247–62.

WSDOT (Washington State Department of Transportation) (2006), *Measures, Markers and Mileposts*, Olympia, WA: WSDOT.

Websites

http://www.cato.org/pubs/regulation/regv23n1/poole.pdf.
http://www.itsdocs.fhwa.dot.gov/JPODOCS/REPTS_TE/hot/chapter_1.htm.

17. The US context for highway congestion pricing

Bumsoo Lee and Peter Gordon

1 INTRODUCTION

If price does not ration, something else will. We also know that auto ownership and use respond to rising income and that congestion has become the default rationing mechanism on most of the world's roads and highways. Economists and others have pointed out that this is increasingly wasteful and have argued that time-of-day pricing should be implemented (see, for example, the recent collection of essays edited by Roth, 2006). Modern monitoring and collection technologies suggest that this can now be done at low cost – although that assertion is challenged in a recent examination of the Stockholm road pricing trial, by Prud'homme and Kopp (2006).

Policy makers in the US, however, have for the most part been reluctant to go along, fearing the prospect (or the appearance) of regressive impacts – even though they are thereby forgoing a new and considerable revenue source. In Chapter 19 of this volume, King et al. argue that improved revenue targeting and sharing schemes could develop greater political support.

The world's best-known experiments with road pricing have been the area-pricing programs in Singapore (since 1975)[1] and London (since 2004). On a smaller scale, there have been scattered cases around various cities of the developed countries with moderately scaled pricing experiments on specific areas or on specific stretches of highways.

Recently, some writers have suggested that the US is now near a tipping-point, and that many more road pricing projects will soon be implemented (Poole and Orski, 2006). What do we know about the modern US urban transportation context? What does it suggest for further pricing projects in this country? This chapter is a survey of recently accumulated descriptive research findings and attempts to answer both questions.

There are two key results that emerge. First, in most major US cities, trip origins and destinations are dispersed. This makes London- or Singapore-style area-pricing impractical. A HOV- to HOT-lane conversion plan may be more appropriate. Second, the growth of non-work travel – many trips which are probably more responsive to dollar cost increases than commutes – at *all* times of the day makes the pricing alternative more attractive than had heretofore been thought. A look at the online TDM encyclopedia,[2] which includes a summary of recently estimated transportation demand elasticities, suggests that more specific knowledge in this area would be helpful. Our study of non-work travel also depicts the growth of *chained tours* that combine work trips with non-work trips. These are not likely to be done via transit or carpools. This further strengthens the case for the HOT-lane approach.

2 DATA

Cities have been decentralizing for many years. In the US, the group of 75 largest cities gained population share until about 1940, but have been losing their proportionate importance ever since. The suburbanization of origins and destinations has been used to explain relatively benign commuting times – in spite of the absence of road pricing (Gordon et al., 1989; Crane and Chatman, 2003). In fact, *average* travel speeds on US roads had been *increasing* to 1995. Less benign travel speed and time trends since 1995 have been explained by the prosperity of the late 1990s (more income, more cars, more errands by car, see below) coupled with a decline in road construction (Gordon et al., 2004).

Our analysis begins where these well-known facts leave off, by considering trends for the smallest spatial units for which data are available to us and by considering employment as well as population locations.

We relied on two datasets. One was the 2000 Census Transportation Planning Package (CTPP) data, drawn from the decennial census journey-to-work survey. The CTPP is one of the few sources of employment data by place of work for small geographical units such as census tracts or traffic analysis zones (TAZs). It provides tabulations of households, persons and workers by place of residence, by place of work, and for journeys to work. This information is all-important for grasping the extent of decentralization and dispersion of population and jobs in US metropolitan areas.

We also worked with data from the 1990 and 1995 Nationwide Personal Transportation Surveys (NPTS) and the 2001 National Household Travel Survey (NHTS) for the study of work and non-work travel patterns. In the 11-year interval, the US population grew by 16 percent while the number of workers grew by 22 percent, household vehicles increased by 23 percent, person-trips by 34 percent and person-miles by 40 percent. Constant dollar per capita income over the 11 years grew by 21 percent.

The surveys, initiated in 1969 by the US Department of Transportation (USDOT), provide detailed data on households, people, vehicles and travel for all purposes by all modes. Thus, the NPTS/NHTS data series are one of the best data sources for the analysis of nationwide travel trends. Nevertheless, there are some comparability issues from one survey to another because the survey techniques changed between survey years. In particular, a travel diary (replacing memory recall) and household rosters have been used only since the 1995 survey. These changes have significantly improved interview responses. Hu and Young (1999) provide a method to adjust 1990 data for comparison with 1995 and 2001 results by estimating the impact of the two new techniques that had been used in the 1990 survey. Another problem is that, given what is known of work-trip trends in the 1990s, the 1995 survey is believed to overestimate work trips. For these reasons, our trip-level analysis relies on the 1990 and 2001 data only. However, we did use 1995 and 2001 data for tour-level analyses because there was no information on trip-chaining behavior in the 1990 data.

3 CENTRALIZATION, DISPERSION AND DECENTRALIZATION

London- or Singapore-style area pricing schemes are effective in cities with highly centralized employment or sizable central business districts (CBDs). For instance, the original congestion charging zone in central London[3] – an eight-square mile area inside the inner ring road – contained about 1.1 million jobs, or 27 percent of total employment in Greater London as of 2005 (Santos and Fraser, 2006). Singapore's Restricted Zone covered an area of about 2.8 square miles including the CBD and 315,000 jobs, or about 20 percent of the city state's total employment were centralized in the zone as of 1990 (McCarthy and Tay, 1993).

However, US metropolitan areas are much more decentralized and dispersed. We were able to define and identify major employment centers and subcenters. In general, urban employment centers are defined as clusters of zones that have higher employment density than the surrounding areas. Two types of procedure have been popularly used in identifying density peaks, a minimum density procedure (Giuliano and Small, 1991) and a nonparametric method (McMillen, 2001). In a previous study (Lee, 2007), we found that employment centers defined by the latter best reflect downtowns as we know them.

We identified CBDs and subcenters in the 79 largest US metropolitan areas using a modified version of McMillen's geographically weighted regression (GWR) procedure. Whereas he identified TAZs that have higher actual density than the estimated GWR surface, we compared two estimated employment density surfaces – one with a small window size (10 neighboring census tracts) and the other with a large window size (100 census tracts) (for detailed descriptions of the procedure, see Lee, 2007). Among the identified employment density peaks, we qualified only those with more than 10,000 jobs as employment centers.

Table 17.1 shows CBD size and employment share by location type in the largest US metropolitan areas and average values in each metro size class. In every one of the major metro areas (those with population over 3 million), the largest share of employment was dispersed, outside of any identifiable center. Also in every one of the largest metro group, the CBD accounts for less than 10 percent of total metro employment. The table also reveals similar tendencies for the smaller metros in our sample. Agglomeration opportunities come in many shapes and sizes but there are now many more Silicon Valleys than Manhattans. The dispersion tendency, of course, takes some congestion pressures away from centers but it also decreases the possible usefulness of the area pricing approach.

Table 17.2 presents (one-way) commute times in 2000, by workplace location type. These data are for drive-alone commuters only so that mode mix changes do not perturb the results. The table shows that the shortest commutes are for workers with destinations at dispersed locations. CBD workers spend significantly more time in commuting than other metro commuters, especially in largest metro areas. CBD workers' commute time is almost twice as long as the metropolitan average in New York and they spend 40 to 53 percent extra commute time in Philadelphia, Chicago and Boston, which have relatively large CBDs. These older CBDs can still offer enough agglomeration economies to fund the wages that offset the longer commutes.

We were also able to describe changes in employment decentralization and dispersion over the last one or two decades. But, unfortunately, census tract level employment data conversion between census years was possible only for six metro areas. Los Angeles and San Francisco are more polycentric than the other four metropolitan areas in relative terms.

Employment decentralization occurred for the years shown in all six of these metro areas. The CBD's employment share shrank in each metro. Second, subcenters' employment share also fell in New York, Boston, Philadelphia and Portland while significant subcentering occurred in the two western polycentric metros. It is interesting to find that more centralized places experienced increased dispersion. Nevertheless, either version of these spatial evolutions makes area congestion charging schemes even less attractive.

Table 17.1 Employment share by location type and CBD size in the largest US metropolitan areas, 2000

Metro name	Population (thousands)	Employment (thousands)	CBD		No. of sub-centers	Share of employment (%)		
			Emp. (thousands)	Area (sq.miles)		CBD	Sub-centers	Dis-persed
New York	21,200	9,418	937.1	1.9	33	9.9	11.2	78.8
Los Angeles	16,370	6,717	190.1	2.5	53	2.8	28.8	68.4
Chicago	9,158	4,248	297.8	1.2	17	7.0	11.9	81.1
Washington	7,608	3,815	283.3	1.6	16	7.4	11.8	80.8
San Francisco	7,039	3,513	205.6	0.8	22	5.9	24.2	70.0
Philadelphia	6,188	2,781	239.7	2.4	6	8.6	4.5	86.9
Boston	5,829	2,974	238.1	1.7	12	8.0	8.0	84.0
Detroit	5,456	2,509	129.8	4.7	22	5.2	22.2	72.6
Dallas	5,222	2,566	126.0	4.7	10	4.9	15.8	79.3
Houston	4,670	2,076	165.5	4.1	14	8.0	20.8	71.2
Atlanta	4,112	2,088	166.9	3.9	6	8.0	10.7	81.3
Miami	3,876	1,624	121.0	4.0	6	7.5	15.0	77.5
Seattle	3,555	1,745	163.1	1.7	7	9.3	11.9	78.8
Phoenix	3,252	1,464	104.4	5.5	9	7.1	12.9	79.9
Averages by metropolitan area population size group								
3 million +					17.0	7.1	15.0	77.9
1–3 million					2.6	10.8	7.0	82.2
0.5–1 million					0.9	12.2	5.2	82.6

Source: modification from Lee (2006).

Table 17.2 Commute time by drive alone mode by location type, 2000

Metro name	Population (thousands)	Employment (thousands)	2000 commute time by drive alone mode (min)			
			Metro	CBD	Subcenters	Dispersed
New York	21,200	9,418	28.5	55.6	30.2	27.8
Los Angeles	16,370	6,717	27.8	36.6	28.9	27.0
Chicago	9,158	4,248	28.9	41.8	32.1	28.0
Washington	7,608	3,815	30.3	40.2	30.2	29.8
San Francisco	7,039	3,513	28.4	39.3	29.3	27.8
Philadelphia	6,188	2,781	26.1	36.6	26.1	25.7
Boston	5,829	2,974	27.1	41.6	25.9	26.7
Detroit	5,456	2,509	26.2	31.0	27.7	25.4
Dallas	5,222	2,566	27.4	31.5	28.0	27.1
Houston	4,670	2,076	28.1	32.9	28.9	27.3
Atlanta	4,112	2,088	30.9	36.0	31.4	30.3
Miami	3,876	1,624	27.9	33.8	28.9	27.1
Seattle	3,555	1,745	26.2	30.7	26.3	25.8
Phoenix	3,252	1,464	25.4	31.1	24.7	25.0
Averages by metropolitan area population size group						
3 million+			27.8	37.1	28.5	27.2
1–3 million			24.1	26.9	23.4	23.8
0.5–1 million			22.3	23.3	21.7	22.2

Source: Modification from Lee (2006).

4 NON-WORK TRIPS

Much of the discussion of the 'urban transportation problem' focuses on commuting, and commuting is thought to be a peak-hour problem. Both thoughts require some re-examination. In recent research, we found that most travel by Americans does not involve commuting. In fact, a large majority of peak period travel is not work related.

In a recent paper, we investigated work and non-work travel patterns in terms of temporal variation (Lee et al., 2006). All trips in 1990 and 2001 were grouped by 10 distinct periods of the week according to their departure time (Table 17.3). Non-work trips accounted for more than four-fifths of all trips in each year of the surveys, and were a sizable majority in every one of the 10 time-of-week periods including peak-hour periods (Table 17.4). They also grew more quickly between the 1990 and 2001 survey years

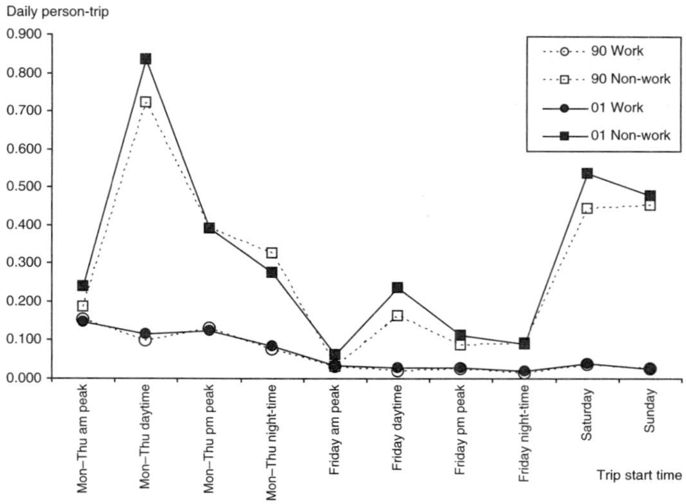

Source: Lee et al. (2006).

Figure 17.1 *Average daily person-trips per person by trip purpose and by time of week, 1990 to 2001*

Table 17.3 *Definitions of 10 periods of the week*

Time of day/week	Week	Departure time
Mon.–Thu. am peak	Mon.–Thu.	6:00am–8:59am
Mon.–Thu. day off-peak	Mon.–Thu.	9:00am–3:59pm
Mon.–Thu. pm peak	Mon.–Thu.	4:00pm–6:59pm
Mon.–Thu. night off-peak	Mon.–Thu.	7:00pm–5:59am
Friday am peak	Friday	6:00am–8:59am
Friday day off-peak	Friday	9:00am–3:59pm
Friday pm peak	Friday	4:00pm–6:59pm
Friday night off-peak	Friday	7:00pm–5:59am
Saturday	Saturday	0:00am–12:59pm
Sunday	Sunday	0:00am–12:59pm

than work trips (by 30 percent as opposed to 23 percent, while the US population grew by 15.8 percent).

The Monday–Thursday am peaks included the largest number and share of work trips, but these work trips were never the majority trip type, and their share even fell significantly between survey years. The Friday am peak

Table 17.4 Annual person-trips by trip purpose and by time of week, 1990 to 2001

	All	(%)	Work	(%)	Non-work	(%)
1990 All (millions)	284,551	100	49,327	17.3	235,224	82.7
Mon–Thu am peak	27,272	100	12,227	44.8	15,045	55.2
Mon–Thu off-peak day	66,526	100	7,906	11.9	58,620	88.1
Mon–Thu pm peak	42,259	100	10,495	24.8	31,764	75.2
Mon–Thu off-peak night	32,709	100	6,152	18.8	26,557	81.2
Friday am peak	5,068	100	2,536	50.0	2,532	50.0
Friday off-peak day	14,890	100	1,655	11.1	13,235	88.9
Friday pm peak	9,094	100	2,032	22.3	7,062	77.7
Friday off-peak night	8,723	100	1,233	14.1	7,489	85.9
Saturday all day	39,108	100	2,982	7.6	36,127	92.4
Sunday all day	38,902	100	2,109	5.4	36,793	94.6
2001 All (millions)	366,458	100	60,651	16.6	305,807	83.4
Mon–Thu am peak	36,121	100	13,683	37.9	22,438	62.1
Mon–Thu off-peak day	89,124	100	10,724	12.0	78,400	88.0
Mon–Thu pm peak	48,367	100	11,712	24.2	36,655	75.8
Mon–Thu off-peak night	33,750	100	7,818	23.2	25,932	76.8
Friday am peak	9,136	100	3,270	35.8	5,866	64.2
Friday off-peak day	24,927	100	2,712	10.9	22,215	89.1
Friday pm peak	13,240	100	2,679	20.2	10,561	79.8
Friday off-peak night	10,180	100	1,815	17.8	8,365	82.2
Saturday all day	54,218	100	3,786	7.0	50,431	93.0
Sunday all day	47,395	100	2,452	5.2	44,943	94.8
Growth 1990–2001 (%)	28.8		23.0		30.0	
Mon–Thu am peak	32.4		11.9		49.1	
Mon–Thu off-peak day	34.0		35.6		33.7	
Mon–Thu pm peak	14.5		11.6		15.4	
Mon–Thu off-peak night	3.2		27.1		−2.4	
Friday am peak	80.2		28.9		131.7	
Friday off-peak day	67.4		63.9		67.9	
Friday pm peak	45.6		31.8		49.5	
Friday off-peak night	16.7		47.1		11.7	
Saturday all day	38.6		27.0		39.6	
Sunday all day	21.8		16.3		22.2	

Notes:

1. 1990 data are adjusted to be comparable with 2001 data because new survey techniques such as travel diary and household rostering have been used since 1995 NPTS (Hu and Young, 1999).
2. Persons of age 0 to 4 are excluded from 2001 data because they were not surveyed in the 1990 survey.

Family/personal	(%)	School/church	(%)	Social/recreation	(%)
130,770	46.0	27,848	9.8	76,605	26.9
6,700	24.6	6,968	25.5	1,377	5.0
40,296	60.6	7,189	10.8	11,135	16.7
19,240	45.5	2,153	5.1	10,371	24.5
11,897	36.4	1,853	5.7	12,807	39.2
1,198	23.6	1,113	22.0	221	4.4
9,268	62.2	1,235	8.3	2,731	18.3
4,199	46.2	191	2.1	2,672	29.4
2,957	33.9	184	2.1	4,349	49.9
19,646	50.2	752	1.9	15,728	40.2
15,368	39.5	6,211	16.0	15,214	39.1
168,438	46.0	37,659	10.3	99,711	27.2
11,177	30.9	8,328	23.1	2,934	8.1
53,182	59.7	8,589	9.6	16,629	18.7
19,648	40.6	3,573	7.4	13,434	27.8
10,806	32.0	2,204	6.5	12,923	38.3
3,043	33.3	2,028	22.2	794	8.7
15,333	61.5	1,898	7.6	4,984	20.0
5,745	43.4	625	4.7	4,191	31.7
3,192	31.4	331	3.3	4,842	47.6
27,420	50.6	1,686	3.1	21,325	39.3
18,891	39.9	8,397	17.7	17,655	37.3
28.8		35.2		30.2	
66.8		19.5		113.1	
32.0		19.5		49.3	
2.1		66.0		29.5	
−9.2		18.9		0.9	
154.1		82.2		258.9	
65.4		53.7		82.5	
36.8		227.2		56.9	
8.0		80.3		11.3	
39.6		124.3		35.6	
22.9		35.2		16.0	

3. Trips for which day of week or time of day are unknown are excluded.
4. The column of all trips does not equal to total person-trips because it excludes trips for such purposes as work related, pleasure driving and vacation.

Source: Lee et al. (2006).

Table 17.5 Direct and chained tours by period of the week, 1995 and 2001

	Commute				Non-commute				Other	All
		(%)	Direct	Chain		(%)	Direct	Chain		
1995 All (millions)	58,681	(21.5)	45,868	12,813	205,870	(75.5)	168,193	37,677	8,248	272,799
Mon–Thu am peak	13,882	(47.9)	11,221	2,660	14,313	(49.4)	12,001	2,312	785	28,979
Mon–Thu off-peak day	10,665	(16.5)	7,644	3,021	51,006	(78.9)	39,671	11,335	2,975	64,646
Mon–Thu pm peak	11,519	(31.9)	8,425	3,094	23,338	(64.7)	19,434	3,905	1,234	36,091
Mon–Thu off-peak night	6,199	(23.2)	5,490	709	19,700	(73.8)	17,370	2,330	805	26,704
Friday am peak	3,305	(45.7)	2,686	619	3,778	(52.3)	3,094	684	144	7,226
Friday off-peak day	2,684	(15.2)	1,881	803	14,192	(80.4)	10,960	3,232	770	17,646
Friday pm peak	2,553	(26.5)	1,897	656	6,778	(70.4)	5,569	1,209	296	9,626
Friday off-peak night	1,434	(17.4)	1,251	183	6,620	(80.4)	5,766	854	176	8,229
Saturday all day	3,875	(10.1)	3,198	677	33,976	(88.2)	27,208	6,768	672	38,523
Sunday all day	2,567	(7.3)	2,176	391	32,168	(91.6)	27,121	5,048	393	35,128
2001 All (millions)	56,903	(20.7)	43,162	13,740	213,827	(77.7)	174,461	39,366	4,497	275,226
Mon–Thu am peak	13,519	(45.1)	10,440	3,079	15,835	(52.9)	13,337	2,498	590	29,943
Mon–Thu off-peak day	10,793	(16.7)	7,613	3,180	53,041	(82.1)	41,250	11,792	762	64,596
Mon–Thu pm peak	11,198	(31.8)	8,059	3,139	23,470	(66.6)	19,672	3,798	594	35,262
Mon–Thu off-peak night	5,913	(22.8)	5,097	816	18,905	(72.8)	16,599	2,305	1,157	25,974
Friday am peak	3,266	(43.0)	2,544	722	4,218	(55.5)	3,580	639	110	7,595
Friday off peak day	2,749	(15.6)	1,927	822	14,744	(83.5)	11,304	3,440	171	17,664
Friday pm peak	2,476	(26.4)	1,761	715	6,725	(71.7)	5,504	1,221	175	9,375
Friday off-peak night	1,298	(16.3)	1,077	221	6,317	(79.3)	5,595	722	353	7,968
Saturday all day	3,443	(8.6)	2,813	630	36,016	(90.5)	28,563	7,453	356	39,815
Sunday all day	2,247	(6.1)	1,831	416	34,558	(93.3)	29,059	5,498	229	37,034

Source: Lee et al. (2006).

showed a larger and increasing proportion of non-work trips. The only period showing a large increase in the proportion of work trips was the Monday–Thursday night off-peak period.

We found a stark contrast among growth patterns for work and non-work trips in terms of their temporal distribution across weekly periods. Whereas work trips became more spread out, extending to off-peaks, non-work trips grew faster in the morning peak. The spreading of work trips may be attributed to increasingly flexible work schedules while the growth in morning-peak non-work trips reflects the increased frequency of non-work trip-chaining into commute tours (see below). Both tendencies, increasing flexibility in work schedules and the prevalence of non-work trips in peak hours, make peak-hour pricing more attractive.

Trips for family or personal business (including shopping and doctor visits) accounted for the majority of non-work trips. Yet, there was also considerable growth in the school/church trips and the social/recreation trips categories. Non-work trip frequencies grew most in the Friday am peak period, perhaps the result of a trend towards early weekends.

This point is sharpened when we introduce a tour-level analysis which highlights the non-work trips that are a part of many commutes (Table 17.5). The tour analysis is the more interesting because it accounts for some of the growth in non-work travel. It makes sense that a growing labor force participation rate, especially among women, causes more errands to be included in tours to and from work.

The Federal Highway Administration (FHWA) defines a trip chain as 'a sequence of trips bounded by stops of 30 minutes or less' (McGuckin and Nakamoto, 2004, p. 1). Any stop of more than 30 minutes becomes either the origin or the destination of a tour. Thus, a tour denotes a single trip or chained trips bounded by two anchor destinations (of more than 30-minute dwell time). Unlike in previous research, the FHWA definition includes places other than home and workplace as anchor destinations that constitute either end of a tour. Thus, trip chain datasets for 1995 NPTS and 2001 NHTS classify all tours into nine tour types according to origin and destination place types: (i) home-to-home, (ii) home-to-other, (iii) home-to-work, (iv) other-to-home, (v) other-to-other, (vi) other-to-work, (vii) work-to-home, (viii) work-to-home, and (ix) work-to-work. The home-to-work and work-to-home tours are apparently commute tours, whether direct or chained.

However, a commute tour in the general sense can be much more complex, possibly involving intervening stops of more than 30 minutes, such as a visit to a fitness center. To distinguish these kinds of commutes, we identified commutes with a stop of more than 30 minutes by connecting two pairs of continuing FHWA-defined tours in the categories home-to-other and other-to-work; and in the categories work-to-other and other-to-home. If, however, there are two or more intervening stops of more than 30 minutes en route to or from the workplace, we do not count the tour as a commute.

Our new definitions add to our point about the dominance of non-work travel because they emphasize the fact that not only are non-work trips the vast majority in each peak period but, once we re-define commute tours, we find that many of them (23 percent in the Monday–Thursday am peak and 28 percent in the Monday–Thursday pm peak) also involve non-work trips. The proportion of chained commutes during both am and pm peaks reflects a significantly increased trip-chaining tendency between survey years. This increase of chained tours in the morning peak may be an important factor behind the increased road congestion found for the late 1990s (Gordon et al., 2004).

Trip-chaining is an individual level strategy to economize on travel times by combining multiple trips on various purposes into a tour and can be done most easily by car (Hensher and Reyes, 2000; Lee et al., 2006). Therefore, the increasing tendency toward trip-chaining further strengthens the case for the HOT-lane approach over other approaches for coping with road congestion.

5 DISCUSSION

Unpriced access to busy roads and highways has long served as a textbook example of a market failure. Actually, as new technologies make toll collection and road monitoring costs cheaper, the widespread lack of road pricing can be seen as a policy failure.

The existence of growing networks of HOV lanes on the freeways of major US metro areas provides an interesting opportunity for policy makers to implement pricing without major disruption because most HOVs are presently underutilized. Where they exist, they occupy 25 percent of the road space (one of four lanes) but can accommodate just 7 percent of the vehicles (those estimated to carry two or more passengers). The availability of HOVs was supposed to increase carpooling but this has not happened. California law was recently changed to allow hybrid vehicles onto the state's underutilized HOVs, no matter what the vehicle occupancy.

The HOT-lane proposal (to convert HOV lanes to HOT lanes) is summarized in a recent paper by Poole and Orski (2006, pp. 453–4). Note that they describe it as a transit as well as a highway policy.

> By changing the access requirement from vehicle occupancy to willingness to pay a market price (for cars) but allowing super high-occupancy vehicles (buses

Table 17.6 Comparative throughput of HOV lanes and HOT network

	Typical HOV-2	Typical HOV-3	Ideal HOV-3	HOT Network
SOVs (average 1.1 person/veh.)	0	0	0	1100
HOV-2s (average 2.1 person/veh.)	788	0	0	300
HOV-3s (average 3.2 person/veh.)	150	350	1200	200
Vanpool (average 7.0 person/veh.)	10	20	20	60
Express bus (average 35 persons/veh.)	2	3	40	40
Vehicles/hour	950	373	1260	1700
Persons/hour	2275	1365	5380	4300

Sources: Table 19.2 in Poole and Orski (2006).

> and vanpools) to use the lanes at no charge, we can accomplish three important goals: 1. Generate sufficient new revenue to building out today's fragmented HOV lanes into a seamless network; 2. Provide a congestion-free alternative for motorists on every congested freeway in the same metro area; and 3. Provide a congestion-free guideway for bus Rapid Transit service that can make this form of transit significantly more competitive with driving.

The authors show that converting HOV-2 lanes (those that allow two-person carpools) to HOT lanes would approximately double vehicle-per-hour and person-per-hour throughput (Table 17.6). They also estimate the costs of implementing their system in eight of the major metro areas of the US. The estimates range from $2.7 billion (Miami) to $10.8 billion (Los Angeles) but dedicated revenue bonds would cover two-thirds of these costs. In the light of the high proportion of costs that can be met in this way, private capital and private management become plausible. This is an added attraction of the proposal.

Interestingly, there are two HOT lanes currently in operation in Southern California. They have each been in operation for over 10 years and are described by Sullivan (2006):

> The two Southern California projects are applications of 'value pricing,' described in a U.S. DOT report to Congress as 'a market-based approach to traffic management which involves charging higher prices for travel on roadways during periods of peak demand. Also known as congestion pricing or road pricing, value pricing is designed to make better use of existing highway capacity by encouraging some travelers to shift to alternative times, routes, or modes of transportation.' . . . The Interstate 15 project uses dynamic value pricing where the toll can change in real time to adapt to unusual changes in demand. However, a schedule of typical daily tolls is also published. The State Route 91

project sticks to a published toll schedule, based on established patterns of daily demand.

These projects have enjoyed substantial public acceptance, in part because they have been marketed as a kinder and gentler form of congestion pricing, in which innovative pricing is used to create a new product – a congestion-free travel option in an otherwise congested commute corridor. Travelers are free to use or avoid the value priced facilities as they see fit, since the original congested travel options remain available. This approach stands in sharp contrast to mandatory pricing of all private vehicle trips at targeted locations and times, which some regard as ideal congestion pricing. (pp. 189–90)

The impact studies have shown that the value-priced toll facilities, where travelers can bypass congestion for a price, are associated with significant and systematic responses in travel behavior. This suggests that demand-dependent pricing can be a powerful tool for managing highway traffic and providing more choices to the traveling public in similar corridors elsewhere. (p. 214)

The success of these facilities is not surprising because they exist and operate in the context described in our analysis of the US data. Origins and destinations are dispersed and travel patterns are most amenable to the use of singly-operated motor vehicles.

6 CONCLUSIONS

Our findings complement and elaborate the recommendations of Poole and Orski (2006). Converting HOV lanes to HOT lanes and redirecting current planning away from more HOV lane development (as well as from conventional transit planning) towards their suggested plan is the way to go in the light of what we know of US settlement and travel trends. Dispersed origins and destinations are unlikely to be well served by conventional transit or by carpooling. And the increasing tendency to combine work trips with non-work trips reflects this and also favors the HOT-lanes policy.

NOTES

1. In 1998, Singapore replaced the manual area licensing scheme by an electronic road pricing scheme, which charges tolls per entry at varying prices at different times of the day (Phang and Toh, 2004).
2. See http://www.vtpi.org/tdm/.
3. They extended the London Congestion Charging Scheme to include the Royal Borough of Kensington and Chelsea (the so-called 'western extension') effective from February 2007.

REFERENCES

Crane, Randall and Daniel Chatman (2003), 'Traffic and sprawl: evidence from U.S. commuting, 1985 to 1997', *Planning and Markets*, **6**, 14–22.

Giuliano, Genevieve and Kenneth A. Small (1991), 'Subcenters in the Los Angeles region', *Regional Science and Urban Economics*, **21**, 163–82.

Gordon, Peter, Ajay Kumar and Harry W. Richardson (1989), 'The influence of metropolitan spatial structure on commuting time', *Journal of Urban Economics*, **26**, 138–51.

Gordon, Peter, Bumsoo Lee and Harry W. Richardson (2004), 'Travel trends in U.S. cities: explaining the 2000 census commuting results', Working Paper 2004-1007, Lusk Center for Real Estate, University of Southern California.

Hensher, David A. and April J. Reyes (2000), 'Trip chaining as a barrier to the propensity to use public transport', *Transportation*, **27**, 341–61.

Hu, Patricia S. and Jennifer R. Young (1999), 'Summary of Travel Trends: 1995 Nationwide Personal Transportation Survey', Federal Highway Administration, US Department of Transportation, Washington, DC.

Lee, Bumsoo (2006), 'Urban spatial structure and commuting in US metropolitan areas', Paper presented at the Western Regional Science Association 45th Annual Conference, Santa Fe, New Mexico, February.

Lee, Bumsoo (2007), ' "Edge" or "edgeless cities"? Urban spatial structure in U.S. metropolitan areas, 1980–2000', *Journal of Regional Science*, **47**, 479–515.

Lee, Bumsoo, Peter Gordon, James E. Moore II and Harry W. Richardson (2006), 'Residential location, land use and transportation: the neglected role of non-work travel', Paper presented at the Western Regional Science Association 45th Annual Conference, Santa Fe, New Mexico, February.

McCarthy, Patrick. S. and Richard Tay (1993), 'Pricing road congestion – recent evidence from Singapore', *Policy Studies Journal*, **21**, 296–308.

McGuckin, Nancy and Yukiko Nakamoto (2004), 'Trips, chains, and tours: using an operational definition', Transportation Research Board, Washington, DC.

McMillen, D.P. (2001), 'Nonparametric employment subcenter identification', *Journal of Urban Economics*, **50**, 448–73.

Phang, Sock-Yong and Rex S. Toh (2004), 'Road congestion pricing in Singapore: 1975 to 2003', *Transportation Journal*, **43**, 16–25.

Poole, Robert W. and C. Kenneth Orski (2006), 'HOT networks: a new plan for congestion relief and better transit', in Roth (ed.), pp. 451–99.

Prud'homme, Rémy and Pierre Kopp (2006), 'The Stockholm toll: an economic evaluation', Unpublished manuscript, University of Paris XII, Paris.

Roth, Gabriel (2006), *Street Smart: Competition, Entrepreneurship, and the Future of Roads*, New Brunswick, NJ: Transaction Publishers.

Santos, Georgina and Gordon Fraser (2006), 'Road pricing: lessons from London', *Economic Policy*, **21**, 263–310.

Sullivan, Edward C. (2006), 'HOT lanes in Southern California', in Roth (ed.), pp. 189–223.

18. Expansion of toll lanes or more free lanes? A case study of SR91 in Southern California

Harry W. Richardson, Peter Gordon, James E. Moore II, Sungbin Cho and Qisheng Pan

1 INTRODUCTION

The research reported here builds on our earlier work modeling the regional economic impacts of highway and other infrastructure projects. We are particularly interested in treating the full effects of highway capacity gains and losses, and this application elaborates this work in two important directions. First, we have extended our modeling capability to include highway lanes that are tolled. Second, we apply the new model to an important prototype application, the originally private, now public, lanes in the median of a 10-mile segment of California State Route (SR) 91.

The possible widening of this route with additional tolled or general-purpose lanes has been the subject of considerable controversy. A non-compete provision in the franchise awarded to the California Private Transportation Company (CPTC) had stood in the way of public agencies' efforts to provide additional capacity in the corridor. Our approach sheds light on such controversies and, thereby, may reduce political conflict and misunderstanding. We also show that, whereas congestion tolls are widely presumed to be efficient, the efficiency outcomes are complex when only a part of the network is tolled.

2 CALIFORNIA SR91 EXPRESS LANES

The SR91 express lanes were California's first private toll highway project, which was developed under enabling legislation passed by the California legislature in 1989. A franchise was eventually awarded to the CPTC who financed, built and operated two tolled lanes in each direction along

10 miles of the SR91's unused median strip. Development costs are estimated to have been $135 million. These lanes opened to traffic on 27 December 1995. Drivers pay electronically via windshield-mounted transponders, a widely used Texas Instruments technology called FastTrak that also serves as a California bridge toll standard, and are billed monthly.

The SR91 toll lanes are an example of value pricing, that is, of providing travelers with an opportunity to pay a premium for access to a higher level of service (Small, 2001). This context provides toll facilities with an attractive policy dimension, but introduces a host of questions ranging from modeling to the politics of congestion.

CPTC developed, refined and applied state-of-the-art pricing and photo-enforcement software and hardware and demonstrated that these perform well. Tolls varied from $0.60 to $3.20 in 1998, depending on traffic flow conditions, and the toll schedule was periodically adjusted so that 65 mph average speeds could be maintained. Most recently, peak-hour tolls went as high as $9.50. Given the target 25-minute time savings, this implies a high valuation of travel time for peak-period toll users. Three-person or larger carpools on the express lanes received free access.

Recent volumes on the express lanes were greater than 30,000 trips per day, 14 percent of weekday SR91 corridor use. Peak-hour use was 1,400–1,600 vehicles per hour per lane. Yet, capping a controversy over how corridor capacity should be expanded to respond to growing demand, the lanes were sold to the Orange County Transportation Authority (OCTA) for $207 million, with ownership transfer occurring in early 2003. Now OCTA has the flexibility to add to capacity or change the mix of toll and free lanes. To our knowledge, the type of analysis reported below was not done. Detailed results of user surveys are described at http://www.path.berkeley.edu/~leap/TTM/Demand_Manage/pricing.html. Many of the cited descriptive data are from Caltrans's final evaluation report published in 2000 and available at http://www.ceenve.calpoly.edu/sullivan/SR91/sr91.htm.

3 THE SOUTHERN CALIFORNIA PLANNING MODEL (SCPM)

Regional economists and regional planners often rely on interindustry models. The details of intersectoral linkages in these models are useful for exploring regional economic structure. However, this approach has not permitted an adequate treatment of transportation costs, not all of which are transacted because most roads are publicly provided. This problem has recently been addressed at the national level by the Bureau of

Transportation Statistics' effort to create Transportation Satellite Accounts (Han and Fang, 2000). Such spatial elaborations require explicit treatment of the resources consumed by flows between origin–destination pairs. While explicit representation of the transportation network is usually less important in multiregional approaches, it is another matter at the intrametropolitan level, because congestion dominates line-haul costs.

Along with the explicit treatment of transport costs, we have developed sub-metropolitan area level disaggregations of the regional economic model. Richardson et al. (1993) developed the Southern California Planning Model-1 (SCPM1), combining a metropolitan-level input–output model with a Garin–Lowry model to spatially allocate induced economic impacts. This model operationalized spatial input–output analysis at the intrametropolitan level. SCPM1 could allocate impacts in terms of jobs or the dollar value of output to the area's 308 sub-regional (municipal) zones. It did not treat the transportation network explicitly. Congestion effects were ignored, and transportation flows were exogenous.

SCPM1 and its successors have been developed for the Los Angeles metropolitan region. The study area includes Los Angeles, Orange, Riverside, San Bernardino and Ventura counties. The area covers almost 34,000 square miles. The 2000 population of the five-county area was over 16.3 million.

In 2000, the urbanized portions of the five-county area extended to 1,668 square miles; population density in the urbanized area was about 7,068 people per square mile, the highest in the US. The urbanized area is described in terms of the Southern California Association of Government's (SCAG's) 1527 disaggregated 1990 traffic analysis zones (TAZs; reduced to 1464 TAZs in the exercise that follows for modeling convenience). The regional highway network includes 22,244 links.

Table 18.1 provides some recent aggregate data describing the region. The total number of households in the area was 5.35 million in 2000. The nonfarm employment in the SCAG region was over 5.11 million in 1999. The employment distribution across industry sectors was: 34.3 percent in services, 16.4 percent in manufacturing, 13.3 percent in government, 9.6 percent in retail and 7.3 percent in Finance, Insurance and Real Estate (FIRE). International exports from the five-county area have been reported to be $35.7 billion in 1996 (Exporter Location Series, US Bureau of the Census); our analysis suggests, however, that this may be a significant underestimate.

Integrating a transportation network into SCPM1 provides important opportunities. Distance decay relationships (destination choice) can be endogenized, permitting an improved spatial allocation of indirect and induced economic impacts. Also, this integration makes it possible to better

Table 18.1 Socio-economic profile, Los Angeles five-county area

County	Pop. 2000	House-holds 2000	Private non-farm emp. 1999	Median income 1999	Mean commute time* (mins) 2000	Land area (sq. mi.) –
Los Angeles	9,519.3	3,133.8	3,747.8	\$42,189	29.4	4,061
Orange	2,846.3	935.3	1,331.0	\$58,820	27.2	789
Riverside	1,545.4	506.2	366.4	\$42,887	31.2	7,207
San Bernardino	1,709.4	528.6	441.0	\$42,066	31.0	20,052
Ventura	753.2	243.2	224.8	\$59,666	25.4	1,845
Five-county area	16,373.6	5,347.1	6,110.8	\$45,957	29.1	33,954

Note: * Weighted mean; median not available.

Source: See http://quickfacts.census.gov/qfd/.

account for the economic consequences of changes in transportation network capacity, such as the prospect of widening portions of SR91.

Recent work resulted in the development of SCPM2, which treats the transportation network explicitly, endogenizing otherwise exogenous matrices describing the travel behavior of households, achieving consistency across network costs and origin–destination requirements, and endogenizing indirect and induced economic deliveries and arrivals (impacts) over zones.

The current research extended the model so that network segments that are tolled can be highlighted and examined. The current version of the model, SCPM2.5, relies on a constrained optimization that combines traffic assignment and trip distribution. A path-flow version of this model is:

$$Z = \min Z(\mathbf{x}, \mathbf{q}) = \sum_a \int_0^{x_a} t_a(w)\, dw + \sum_p \left[\frac{1}{\beta^p} \sum_r \sum_s q_{rs}^p \cdot \ln(q_{rs}^p) \right] \quad (18.1)$$

subject to:

$$f_{rs}^{pk} \geq 0 \quad \forall p, k, r, s \quad (18.2)$$

$$q_{rs}^p \geq 0 \quad \forall p, r, s \quad (18.3)$$

$$\sum_k f_{rs}^{pk} = q_{rs}^p \quad \forall p,r,s \tag{18.4}$$

$$x_a = \sum_p \sum_{rs} \sum_k f_{rs}^{pk} \cdot \delta_{a,k}^{rs} \quad \forall a \tag{18.5}$$

$$q_{rs}^p = O_r^p \cdot D_s^p \cdot K_{rs}^p \cdot \exp(\alpha^p + \beta^p u_{rs}) \quad \forall p, r, s \tag{18.6}$$

$$O_r^n = \mathbf{O}_r^n + \sum_j b_{ij} \cdot V_r^n, \; D_s^n = \mathbf{D}_s^n + \sum_j a_{ij} \cdot V_s^n \tag{18.7}$$

where:

t_a	=	link performance function of link a;
x_a	=	flow on link a;
q_{rs}^p	=	trip rate of type p between OD pair r–s;
f_{rs}^{pk}	=	flow of trip type p on path k connecting OD pair r–s;
u_{rs}	=	travel time between OD pair r–s;
$\delta_{a,k}^{rs}$	=	1 if link a is on path k between OD pair r–s, 0 otherwise;
O_r^n	=	freight trip generated from industry n in zone r;
D_s^n	=	freight trip destined to industry n in zone s, and its baseline;
$\mathbf{O}_r^n$	=	baseline trip generation from industry n in zone r;
$\mathbf{D}_s^n$	=	baseline trip destination to industry n in zone s;
α^p, β^p, K_{rs}^p	=	calibration parameters; and
p	=	number of trip types, $p \in \mathrm{P} = \{m, n\}, m = 1 \ldots 9, n = 1 \ldots 5$.

Trips are disaggregated into nine types of personal trips (m), and freight trips generated by four industrial sectors (n):

m	1 = home-to-work, 2 = work-to-home, 3 = home-to-shop, 4 = shop-to-home, 5 = home-to-other, 6 = other-to-home, 7 = work-to-other, 8 = other-to-work, 9 = other-to-other.
n	1 = Mining, 2 = Manufacturing (non-durable), 3 = Manufacturing (durable), 4 = Transportation, 5 = Communication & Utilities.
V_r^n	Economic impact on industry n in zone r:

$$V_r^n = V_r^{n,D} + V_r^{n,I} + V_r^{n,U},$$

$V_r^{n,D}$ ≠ Net direct impact on industry n in zone r;

$V_r^{n,I}$ ≠ net indirect impact on industry n in zone r:

$$V_r^{n,I} = \frac{\exp(\alpha^n + \beta^n \cdot u_{sr})}{\sum_z \exp(\alpha^n + \beta^n \cdot u_{sz})} \cdot V^{n,I} \cdot \left(\frac{V_s^{n,D}}{V^{n,D}}\right);$$

$V_r^{n,U}$ net induced impact on industry n in zone r:

$$V_r^{n,U} = JHS \cdot (JHW)^U \cdot \frac{\exp(\alpha^n + \beta^n \cdot u_{sr})}{\sum_z \exp(\alpha^n + \beta^n \cdot u_{sz})} \cdot V^{n,U} \cdot \left(\frac{V_s^{n,D}}{V^{n,D}}\right),$$

$$JHW = \frac{q_{rs}^{m=1}}{\sum_r q_{rs}^{m=1}},\ q_{rs}^{m=1} = \text{journey home-to-work matrix},$$

$$JHS = \frac{q_{rs}^{m=3}}{\sum_s q_{rs}^{m=3}},\ q_{rs}^{m=3} = \text{journey home-to-shop matrix}.$$

4 DATA INPUTS

Spatially disaggregated modeling requires the preparation of input data at fine levels of geographic detail. Most model input data were collected at the census block level, the base unit for all census data. We assembled detailed spatial and aspatial information for the study area, including Census 2000 population data, SCAG 1997 employment data providing employment by four-digit SIC category by street address, United States Geological Survey (USGS) 1-meter resolution air photos and similar sources. These data are available from the authors.

Because the expansion of SR91 is still a hypothetical project, there are no data on the exact expansion boundaries. However, it is possible to identify the housing and business units likely to be along the freeway alignment by referring to USGS air photos. Unfortunately, the available photos were taken several years ago and do not provide up-to-date land-use information. It was also problematic to match the air photos with the alignment of the freeway because they are represented in different projection systems. To obtain up-to-date information on the land uses most likely to be impacted, we relied on field inspections to update the land uses shown by the USGS air photos. From our field inspections it was determined that 266 housing units are likely to be impacted by the hypothetical freeway expansion project. Most of these are located in low-density residential areas. All of them are in the city of Anaheim. No businesses were found to be located in the likely impact area of the freeway expansion because all existing businesses are set back from the alignment.

Predicting the destination settlements of the relocating households involved two steps. First, an empirically established distribution function was used to generate moving distances (Clark et al., 2002). The mean move was estimated to be 6.28 miles. Second, most likely move-in locations for each impacted household were determined by identifying the center of the census block with average housing unit price closest to that of the census block from which the household would move. This means that each move-out household is relocated to a place with a housing unit price similar to the original residence. Some households might decide to trade up or down. Others might decide to move out of the region. There are no data available on this possibility, so we assumed and modeled a quasi-equilibrium response.

The number of move-out and move-in households in each census block was used together with county, city, TAZ, congressional district and school district information to generate input data for the SCPM runs. As households are relocated, their expenditures, including property and sales taxes, are also relocated. Based on available data, it is possible to determine the median housing value, household income, sales tax rate, property tax rate and other inputs to SCPM. Detailed input data development procedures are available from the authors.

5 MODEL RESULTS: HYPOTHETICAL EXPANSION OF SR91 CAPACITY

Household Relocation Effects

The aggregate regional effects of household relocation are minor (see Table 18.2). Approximately $24 million in annual household expenditures are removed from the path of the highway expansion and are relocated

Table 18.2 Summary economic impacts of residential relocation, one-lane expansion of SR91 in each direction (1999 $1,000s)

$1000	Positive	Negative	Net
Direct	14,405	−14,421	−16
Indirect	3,621	−3,635	−15
Induced	5,834	−5,823	12
Total	23,860	−23,879	−19

Note: Negative impacts are generated at residents' move-out relocations. Positive impacts are generated at residents' move-in relocations.

throughout the region. Aggregate impacts are small. The importance of these calculations is to show spatial redistribution effects. We have been able to calculate the direct, indirect and induced impacts of the relocations by traffic analysis zone (maps available from the authors).

Direct impacts result from displaced households. Negative impacts occur in locations adjacent to the freeway right of way. Positive impacts are more widely distributed as households relocate (Clark et al., 2002). Indirect, induced and total impacts are much more widely dispersed. SCPM2 has the unique capability to estimate these complex spatial effects. SCPM2 results are sector specific, but are reported in terms of total dollars.

Highway Network Effects

Network flow estimates are shown in Tables 18.3 and 18.4. Table 18.3 provides baseline and scenario results, reporting network delay costs and tolls paid. Tolls paid are a transfer from users to owners. However, total user costs are the sum of delay costs and tolls, and it is this total cost to which users respond. About 17 percent of the estimated baseline passenger-car-equivalents on the network consist of freight shipments.

We consider and report on two facility expansion scenarios combined with six operating options for a total of 12 scenarios, plus five operating options for the existing facility. Tolls on SR91 vary by time of day. SCPM2 results are scaled to 24-hour periods. The analysis relies on a composite SR91 toll that approximates a weighted average charge across 24 hours. Ideally, we would also model flow adjustments across time of day. This would increase model complexity very substantially. We account for adjustments in origins and destinations and route choice for multiple modes. Accounting for time-of-day adjustments is a future research activity.

The two facility expansion scenarios are:

- add a toll lane in each direction, providing four general-purpose lanes and three toll lanes in each direction (the 4 + 3 scenario), or
- add a general-purpose lane in each direction, providing five general-purpose lanes and two toll lanes in each direction (the 5 + 2 scenario).

The operating options are defined by varying the composite toll charged on the tolled lanes from values of \$1 to \$11 (as pointed out above, the peak toll is currently \$9.25). Table 18.4 shows changes in network delay costs and tolls collected under alternative scenarios. Turning to the dollar values of impacts, and assuming no shifts in origin–destination demand, annual reductions in travel costs over the network range between \$13.3 and \$14.7 million with high tolls (\$7–11) in the 4 + 3 (toll road added) combination.

Table 18.3 Annual network user costs ($1,000s)

Scenarios		Passenger travel time cost		Toll paid		Truck travel time cost	Sum		
	Toll	Drive-alone	HOV	Drive-alone	HOV		Delay cost	Toll paid	Total
4 + 2	$1	12,884,798	8,512,183	7,538	1,567	27,564,586	48,961,566	9,105	48,970,671
	$3	12,871,093	8,508,088	19,128	4,449	27,552,010	48,931,191	23,577	48,954,768
	$5	12,871,840	8,506,040	19,053	6,512	27,527,183	48,905,062	25,566	48,930,627
	$7	12,882,848	8,507,591	7,167	4,483	27,526,280	48,916,718	11,649	48,928,367
	$9	12,889,804	8,511,973	–	–	27,525,491	48,927,268	–	48,927,268
	$11	12,890,056	8,512,098	–	–	27,526,420	48,928,575	–	48,928,575
4 + 3	$1	12,883,933	8,512,319	8,471	1,722	27,568,879	48,965,131	10,193	48,975,324
	$3	12,866,602	8,507,884	23,543	4,650	27,561,814	48,936,299	28,193	48,964,492
	$5	12,865,310	8,505,624	25,152	6,851	27,555,592	48,926,526	32,002	48,958,529
	$7	12,882,175	8,507,708	7,737	4,283	27,526,734	48,916,617	12,019	48,928,636
	$9	12,889,804	8,511,970	–	–	27,525,458	48,927,231	–	48,927,231
	$11	12,890,020	8,512,077	–	–	27,526,283	48,928,380	–	48,928,380
5 + 2	$1	12,884,852	8,512,261	7,514	1,574	27,564,714	48,961,827	9,088	48,970,914
	$3	12,870,826	8,508,050	19,301	4,496	27,554,968	48,933,844	23,797	48,957,641
	$5	12,871,044	8,505,171	18,758	7,069	27,554,310	48,930,525	25,827	48,956,352
	$7	12,888,740	8,511,771	959	431	27,553,897	48,954,408	1,390	48,955,798
	$9	12,890,013	8,512,339	–	–	27,555,199	48,957,550	–	48,957,550
	$11	12,889,953	8,512,321	–	–	27,554,505	48,956,779	–	48,956,779

Note: Values of time = $13.0/hour for drive-alone, $28.9/hour for carpool, and $71.0/hour for trucks. On average $29.44/hour.

Table 18.4 Annual changes in network user costs ($1,000s)

Scenarios		Passenger travel time cost		Toll paid		Truck travel time cost	Sum		
	Toll	Drive-alone	HOV	Drive-alone	HOV		Delay cost	Toll paid	Total
4+3	$1	14,335	5,545	−12,291	−3,916	29,682	49,562	−16,208	33,355
	$3	−2,997	1,110	2,781	−988	22,617	20,731	1,793	22,523
	$5	−4,288	−1,149	4,389	1,212	16,395	10,958	5,602	16,560
	$7	12,576	934	−13,026	−1,356	−12,462	1,048	−14,381	−13,333
	$9	20,205	5,196	−20,762	−5,638	−13,738	11,663	−26,401	−14,738
	$11	20,421	5,304	−20,762	−5,638	−12,914	12,811	−26,401	−13,589
5+2	$1	15,253	5,487	−13,248	−4,065	25,518	46,258	−17,313	28,945
	$3	1,228	1,276	−1,461	−1,143	15,771	18,276	−2,603	15,672
	$5	1,446	−1,603	−2,004	1,430	15,113	14,956	−574	14,383
	$7	19,141	4,997	−19,803	−5,208	14,700	38,839	−25,010	13,829
	$9	20,414	5,565	−20,762	−5,638	16,002	41,981	−26,401	15,581
	$11	20,354	5,547	−20,762	−5,638	15,309	41,211	−26,401	14,810

Note: Values of time = $13.0/hour for drive-alone, $28.9/hour for carpool, and $71.0/hour for trucks. On average $29.44/hour. Value of travel time assumptions are controversial. The Caltrans (2000) SR91 report notes that $6 to $14 per hour values were inferred from patterns of tollway use on the SR91. See also Small and Yan (2001).

All other scenarios (lower tolls in the 4 + 3 scenarios and all tolls in the 5 + 2 scenarios) result in higher travel costs in the $14.4 to $33.4 million range. The results vary with different values of travel time. They are somewhat counterintuitive because although congestion tolls should reduce congestion overall, in the adjacent toll-free road cases while tolls improve the level of service on the tolled facilities, high tolls prompt drivers to move to the free lanes and intensify demand there. One complication of our results is that we include tolls in total travel costs.

The key to interpreting our results is that from a *network* perspective, adding tolls on selected facilities has indeterminate effects on system performance. Tolls, while allowing for more efficient use of the tolled segment, also have the effect of diverting traffic to other parts of the network, resulting in other links carrying greater volumes. Then, the result depends on the magnitude of the tolls and the degree of congestion on alternative routes.

Partial equilibrium effects often differ from general equilibrium effects. System-wide tolls can be set to maximize net revenues and throughput, or to minimize travel delay. Minimizing travel delay delivers efficiency improvements if reductions in total delay are sufficient to offset the administrative cost of collecting the tolls. Limited tolling may create efficiencies along a link, but unless all segments of the network are tolled, it is not clear that such a limited toll strategy will increase network efficiency. Traffic may be shifted to other routes, often with unpredictable results. Indeed, this was one of the sources of the political controversy over the California Department of Transportation's non-compete agreement with CPTC. This agreement precluded expansion of the SR91 general-purpose lanes.

Most of the early academic literature on congestion tolls concludes that they are efficient, with modest attention to imperfections. In recent years, considerable theoretical attention has been directed to the question of second-best toll strategies that are consistent with value-pricing options. In these circumstances, tolls are introduced incrementally on new or existing facilities competing with untolled links. This makes the level of congestion on untolled lanes an important variable or parameter, depending on whether the system of interest consists of the tolled facilities or the network (Verhoef et al., 1996; Small and Yan, 2001).

The standard theoretical discussion examines various public and private objectives given a simple hybrid system consisting of an origin–destination pair, and a toll facility competing with a single, parallel, untolled path. This approach makes it possible to investigate general principles and strategies.

Our examination of the SR91 toll facilities treats these links in the context of the real-world SCPM network described above. While we are able to simulate flows and changes in flows, we do not identify optimal tolls. Nevertheless, similar to the cited literature, our analysis of network flows suggests that efficiency gains will not necessarily be achieved by selective tolling. Selective tolls again produce a second-best result whereby more efficient use of a particular link occurs at the expense of performance throughout the rest of the system. The more congestion there is on untolled facilities, the greater the possible efficiency loss from value tolls. This has substantial policy relevance, because tolls are inevitably introduced on a facility-by-facility basis. Further, private interests have the greatest incentive to risk their capital on the construction of new facilities when congestion on competing routes is high. The modeling consequences highlighted here include the importance of being able to compare the system effects of adding tolled versus untolled capacity. Our results show that large-scale facility investment decisions require that this be done in the context provided by a model of the complete network examining alternative tolls and alternative road configurations.

The network simulations with our model involve modifying the volume delay functions for network links to include tolls. System cost is the product of link volumes and corresponding travel times summed over all the links in the network. Consider the simplified example provided in Figure 18.1. A fixed demand for travel is allocated over competing links A and B.

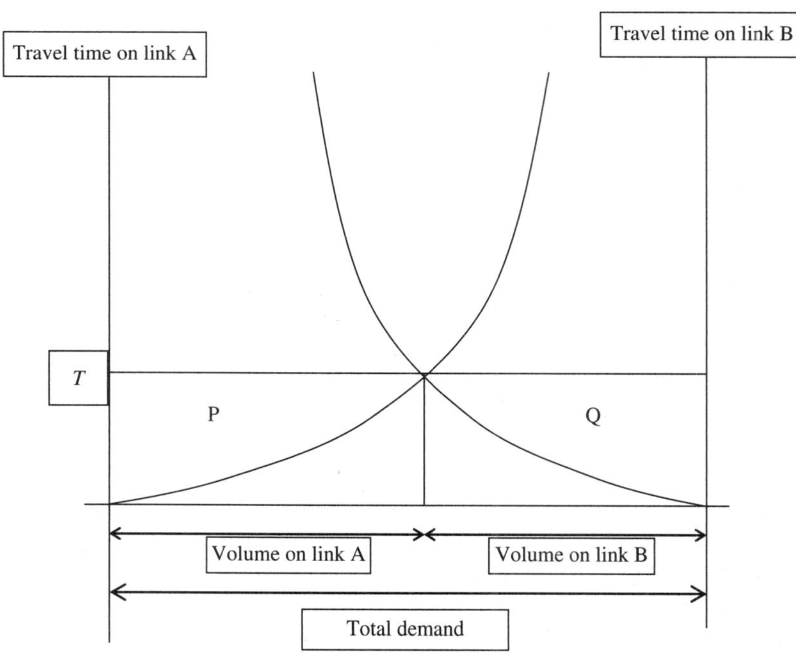

Figure 18.1 Travel demand on competing links

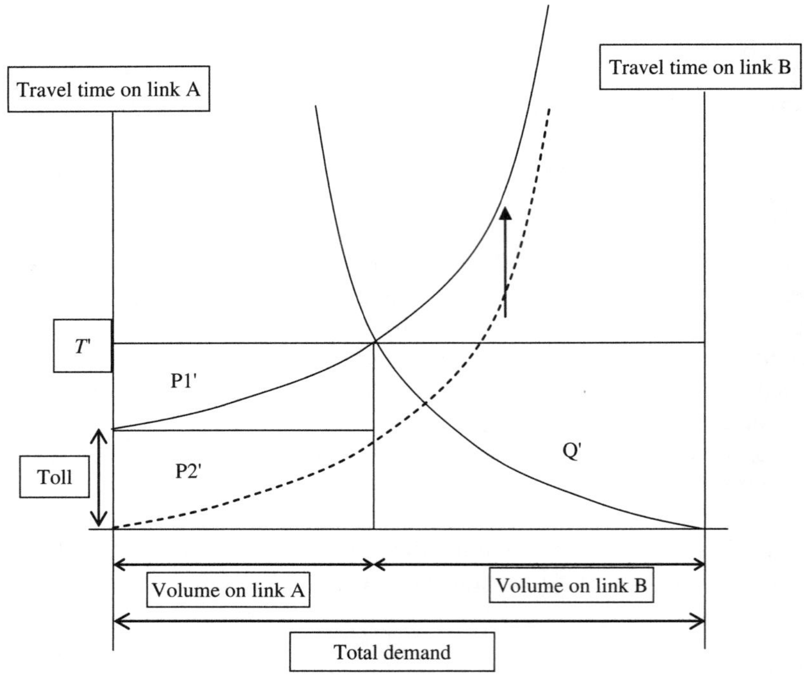

Figure 18.2 The effect of cost increases

Equilibrium flows occur at congested travel time *T*. In the untolled case, system cost is the sum of areas P and Q.

Time has value. For modeling purposes, imposing a toll on a link is equivalent to modifying the volume-delay function for the link so that all vehicles on the link experience a corresponding increase in travel time. In the case of an ideal toll, this resets average cost to marginal cost. If a toll is applied to link A, the existing volume delay function shifts upward by a value corresponding to the toll.

In this example, the toll is set to keep the congestion on link A low with no consideration given to congestion impacts on link B. If travel demand is fixed, the new equilibrium travel time *T*′ is greater than *T*. As a result, the system cost increases relative to the untolled case (see Figure 18.2). P1′ is the system cost due to congestion on link A. P2′ is the revenue provided to the owner of the toll road, expressed in units of time.

Applying SCPM makes it possible to support a detailed cost–benefit analysis of individual projects. We have already calculated the annual network net benefits of various capacity options. The model can also be used to calculate the spatial and sector incidences of construction expenditures, as well as those of alternative approaches to financing the project. This joint treatment of benefits and costs in substantial spatial and sectoral detail allows a discussion of equity as well as efficiency consequences.

6 CONCLUSIONS

What have we accomplished?

1. We have elaborated a network model to account for the effects of tolls on selected freeway lanes.
2. We have integrated the network model with a spatially detailed economic model of the regional economy.
3. We have applied the model to the hypothetical case of highway widening on a segment of California's SR91, comparing the network-wide effects of adding tolled versus non-tolled lanes. The application included substantial data gathering and analysis so that a single integrated model can be used to analyse: (a) the spatial economic effects of household displacement, and (b) the network effects of various highway widening and tolling alternatives.
4. We have found that system-wide network effects of adding tolled lanes on just a small link of that network reveals a complex set of results. Most research on value pricing has necessarily been of a partial equilibrium nature, and does not consider the full network effects. Substantially more research should be done at the network level.

On the premise that all politics are local, we suggest that our analysis of distributional impacts is useful to policy analysts. While we are able to identify costs and benefits, we are also able to estimate which communities bear the costs and which gain the benefits. Whereas network studies in regional highway analyses are common, none to our knowledge includes the comprehensive results from an integrated model shown here.

REFERENCES

Clark, William A.V., Youqin Huang and Suzanne Davies Withers (2002), 'Does commuting distance matter? Commuting tolerance and residential change', *Regional Science and Urban Economics*, **33**, 199–221.

Han, Xiaoli and Bingsong Fang (2000), 'Four measures of transportation's economic importance', *Journal of Transportation Statistics*, **3**, 15–30.

Richardson, Harry W., Peter Gordon, Myun-Jin Jun and Moon H. Kim (1993),

'PRIDE and prejudice: the economic impacts of growth controls in Pasadena', *Environment and Planning A*, **25A**, 987–1002.
Small, Kenneth A. (2001), 'The value of value pricing', *Access*, No. 18, 23–7.
Small, Kenneth A. and Jia Yan (2001), 'The value of "value pricing" of roads: second-best pricing and product differentiation', *Journal of Urban Economics*, **49**, 310–36.
Verhoef, Erik T., Peter Nijkamp and Piet Reitveld (1996), 'Second-best congestion pricing: the case of an untolled alternative', *Journal of Urban Economics*, **40**, 279–302.

19. The political calculus of congestion pricing

David King, Michael Manville and Donald Shoup

It has been a commonplace event for transportation economists to put the conventional [*congestion theory*] diagram on the board, note the self-evident optimality of pricing solutions, and then sit down waiting for the world to adopt this obviously correct solution. Well, we have been waiting for seventy years now, and it's worth asking what are the facets of the problem we have been missing. Why is the world reluctant to do the obvious?

(Charles Lave)

1 INTRODUCTION

Transportation planners and economists generally agree that congestion pricing is the best way, and perhaps the only way, to significantly reduce traffic congestion, but most policy makers do not agree. Few elected officials with a sense of self-preservation will endorse a program that places new charges on a majority of their constituents. Pricing proponents often respond by arguing that although congestion pricing may be politically unpopular now, once it is implemented the public will understand its benefits, and its political problems will disappear. Implementation, however, will not solve the political problem, because implementation *is* the political problem. The political difficulty with congestion pricing is persuading people to do it in the first place, not in convincing them of its value after the fact.

Congestion pricing has broadly distributed costs (most people end up paying tolls) and broadly distributed benefits (drivers suffer less congestion and the tolls can pay for added public services). What pricing lacks is a constituency that will derive *concentrated* benefits that exceed their costs. The high political cost of supporting road pricing falls entirely on those who spend their time, money and political capital trying to implement tolling. Unless new tolls offer someone benefits that exceed these political costs, no one will take action.

This chapter was previously published in *Transport Policy*, **14**(2), 2007, 111–23.

Congestion pricing suffers, therefore, from an absence of strong advocates. 'There is nothing more difficult to take in hand, or more uncertain in its success', Machiavelli wrote in *The Prince*, 'than to take the lead in the introduction of a new order of things. Because the innovator has for enemies all those who have done well under the old order of things, and lukewarm defenders in those who may do well under the new'.

Machiavelli wrote those words in 1532. In 1993, the University of California's Martin Wachs made the same point, albeit in less florid prose, when he summarized the political dilemma that faced congestion pricing: 'In addition to professors of transportation economics and policy, who hardly constitute a potent political force, I can think of few constituencies who would willingly and vigorously fight for the concept'.

2 A CONSTITUENCY FOR CONGESTION TOLLS: CITIES

In this chapter we propose a new way to create political support for congestion pricing on urban freeways: distribute the toll revenue to cities, and particularly to the cities through which the freeways pass. With the revenue as a prize, cities can become the champions of congestion pricing; the benefit to public officials in these cities can be worth far more than the costs of fighting for its passage.

Policy proposals often succeed not because (or not only because) they benefit the public interest, but because they benefit *particular* interests, and these interests organize to champion the policies. Yet when transportation planners recommend tolls to reduce traffic, they tend to focus on the widespread economic benefits of congestion relief, rather than on the political benefits. But as Philip Goodwin (1997: 2) says, 'discussion of pricing without explicit attention to the use of revenue streams is inherently unlikely to be able to command a consensus in its support. I treat this as an axiom of contemporary transport policy'. Rather than spend the revenue to reduce drivers' opposition to congestion pricing, we propose distributing the revenue to increase local political leaders' support for congestion pricing. In economic jargon, we propose creating politically influential residual claimants for the toll revenue: a group entitled to the net revenue from the priced roads. James Q. Wilson (1980) posited a theory of 'client politics' that provides the framework for this argument. Wilson contends that policies with concentrated benefits and widely dispersed costs are likely to succeed:

> When the benefits of a prospective policy are concentrated but the costs widely distributed, client politics is likely the result. Some small easily organized group

> will benefit and thus has a powerful incentive to organize and lobby; the costs of the benefit are distributed at a low per capita rate over a large number of people, and hence they have little incentive to organize in opposition. (p. 369)

If toll revenues are given to cities with freeways, elected officials from these cities stand to gain considerably, and have a strong incentive to lobby for the tolls. The drivers who pay the tolls, in contrast, will each lose only moderately (and any loss will be at least partially offset by reduced congestion).

Four basic conditions for the political approval of congestion pricing help explain why cities are the appropriate claimant for the revenue. We have just discussed the first condition, that the potential gains to revenue claimants must be obvious. Second, the claimants must be organized and politically powerful. Third, the claimants must have some defensible claim to the revenue. And fourth, the gain must be *concentrated.* There cannot be so many claimants that no one gains enough to make political action worthwhile. Drivers who pay the tolls might have a defensible claim to the revenue, but nevertheless are not suitable political claimants because they are many and dispersed which makes them unlikely to generate political power.

Freeways have regional benefits, so it might seem sensible to allocate the money to some regional authority – a public transportation or highway agency, for example. But here we can introduce another condition for receiving congestion toll revenue: the recipient must have a claim to the revenue that is both economic *and* political, with the political claim being more important. And while there are some good arguments for giving the toll revenue to regional agencies, they are not *political* arguments. A regional agency would be hard-pressed to produce a public service that the region's residents considered a reasonable compensation for the loss of free access to the freeways. Even spending the revenue on regional transit improvements may do little to improve the prospects for pricing. In the United States a politically weak and unorganized minority travels by public transport, which dims the chance of effective political support.

Individual cities, however, could conceivably arrive at a mix of public goods and services that would create support for congestion pricing. Dividing the toll revenue among cities would allow each community to choose its preferred mix of public goods and services; the gains to individuals in their roles as residents, when combined with time-savings from the tolls, should outweigh the losses to individuals in their roles as motorists. Instead of a regional agency profiting at the expense of all drivers, citizens of each community would benefit from tolls levied on motorists from outside their borders. While that distinction is more one of perception than reality (motorists from neighboring cities would just end up subsidizing each other's public goods) the way that choices and policies are framed matters tremendously in political decision making.[1] Similarly, distributing the money to cities means that the toll revenue would be spent locally but collected region-wide, allowing local officials to claim credit for providing new benefits while shielding them from the resentment attached to congestion pricing's costs.

We do not rule out other claimants to the revenue, nor do we argue that political considerations are the only ones that matter. Once cities are mobilized, logrolling and vote-trading will doubtless occur en route to pricing's approval, and the cities may well have to share the toll revenue with transportation agencies to gain their support or at least quell their opposition. And depending on the specific context of each region, various extensions and adjustments can be made to the revenue distribution, based on equity or planning concerns. But political support for congestion pricing will depend on *who* gets the toll revenue, and no one will receive any revenue until congestion pricing is adopted.

A final note before we move on: if our proposal sounds like rent seeking, it is. City governments will lobby for a regulation (congestion pricing) because it will deliver them a revenue windfall. The term 'rent seeking' is usually employed pejoratively, and for that matter so too is 'client politics' – Wilson coined the term but he did not write approvingly of the practice. The individuals or groups who seek rents generally do so to shelter themselves from the discipline of the market. Competing by regulation, rather than innovation, dissipates otherwise productive resources and stifles industrial development; the company that spends its money lobbying for a protective tariff rather than improving its products is a drag, not a boon, to the larger economy.[2] In this case, however, cities that rent seek (or 'toll seek') will be introducing – rather than curtailing – a market mechanism. Congestion pricing can be ushered into existence through efficient rent seeking.

In the remainder of this chapter we first situate our revenue-distribution proposal in political and economic theory, using the concepts of client politics and loss aversion (Section 3). We then evaluate other proposed claimants for the toll revenue in the light of these theories (Section 4). Next we outline the reasons for giving the revenue to cities, and then illustrate how such a distribution program might work (Section 5). In Los Angeles, distributing the money to cities with freeways would be both progressive and politically expedient (Section 6). In the Twin Cities we suggest a distribution that reflects the region's existing commitment to regional redistribution (Section 7). The important point is that coalitions of local governments would have the power and incentive to create political momentum for congestion pricing. Section 8 presents a brief discussion of two other case studies, New York and San Francisco. Section 9 concludes.

3 THE POLITICS OF CONGESTION PRICING

Congestion pricing will do two things: reduce congestion and raise revenue. We thus cannot predict who will receive a net benefit from congestion pricing until we know how the toll revenue is used. In a study of the congestion pricing pilot program in Stockholm, Eliasson and Mattsson (2006) estimated that the toll revenue was about three times the net benefits of reducing congestion. That is, motorists pay \$3 in tolls for every \$1 of benefit they receive from congestion relief. To achieve equity, the distribution of the toll revenue is thus more important than the distribution of congestion-relief benefits.

Even before any distribution of the revenue, congestion pricing will create a net benefit for two groups because of improved traffic flow:

1. drivers whose time saved is worth more than the tolls they pay; and
2. people who already use transit and will not pay tolls but will travel faster.

Again before considering the use of the revenue, congestion pricing will create a net loss for three other groups:

3. drivers whose time saved is worth less than the tolls they pay;
4. drivers who switch to a less convenient route to avoid the tolls; and
5. people on non-tolled routes whose traffic increases when drivers from group 4 switch to their roads.

Members of groups 1 and 2 are better off regardless of whether they receive any benefits from the toll revenue, while members of groups 3, 4 and 5 are better off only if they receive benefits from the toll revenue that outweigh the tolls they pay.

If we focus only on how congestion pricing affects *drivers*, and if we neglect the potentially large number of people who will benefit from the toll revenue in their role as residents of the cities receiving the revenue, the losers almost certainly outnumber the winners. But if we also consider the benefits to residents from the public services (or tax reductions) financed by the toll revenue, congestion pricing can produce many more winners than losers. Although pricing may harm most drivers, no one is *only* a driver. That is, many people in groups 3, 4 and 5 may gain more in their role as residents who receive the added public services than they lose in their role as drivers. The political results of congestion pricing thus depend crucially on how the toll revenue is spent.

For example, consider the possible outcomes for the members of group 3. Suppose they each pay \$100 a month in tolls but save time that is worth only $60 a month. The tolls impose a net cost of $40 a month on these people in their role as *drivers*. Now suppose the toll revenue pays for added public services these drivers think are worth $50 a month in their role as *residents*. The outcome of the congestion tolls and the added services financed by the toll revenues is a net benefit worth $10 a month for all members of group 3, even though the tolls make them worse off in their role as drivers. When we consider the value of added public services financed by the tolls, congestion pricing can similarly make the members of groups 4 and 5 better off. And the members of groups 1 and 2 are also better off even without the added public services. In Stockholm, for example, Eliasson and Mattsson (2006: 618) estimate that congestion pricing would create a net cost per resident of 482 Swedish kroner (SEK) a year before considering the use of the revenue, but a net benefit of SEK 222 a year after considering the use of the toll revenue of SEK 704 a year.

Consider the prospects for congestion pricing in Los Angeles County, which has the worst traffic congestion in the United States (TTI, 2005). Giuliano (1992) argues that in auto-dependent regions such as Los Angeles, congestion pricing will initially make many drivers worse off. The demand for driving in Los Angeles (as in most other urban areas in the US) is highly inelastic, so most people confronted with congestion pricing will end up paying the tolls or driving a less convenient route instead of switching to another travel mode or time. Los Angeles, in other words, has a disproportionate number of people in groups 3, 4 and 5. If we neglect the benefits of the added public services financed by the toll revenue, congestion pricing makes these people worse off. A study of congestion pricing's likely impacts in the Twin Cities made a similar point: for all but two small groups – transit users and affluent drivers – the tolls would exceed the time savings (Anderson and Mohring, 1997). And survey evidence from John Calfee and Clifford Winston (1998) suggests that most drivers in the United States do not value time savings enough to benefit from the reduction in congestion that tolls will achieve.[3]

Conventional wisdom holds that congestion pricing will be adopted only if at the outset it seems to create more winners than losers, and the winners did outnumber the losers in London, Singapore and Stockholm, which have three of the world's most prominent congestion pricing programs. When Singapore introduced congestion pricing in 1975, it had only one car per 16 people, so the tolls fell on a small minority of the population (Cervero, 1998: 171). When London introduced congestion pricing in 2003, only 12 percent of all commuting into the cordoned area was by private car (Transport for London, 2003). Before Stockholm began its trial of congestion pricing in 2006, only 33 percent of the household travel into the toll zone was by car, and 59 percent was by public transit (Armelius and

Hultkrantz, 2006: 167). Because all three cities used the toll revenue to improve public transport, the toll burden fell on the motoring minority while the benefits accrued to the transit-riding majority.

Because motorists are a minority in many cities – especially in developing countries where traffic congestion impedes public transport used by the majority – it might seem surprising that congestion pricing has not been widely adopted; a nonmotoring majority will help to adopt congestion tolls, but it is clearly not sufficient. And because motorists are a majority in the United States, it might seem even more surprising that congestion pricing would ever be adopted. To explain both the absence of congestion pricing in congested cities with a minority of motorists and the prospects of congestion pricing in cities with a majority of motorists, we shall discuss two important political barriers to congestion pricing: *loss aversion* and the *free-rider problem*. We shall then propose ways to overcome these barriers.

Loss Aversion

One explanation for the unpopularity of congestion pricing is that its practical advantages are also political liabilities: tolling is both local and transparent. On a priced road, as drivers pay the tolls they alter their behavior because they face new costs. As the manager of Singapore's system told a journalist, road pricing only works because drivers 'feel the pain' (Baum, 2001). The transparency of congestion pricing makes it prone to loss aversion. Loss aversion is the reluctance to part with a benefit one already has, and the tendency to view a new benefit – even one of equal or greater value – as less desirable than one given up. If avoiding loss is more important than acquiring gain, the phenomenon of loss aversion leads individuals to pay more to keep something they have than they would pay to buy it in the first place, and to fight more to protect an existing benefit than to gain a new one of commensurate value. 'The disutility associated with losing a benefit', as Kahneman et al. (1991: 194) explained, 'is greater than the utility associated with acquiring it'. Or, to quote Adam Smith (1759 III, ii, 176–7) in the *Theory of Moral Sentiments*, 'Pain is . . . in almost all cases, a more pungent sensation than the opposite and corresponding pleasure'.

In the context of congestion pricing, loss aversion suggests that efforts to placate drivers by returning the toll revenue to them will not work. The loss of free access to the roads will weigh more heavily than any benefits of the toll revenue. Even if all the revenue were returned to drivers in the form of lower vehicle registration fees or lower gas taxes, most drivers would probably still view congestion pricing as a loss. What economists consider an acceptable trade is instead rejected as intolerable and unfair.[4]

A toll is a visible and repetitive new cost, while a rebate on a registration fee is an infrequent and hidden benefit – it happens once a year and is buried in the minutiae of a large bill that most people rarely examine. The same could be said about a reduction in the gas tax. Compared to the daily task of paying for road access, a slight decrease in the gas tax would seem like no compensation at all, even if market fluctuations in the price of gas did not swamp any price reduction that results from the tax cut.

Loss aversion helps explain why a majority of people are unlikely to support congestion pricing at the outset. But the loss-aversion literature also suggests that initial resistance is likely to be much stronger than subsequent opposition. Individuals will pay much more in time or energy to keep a benefit than they will to regain that benefit once it is lost.[5] The primary political challenge for congestion pricing is thus not to maximize the number of winners, but rather to overcome initial antagonism to the idea. Once pricing becomes the status quo, its political problems will steadily diminish, because it will benefit from the same political inertia that now works against it.

The Free-rider Problem

In their research on the politics of congestion tolls, Deakin and Harvey (1996: 5–15) note: 'the beneficiaries of pricing often will be harder to mobilize politically than the losers; for example, those who would share the benefits of toll revenues may be a large group but individual benefits may be fairly small'. Loss aversion often prevents drivers from understanding that they could gain (or at least not lose) from congestion pricing. But even when the gains are understood, they are often not large enough to convince individuals to mobilize and lobby for tolls. A free-rider problem emerges: even if most drivers think they would be better off with congestion tolls, no one will be so much better off that they will take the lead to implement the program.

In *The Logic of Collective Action*, Olson (1963) explained the paradox that widespread individual interest does not necessarily lead to group action. 'It does *not* follow', Olson wrote, '[that] because all the individuals in a group would gain if they achieved their group objective, that they would act to achieve that objective, even if they were all rational and self-interested' (1963: 2, emphasis as original). Olson further argued that as a group gets larger, the chances of its engaging in collective action decline, because the average rewards to individual members decline as well. Frozen by free riding, group members pursue their individual interests at the expense of their collective interests.

The inertia of large groups opens the door for what Olson calls 'the exploitation of the great by the small' (p. 73). Small groups are less prone to free riding, are easier to organize, and have a greater incentive to engage in political action because it yields larger rewards to the group's individual members. Thus policies can be adopted when a small but well-organized group of supporters outmaneuvers a large but poorly organized group of opponents.

Client Politics

Wilson's theory of client politics extends Olson's work. The insight of client politics is that small groups can mobilize and triumph politically only when they have a strong incentive to win. Success is determined not by the absolute number of winners and losers, but by the relative ease of collective action, and the extent to which the winners win.[6] Such is the well-documented calculus of light-rail politics. Many rail projects are politically viable in part because their benefits are concentrated among contractors, unions and local politicians, while a large share of their cost is spread widely over all federal taxpayers (Altshuler and Luberoff, 2003; Castelazo and Garret, 2004; Richmond, 2004). The local beneficiaries from a federally subsidized rail project have an incentive to fight for it, while those who pay have little incentive to fight against it, and indeed may not even know they are paying.

Congestion pricing will never enjoy all of urban rail's political advantages, of course, because the costs of pricing are transparent while the costs of rail can be hidden. Drivers on priced roads, unlike the taxpayers who pay for rail transit, will always know how much they are paying. But drivers, like the taxpayers who pay for rail, can be difficult to organize. The same free-rider problem that inhibits drivers from supporting congestion pricing can also forestall their rallying against it. The key to political success for congestion pricing does not lie in turning dispersed costs into dispersed benefits, or in other efforts to engineer widespread support. Congestion pricing will be politically viable when it has well-organized winners who see massive gains, and these massive gains are to be found in the toll revenue.

Previous Revenue Proposals

Goodwin (1989) and Small (1992) have both offered proposals to spend congestion toll revenue in ways designed to maximize political support. Although similar in some respects, their proposals do not share the same logic. Where Goodwin's approach is intended to create constituencies that would benefit from pricing, Small's is intended to prevent opposition from those who would lose.

Goodwin argues that congestion pricing does not suffer from a lack of proponents, but that it does suffer from a perception that the proponents are mutually exclusive of one another. Proponents want the tolls implemented *their* way, which is another way of saying that they will support pricing only if they get the revenue. It follows that pricing loses support as it moves closer to reality, because as potential candidates for the revenue are eliminated, the number of interest groups willing to support it declines.

Goodwin's solution to this dilemma is his 'Rule of Three', which calls for distributing toll revenues in a manner that retains the broadest possible group of supporters. He proposed that a third of the toll revenue be put toward road improvements, a third toward public transport, and a third toward the general fund of the city or state. The Rule of Three is thus intended both to create political beneficiaries and to compensate the travelers who pay the tolls.

Small objected to Goodwin's proposal on the grounds that it devotes too much money to roads and public transportation. Small proposed his own three-way distribution of the revenue: one-third to 'travelers as a group'; another third to reduce general taxes that fund transportation; and the last third put toward new transport services, be they public or private. Specific steps to meet these goals might include lower vehicle license fees and gasoline taxes; reducing the sales and property taxes dedicated to transportation; and the provision of commuting allowances.

Small's plan is at odds with what we know about loss aversion; a variation of his proposed distribution was attempted in 1984, and failed. In 1984 the government of Hong Kong tried to sell a congestion-pricing program by assuring the Hong Kong Automobile Association (HKAA) that tolling would be accompanied by a commensurate reduction in vehicle license fees. But the promise of revenue neutrality convinced neither the HKAA nor the public at large. The HKAA, which is a reasonable proxy for 'drivers as a group', rejected the plan, and Hong Kong did not adopt congestion pricing (Borins, 1988).

Other proposals to allocate toll revenue directly to drivers address the problem of loss aversion, but fail to address the free-rider problem. To spread the benefits over the largest group of people, Kockelman and Kalmanje (2005) suggest that toll revenue be allocated as credits to all licensed drivers, and that the credits be used on priced roads. Drivers would pay out-of-pocket for tolls only if they exceeded their credit allowance, and drivers with unused credits could exchange them for cash. (The transaction costs of collecting and distributing the tolls, however, mean that drivers would get back less than they pay.) Anderson and Mohring (1996) also discussed this sort of distribution as part of a congestion-pricing proposal for the Twin Cities region. But even if a giveback program to all drivers were financially viable, and some of Anderson and Mohring's findings suggest it would not be, spreading the toll revenue around would do little to mobilize drivers to fight for pricing's initial implementation. A credit system or other giveback program combines pricing's dispersed costs with dispersed benefits, and dispersed benefits will not create strong advocates for pricing. Strong advocates for pricing will only be forged from the prospect of concentrated gains.

Goodwin's proposed constituencies do have a reasonable claim to the toll revenue. Public transport seems, at first glance, to be a reasonable claimant for toll revenue, particularly since the pricing programs in London and Singapore both pour the bulk of their revenue into regional transit systems. In the United States, however, where fewer than 3.5 percent of all trips are made by public transport, there are simply not enough riders to make it a politically viable claimant for toll revenue.

We are not suggesting that transit agencies have no stake in debates about congestion pricing. Congestion pricing will be a boon for public transport even if none of the money goes to transit agencies. Priced roads will cause some drivers to switch to transit, which, particularly in the US, needs new riders more than it needs new subsidies.[7] Less congested roads will also help buses move faster, improving the quality of transit service and reducing its high time costs. Small (2005) has laid out a scenario where congestion pricing creates a virtuous circle for public transport even if no toll revenue is put toward service upgrades or improvements. He points out that peak-hour automobile tolls will increase transit ridership, and reduced congestion will speed up public transport that shares the roads with cars. The faster public transport will further increase ridership, and the higher speeds will reduce the cost per ride. Higher ridership and lower costs will enable transit providers to increase service frequency, and the lower costs will allow lower fares, both of which will further increase ridership. As more riders are diverted from cars, congestion is reduced and the virtuous circle continues. Small estimates that congestion pricing in a typical US city could increase bus ridership by 30 percent and increase bus speeds by 9 percent; it could also reduce bus fares by 26 percent. But these benefits will accrue to public transport *only* if congestion pricing is approved, and congestion pricing will *not* be approved if the toll revenue is allocated to public transport. With congestion pricing, public transport will gain not through greater subsidies but through greater ridership and efficiency.

Road improvements, like transit, also appear at first glance to be a good candidate for toll revenue. As with the latter, however, congestion pricing can improve road travel even if none of the toll revenue is invested in road improvements. First, taking a few cars off the road can significantly reduce congestion. By increasing speed and flow, tolling during congested periods can thus have the same effect as adding lanes to freeways (Garrison and Ward, 2000). Second, tolls can reveal where road improvements are most justified, thereby making an often profligate investment process more efficient. At specific bottlenecks, the tolls might be extraordinarily high, and these high tolls will provide an excellent guide for highway investment decisions.

In the United States, where roads are financed primarily through gasoline taxes, congestion pricing can make gas tax expenditures more cost effective by showing where expansions in road capacity are most productive. It may be politically wise to set part of the toll revenue aside for road expansion, however, simply to alleviate suspicion that cities will leave bottlenecks in place to extract maximum revenues from them. But there is little political advantage in dedicating a large stream of toll revenue to road improvements. Doing so is unlikely to reduce drivers' opposition, and even if it does, reducing drivers' opposition to pricing is not the same as convincing them to champion it.

4 CITIES AS CLAIMANTS

The last of Goodwin's three claimants for congestion toll revenue is the city or state general fund. Giving the money to the state fails for the same reason that giving the money to a regional authority would fail: it is unlikely that any state program will be valued as highly as unpriced roads, at least *ex ante*. Cities, however, have the advantage of being well-defined entities with established influence and power. They already have lobbyists and officials whose explicit purpose is to promote their interests, and who can be effective advocates at the state and national levels. Los Angeles, for example, is one of the largest lobbyists in California, and intergovernmental lobbying is one of the state's largest categories of lobbying activity.[8] Cities, counties and municipal leagues all lobby actively at higher levels of government, and studies of local officials, such as city managers, show that they function effectively as de facto lobbyists via their job-related contact with officials in other levels of governments.[9] In contrast to millions of dispersed drivers, cities are already organized and their comparatively small numbers will give them high individual payoffs from the toll revenue – a powerful incentive to collective action. Because local governments are limited in their ability to raise new revenue, they will have a strong interest in making road pricing a reality.

City leaders can influence officials at higher levels of government and also bring along the constituents they represent. Local leaders are attuned to the public goods and services their constituents want, and they can allocate their share of the toll revenue to provide those goods and services. At the local level there is a greater chance that these goods and services will be viewed as a reasonable compensation for loss of free access to the freeways. The rich and poor communities of a region would likely never agree on the proper way to spend congestion revenue. But if a rich community could dedicate its funds to more street cleaning or burying power lines, while the poor community could pay for new parks or after-school programs – and each community felt (correctly) that its programs were being funded largely by other cities' drivers – then some of the political opposition to congestion pricing would evaporate.

It follows that the toll revenue should have minimal earmarks on how to spend the money (on the grounds that each jurisdiction will know best how to spend its own money) but strict auditing requirements (to ensure that the revenue is not misappropriated). It also follows that the uses of toll revenue will vary widely, both within and across regions. Such open-endedness is essential to generating local political support. Spending the toll revenue for a regional purpose like public transit, by contrast, would most likely founder on the heterogeneous preferences of the region's residents.

This brings us to a final implication of our proposal: it can overcome the political cooperation problems in fragmented metropolitan areas. Fragmented metropolitan government creates fiscal disparities and makes regional policies difficult. Because small local governments tend to be internally homogeneous, they can reach consensus more easily about how to spend potential toll revenue. Further, a major problem with fragmented regions is that cities do not have the same resources to finance public services. Our proposal distributes additional revenue among cities in a way that it does not threaten existing resources under local control, such as property taxes.

The cities-as-toll-recipients proposal parts company with most transportation research, where fragmentation is often decried as an obstacle to sound regional policy. The evidence seems mixed, but fragmentation, whatever its merits, seems here to stay, and transportation planners might be better served by turning it to their advantage rather than hoping it will disappear. Gómez-Ibáñez (1992) argued that fragmentation could be an obstacle to congestion pricing. Yet he assumed that congestion toll revenue from a fragmented region would be given to the central city, or to a metropolitan transit agency whose riders are disproportionately central city residents. In such instances the tolls could be interpreted as a tax on suburban commuting and a subsidy to a city government that plays little role in most commuters' lives. Gómez-Ibáñez's point is sound, but the lesson to be drawn is not that fragmentation hurts pricing's political prospects. The lesson is that we cannot distribute toll revenue in fragmented regions in the same way we would in areas with few jurisdictions. The suburbs must not only receive money but also be allowed to spend it on services important to their residents. Many suburbanites have little connection to the center cities in their regions, and we cannot pretend that they will share the preferences of central city residents, or happily donate their toll payments to a jurisdiction that little concerns them. Instead we should allow the multiple governments in the region to spend the revenue in multiple ways.

5 A PRECEDENT: SAN DIEGO COUNTY

We have argued that congestion pricing is unlikely to be politically successful unless powerful claimants benefit from the toll revenue. Even in London, where congestion revenue is spent almost entirely on public transportation, the driving force behind congestion pricing was not Transport for London, the city's public transport agency, but Ken Livingstone, the city's larger-than-life mayor.

An example similar to London can be found in Southern California, in the case of the I15 FasTrak corridor in San Diego County. The FasTrak program converted an existing but underused high occupancy vehicle (HOV) lane into a high occupancy toll (HOT) lane. Unlike a HOV lane, which excludes all vehicles that do not have more than one occupant, a HOT lane allows carpools to travel for free, and allows single-occupant vehicles to travel if they pay a toll. The toll varies with the level of congestion and is adjusted every six minutes.

Converting a HOV lane into a HOT lane is not as politically difficult as introducing fully fledged congestion pricing. Loss aversion is not an obstacle; indeed, by being allowed to buy their way into lanes from which they had previously been excluded outright, solo drivers gain rather than lose options. Yet the free-rider problem still looms large: tolling a lane that runs through multiple jurisdictions requires strong incentives to organize and cooperate, because many people oppose tolls – even tolls on a HOV lane. It is worth examining the political support for creating the HOT lane, particularly if, as Gordon Fielding and Dan Klein (1997) argue, HOT lanes can function as stalking horses for fully priced freeways – that we can toll one lane, and then another, until the gradual expansion of HOT lanes gives us 'congestion pricing, one lane at a time'.

Evans et al. (2007) – on whose research this section is based – show that it was the desire for light rail that was, ironically, the political impetus for the I15 HOT lane. The lane's major proponent was Jan Goldsmith, who in 1991 was mayor of the small city of Poway. In 1991 the San Diego Association of Governments (SANDAG), which is San Diego's regional planning agency, allocated money for light rail service to south San Diego County, but not for the northern cities in the county, citing a lack of funds. Goldsmith wanted transit funds for his city ('we had no money for transit and I was making a big deal of it', he said later) but after meeting with SANDAG representatives he became convinced that the agency really did not have additional transit funding. What the agency did have, however, was access to federal funds to test a HOT lane. Goldsmith and the SANDAG planners decided to propose converting the I15's HOV lanes into HOT lanes – essentially sell off excess HOV space – and then dedicate the revenue to public transportation.

Goldsmith's desire for light rail turned him into a champion of congestion pricing. He campaigned aggressively for the HOT lane, and while he devoted considerable effort to selling the idea to the public (through op-eds and public talks) it is telling that most of his politicking was directed at his fellow elected officials:

> I went to all of my colleagues in San Diego County, the mayors of all the cities affected, the County supervisors, and all of the legislators. I had one-on-one meetings and I would bring some traffic planners along to talk about this project. This was in advance of introducing the legislation. By the time we introduced the legislation, we had support from every elected official in San Diego County whose district was affected.

At the end of 1992 Goldsmith was elected to the State Assembly, where he wrote a bill to permit the HOT lane conversion and began shepherding it through the legislature. The bill had a number of powerful opponents, including Bill Lockyer, who was State Senate President ProTem; Richard Katz, the Chairman of the Assembly Transportation Committee; and the Automobile Club of Southern California. Lockyer had previously killed an effort to put congestion tolls on the Bay Bridge in San Francisco. The Auto Club, with the help of Katz, attached a 'poison pill' amendment to Goldsmith's bill, authorizing congestion tolls on *all* of the I15, not just its HOV lane. Goldsmith was able to beat back both Lockyer and the Auto Club because he had already assembled the support of local politicians. The mayors and legislators in northern San Diego knew he had no intention of tolling all the lanes on the I15. And Goldsmith neutralized Lockyer by arguing that the HOT lane was a matter of local prerogative, not ideology. If all the elected officials in his district wanted to toll solo drivers in a carpool lane and put the money toward public services, why should the state government stand in their way?

Faced with this argument, Lockyer agreed not to oppose the bill. The legislature authorized the HOT lane, and the I15 toll revenue now funds an express bus service, the Inland Breeze, that runs along the I15 into downtown

San Diego. Ridership on the Inland Breeze is low, and most of its riders had been using transit previously, meaning that the bus had little direct impact on congestion. Indirectly, however, the bus probably contributed significantly to reducing congestion, because it provided the motivating force that led elected officials to fight for the variably-priced toll lane. Indeed, the bus's greatest contribution to fighting traffic may have been its role in creating the HOT lane. Congestion pricing was, for Goldsmith, a means to an end, with the end being transit. But the transit was also a means to an end, with the end being congestion pricing. The HOT lane made the bus possible just as the bus made congestion pricing possible. What was important, again, was not only *how* the revenue was spent, but also *who* wanted the revenue.

6 LOS ANGELES COUNTY

We can use Los Angeles County to illustrate how our proposal might work for congestion pricing on all freeways, not just HOT lanes. According to the Texas Transportation Institute's *2005 Urban Mobility Study*, Los Angeles has the worst traffic congestion in the United States, and it has five of the 10 most congested freeway interchanges in the US. Seventy percent of the county's commuters drive alone to work, according to the 2000 Census, and only 7 percent use transit. The county is also highly fragmented: it has 88 city governments of varying size and fiscal capacity.

One way to implement our proposal is to charge congestion tolls on the LA freeways and distribute the resulting revenue to the cities with freeways on a per capita basis. Doing this would create a strong claimant coalition of 66 local governments plus the county. The geography of LA's freeways, however, along with the county's population distribution and the fiscal disparities that exist between its local governments, allows us to adjust our proposal. In Los Angeles we can use toll revenue to advance some equity and environmental goals, without sacrificing political support.

Los Angeles County's 882-mile freeway system passes through 66 of its 88 cities, and also through unincorporated territory.[10] Because the freeway cities and the unincorporated area include 97 percent of the county's population, the claimant coalition is large. It is unlikely, of course, that any toll-revenue distribution formula would be so simple. Both federal and state laws would have to be changed to allow pricing, and like much revenue-generating legislation, a road-pricing bill would doubtless emerge with its share of earmarks and a complicated allocation mechanism.[11] For the sake of illustration, however, imagine a simple system where the entire freeway network is priced and all the revenue goes to the cities with freeways.

Estimates of congestion costs in Los Angeles County vary, but the toll revenue would be substantial by any measure. Using a transportation model calibrated for Southern California, Deakin and Harvey (1996) estimated the annual revenue that would result from congestion tolls in the Los Angeles region: $3.2 billion in 1991, rising to $7.3 billion in 2010.[12] Small (1992) estimated that congestion tolls in Los Angeles would have produced $3 billion, net of collection costs, in 1991. The Texas Transportation Institute (2005) estimated that the total costs of traffic congestion in Los Angeles were $8.4 billion in 1991 and $12.8 billion in 2001.[13]

One striking result of the toll revenue distribution in Los Angeles is how progressive it would be. According to the 2000 Census, the average per capita income in LA County was $20,100 a year in the 66 cities with freeways, and $35,100 a year in the 22 cities without them (see Table 19.1). Congestion tolls will thus shift money from richer cities without freeways (like Beverly Hills) to poorer cities with freeways (like Compton). Deakin and Harvey estimated that higher-income motorists will pay most of the tolls, in part because the highest-income quintile own 3.1 times more cars than the lowest-income quintile and drive 3.6 times more vehicle-miles per day.[14] Because higher-income motorists also drive more during the peak hours, the highest-income quintile will actually pay about five times more in tolls than the lowest-income quintile.[15] High-income drivers will pay to provide public services for low-income people.

If we stretch our definition of freeway cities a bit, the revenue distribution is even more progressive. Los Angeles County has four poor, small cities that do not have freeways within their borders (Cudahy, Huntington Park, La Puente and Temple City) but which are bounded closely by freeways on at least one side. It is reasonable to argue that these cities bear harmful freeway externalities. If we include these four cities among our toll-recipients, the per capita income would be $20,000 a year in the 70 toll-recipient cities, and $47,000 a year in the remaining 18 cities.[16]

Because 9.2 million people live in the 70 toll-recipient cities and the unincorporated area, each $1 billion in congestion tolls will produce about $110 per capita in municipal revenue. If the congestion tolls yield $5 billion a year net of collection costs (the 1991 estimate adjusted for inflation to 2005), they will generate about $550 per capita for the recipient cities. The 70 toll-recipient cities' general revenues averaged $577 per capita in 2001, so the tolls will almost double these cities' general revenues, and the poorest cities will gain the most in proportion to their revenues.[17]

The 20 percent of the population who live in the 33 poorest cities receive 12 percent of the county's income but get 21 percent of the toll revenue. In contrast, the 20 percent of the population who live in the 43 richest cities receive 30 percent of the county's income but get only 17 percent of the toll

Table 19.1 Per capita incomes of cities in Los Angeles County ($ per person per year)

66 cities *with* freeways						22 cities *without* freeways	
City	Income/ capita	City	Income/ capita	City	Income/ capita	City	Income/ capita
Agoura Hills	39,700	El Segundo	34,000	Norwalk	14,000	Avalon	21,000
Alhambra	17,500	Gardena	17,300	Palmdale	16,400	Beverly Hills	65,500
Arcadia	28,400	Glendale	22,200	Paramount	11,500	Bradbury	57,700
Artesia	15,800	Glendora	26,000	Pasadena	28,200	Cudahy	8,700
Azusa	13,400	Hawaiian Gardens	10,700	Pico Rivera	13,000	Hermosa Beach	54,200
Baldwin Park	11,600	Hawthorne	15,000	Pomona	13,300	Hidden Hills	94,100
Bell	9,900	Industry	9,900	Redondo Beach	38,300	Huntington Park	9,300
Bell Gardens	8,400	Inglewood	14,800	Rosemead	12,100	La Habra Heights	47,300
Bellflower	16,000	Irwindale	13,100	San Dimas	28,300	La Puente	11,300
Burbank	25,700	La Canada Flintridge	52,800	San Fernando	11,500	Lomita	22,100
Calabasas	48,200	La Mirada	22,400	San Gabriel	16,800	Malibu	74,300
Carson	17,100	La Verne	26,700	Santa Clarita	26,800	Manhattan Beach	61,100
Cerritos	25,200	Lakewood	22,100	Santa Fe Springs	14,500	Palos Verde Estates	69,000
Claremont	28,800	Lancaster	16,900	Santa Monica	42,900	Rancho Palos Verdes	46,300
Commerce	11,100	Lawndale	13,700	Signal Hill	24,400	Rolling Hills	111,000
Compton	10,400	Long Beach	19,100	South El Monte	10,100	Rolling Hills Estates	51,800
Covina	20,200	Los Angeles	20,700	South Gate	10,600	San Marino	59,200
Culver City	29,000	Lynwood	9,500	South Pasadena	32,600	Sierra Madre	41,100
Diamond Bar	25,500	Maywood	8,900	Torrance	28,100	Temple City	20,300
Downey	18,200	Monrovia	21,700	Vernon	17,800	Walnut	25,200
Duarte	19,600	Montebello	15,100	West Covina	19,300	West Hollywood	38,300
El Monte	10,300	Monterey Park	17,700	Westlake Village	49,600	Whittier	21,400
				Average	20,100	Average	35,100

Note: The two groups' average incomes are weighted by the cities' populations.

Source: US Census 2000.

revenue. The 1 percent of the population who live in the eight richest cities receives 4 percent of the county's income and no toll revenue.

Given this distribution, it is reasonable to ask whether high-income motorists, who probably represent the most politically influential segment of the county, would thwart any attempt to price the roads. Of course this is possible, but high-income motorists also have a high value of time. While they may disproportionately pay the tolls, they will also disproportionately benefit from reduced congestion; indeed, the research cited above suggests that high-income motorists are one of the few groups who will benefit immediately after tolling begins. Like all motorists, many affluent drivers will doubtless oppose tolls before they are put in place, but this opposition again points to the need for powerful claimants in the early stages of a political campaign. Once the tolls are operational, it seems unlikely that wealthy drivers will want or be able to derail them.

7 MINNEAPOLIS–ST. PAUL

We can also use Minneapolis–St. Paul to illustrate how distributing toll revenue to cities would affect the political calculus of congestion pricing. The Twin Cities region, which has 13 governments per 100,000 people, is one of the most fragmented metropolitan areas in the United States. Anderson and Mohring (1996) estimated that congestion tolls could generate about $250 million a year in the Twin Cities, or about $90 per capita per year for the 2.7 million residents.[18] Congestion tolls would yield much less revenue in the Twin Cities than in Los Angeles because of the smaller population and lower levels of congestion.[19] And in contrast to Los Angeles, distributing the toll revenue to cities with freeways would not significantly reduce fiscal disparities in the Twin Cities. The average annual income is $26,500 per capita in the 70 cities with freeways and $27,700 in the 112 cities without freeways. The fiscal effects in Los Angeles, where every poor city could receive toll revenue and none of the richest cities would receive anything, would not be repeated in the Twin Cities. Nor would distributing the revenue to cities with freeways create a majority coalition of local governments in support of pricing: many more cities lack freeways than have them.

The Twin Cities region does have, however, an existing system of sharing tax revenue to reduce fiscal disparities, and congestion revenue could be used to augment or replace this existing redistribution mechanism. Under the region's Metropolitan Fiscal Disparities Act, 40 percent of each city's growth in assessed value of commercial and industrial property since 1971 is placed in a seven-county regional pool. The assessed value of the regional tax-base pool is taxed at a uniform rate of 1 percent, and the revenue is distributed to cities according to their population and fiscal capacity. In 2004, the Fiscal Disparities Act transferred $74 million from 51 'contributor' cities to 131 'recipient' cities. The average per capita income was $32,300 in the contributor cities and $23,900 in the recipient cities. The contributor cities paid an average of $79 per capita into the pool but the recipient cities received an average of only $41 per capita because the total population of the recipient cities (1.8 million) was almost twice that of the contributor cities (934,000).[20]

The Fiscal Disparities Act has succeeded in reducing regional fiscal disparities (Hinze and Baker, 2005) but does so by transferring property tax revenue from cities with greater commercial and industrial property growth to cities with less. Congestion tolls, by contrast, can reduce fiscal disparities by leveling up, not down. Rather than taking from one government and giving to another, tolls take money from drivers (to reduce congestion) and give it to tax-poor cities. No city is forced to surrender its existing revenue stream.

Suppose the congestion toll revenue of $250 million a year were used to replace the $74 million a year now redistributed through the Fiscal Disparities Act. The 51 contributor cities would no longer pay $79 per capita into the tax-base pool, yet the 131 recipient cities would receive $136 per capita in congestion tolls, or $95 per capita more than they now receive from the pool.[21] Using congestion tolls to replace tax-base sharing would therefore help all cities in the region, but would help the poorer cities more than the richer ones.

What the Twin Cities example shares with the Los Angeles and I15 examples is the logic of using municipal government as claimants. This logic is the foundation that provides the political support for congestion pricing. The actual congestion-pricing program can be built atop this foundation, incorporating equity or environmental goals that are suitable to the region in question, so long as whatever additions are made do not undermine the political foundation or cause it to collapse. It is unlikely that any two congestion-pricing programs will look alike; what they will share is the initial prospect of enough revenue, for enough cities, to generate support for variably priced roads.

8 NEW YORK AND SAN FRANCISCO

Finally, we can use two other cities as examples to illustrate our proposal. First consider New York City, where transportation economists have long advocated congestion tolls for the bridges and tunnels between the city's five boroughs. Elected officials in the outlying boroughs such as Brooklyn and Queens strongly oppose peak-hour tolls because most of these tolls would be paid by their constituents who drive into Manhattan. 'We look on it as a tax on the other boroughs [outside Manhattan]', said Councilman David Weprin from Queens. The president of the Queens Chamber of Commerce echoed the sentiment: 'Residents and businesses of Queens, Brooklyn, Staten Island and the Bronx . . . would suffer the most from the plan'.[22]

Suppose, however, that the congestion toll revenues are returned to each borough in proportion to the share of the toll revenues paid by its residents. If 35 percent of the New York City residents who pay the congestion tolls live in Queens, for example, 35 percent of the toll revenues will return to Queens for added public spending in Queens. The E-Z Pass electronic toll system or a sample of license plates can determine the borough residence of the toll payers. Because drivers who live outside New York City will also pay congestion tolls that will be divided among the five boroughs, each borough will receive more toll revenue than its residents pay. Long Island residents who drive into Manhattan, for example, will pay congestion tolls that all the boroughs will share.

Each borough can decide how to spend its own toll revenue. Brooklyn might spend some of its money to clean its subway stations, while Staten Island would want to repair its sidewalks. If each borough can spend its toll revenue on the added public services it values the most, returning toll revenues to the five boroughs will create the greatest political support for congestion pricing.

Our last example is congestion pricing on the bridges of the San Francisco Bay Area. The Golden Gate and Bay Bridges are both tolled, but not at a rate that varies with congestion. A logical first step toward congestion pricing in the Bay Area would be to convert the existing bridge tolls into congestion tolls. The suburban communities who supply most of the bridge commuters would doubtless object because the bridge tolls would become, as Gómez-Ibáñez (1992) warned, a penalty on suburban commuting. If, however, each city in the region received all the added toll revenue paid by its own residents, the cities' elected officials might support congestion pricing. Again, FasTrak electronic toll data or a sample of the license plates of the cars paying the tolls would generate an accurate picture of the toll payments from drivers in each Bay Area city. And as in New York, because some portion of the daily traffic also originates from outside the Bay Area, the toll revenues would exceed what the Bay Area residents pay. Because the bridges are already tolled, the collection costs for the new congestion tolls would be minimal, and the system could be implemented quickly.

9 CONCLUSION

'Policy makers do not just happen to create inefficiencies', Winston and Shirley (1998: 68) wrote. 'When economists estimate large welfare losses stemming from public policies as if the losses were simple oversights that officials could correct by paying closer attention to what they were doing, it is the economists, not the officials, who are not paying attention'.

Economists frustrated by congestion pricing's lack of political support should keep Winston and Shirley's admonition in mind. Policy makers' great sin of omission – their failure to price the roads – is not the result of senseless intransigence, or of their inability to 'get it'. Congestion pricing looks good only from an economic perspective. Politically it looks risky and possibly disastrous. We cannot assume that people will vote for congestion pricing simply because it is economically efficient. The solution is not to make drivers want congestion pricing. Good ideas require advocates, and successful advocates are rarely those who pay the costs. Only the prospect of significant rewards will create strong advocates. Most discussions of congestion pricing's political acceptability revolve around using the toll revenue to buy the acquiescence of drivers, but acquiescence will not generate strong political support, and it is in any event highly improbable.

Even if motorists think that pricing will benefit them, they are unlikely to organize and crusade for it. The absence of popular support does not, however, condemn congestion pricing to the fate of being often discussed but rarely tried. The idea that a policy cannot be approved in the absence of popular support is at odds with the way policies are actually advanced. Not every proposed policy lends itself to initial popularity, and some long-standing policies have never been popular at all. But a policy that will not be popular at the outset cannot be marketed as though it will be popular. Congestion pricing cannot be sold as a policy that harms no one, or even as a policy that helps everyone a little. It can, however, be positioned as a policy that will benefit important political actors a lot. Its success depends, to paraphrase Machiavelli, not on convincing those who benefit from the status quo, but on finding others who will 'do well under the new order of things'.

We argue that earmarking the toll revenue can make congestion pricing politically successful. We do not mean conventional earmarks for specific *programs* or *purposes* such as public transit or road improvements. Instead, we mean earmarking the revenue for specific *places* and *people*. We contend that the toll revenue should be earmarked for cities, preferably the cities that are penetrated by the freeways. Cities are well organized and large enough to be powerful, but small enough to engineer consensus among their constituents about how to spend the money. The toll revenue can advance both environmental and equity goals, provided that these goals do not undermine the political incentives for local governments to pursue congestion pricing. In Los Angeles, congestion pricing revenue could be used to compensate cities for the various environmental and public health costs the freeways bring. We believe similar, although probably not identical, strategies could be adopted in other regions.

The overriding factor in our argument, however, is not abstract fairness but political calculation. Arguments can be made, on fairness grounds, for any number of claimants to congestion pricing's revenue. But no one will get the revenue if congestion prices do not exist. Just as the first goal of any politician must be to get elected, the first goal of any toll revenue distribution must be to secure the *initial approval* of congestion pricing. For this reason the path to congestion pricing does not go through transit agencies or highway bureaucracies, and it does not involve efforts to buy off motorists. Rather it involves igniting the self-interest of cities. Only when it offers concentrated benefits to strong political forces will anyone rise to fight for congestion pricing.

NOTES

1. Both experimental and empirical research in behavioral economics shows that minor changes in description or presentation can dramatically alter behavior, even in settings where the consequences of the decision are large and the individuals making the choice are experienced and sophisticated. For an example, see Bertrand et al. (2005).
2. See Tullock et al. (2002). Tullock allows that rent seeking can have social benefits but frequently does not.
3. For a dissenting view, see Small and Kazimi (1995).
4. In formal terms, loss aversion implies a nonreversible indifference curve. If an individual holds benefit x and is indifferent to trading it for benefit y, then theoretically he/she should also be indifferent when holding benefit y to trading it for x. If loss aversion is present this reversibility does not hold, and the indifference curves actually intersect. See Knetsch (1989); Kahneman et al. (1991). Fielding and Klein (1997) discuss loss aversion specifically with respect to congestion pricing. Kahneman et al. do not discuss pricing *per se*, but argue that loss aversion can 'contribute to the extraordinarily high demand for personal compensation for agreeing to the loss of a public good' (1990, p. 1325).
5. This is known as the difference between willingness to pay and willingness to accept. See Kahneman et al. (1991), and Haneman (1991).
6. Essentially, client politics takes into account the varying intensity of preferences. The desire for concentrated gain – or the desire to avoid concentrated loss – can mobilize minority coalitions. The persistence of farm subsidies and affirmative action, as well as the inability of Congress to pass more stringent gun control laws, have all been cited as minority triumphs in American politics.
7. Since the mid-1970s, for instance, rail transit has consistently received half of all public money spent on surface transportation in US urban areas, even though it carries less than 2 percent of all urban passenger movement (and no freight) (Altshuler and Luberoff, 2003).
8. See records from the California Secretary of State's CalAccess program: http://cal-access.ss.ca.gov/Lobbying/Employers/.

9. See Cammisa (1995); Agranoff and McGuire (1998); Marlowe (2003).
10. Like a city, Los Angeles County would receive toll revenue based on the population of the unincorporated area.
11. In practice, the formula for distributing the toll revenue might resemble the federal formulas for distributing gasoline tax revenues to states. It is also unlikely that any congestion-pricing plan would be implemented broadly across a region. Introducing freeway tolls incrementally, perhaps first in high-occupancy vehicle (HOV) lanes, and then across entire freeways would be more realistic (Fielding and Klein, 1997).
12. Deakin and Harvey (1996, Tables 7–14 and 7–18).
13. Small (1992, p. 371). To lend perspective to these estimates, consider that LA County collects $1.2 billion in gas taxes annually. Even if collection costs are high, the tolls can still be worthwhile. Because they reduce traffic congestion, the tolls save time for motorists and they reduce air pollution, accidents and fuel consumption. In contrast, most other sources of public revenue create deadweight losses. Several economists have estimated that each extra $1 raised by taxation *increases* other costs in the economy by about 30 cents (Drèze, 1995, p. 114). Congestion tolls can thus increase efficiency in two ways: first by reducing the cost of transportation, and second by raising enough revenue so that cities can reduce taxes that distort the incentives to work, save and invest.
14. Deakin and Harvey (1996, Tables 8–1 and 8–3). At the national level, in 2002 the highest-income quintile of households owned 2.9 times more cars than the lowest-income quintile (US Bureau of Labor Statistics, 2004, Table 1). It is also worth pointing out that although congestion tolls are regressive, so too are most other proposals for fighting congestion. The gas tax, which finances most highway and some transit spending, is regressive, and sales taxes to increase transit spending are also regressive.
15. Deakin and Harvey (1996, Table 8–6). And because men are more likely than women to drive in congested conditions, men will also pay more in tolls (ibid., Table 8–7).
16. Removing the four poorest cities from the 'without freeways' group sharply increases the per capita income of the 18 remaining cities because the four poorest cities have large populations while most of the richer cities have small populations. Avalon, which would be the poorest remaining city without a freeway, is on Catalina Island – 26 miles off the coast – and would be unaffected by the congestion tolls.
17. The cities' general revenues are taken from the California State Controller's Office, *Cities Annual Report, Fiscal Year 2000–2001*. General revenues are defined as revenues that cannot be associated with any particular expenditure; examples include property taxes, sales taxes and business license fees. General revenues do not include fees and charges for direct services, such as the revenue from municipally owned electric utilities. The population of Los Angeles County is 9.5 million, of whom 990,000 live in unincorporated areas.
18. Anderson and Mohring (1996) were interested in the distribution of the toll revenue and how the revenue could be used to overcome political aversion. They estimated that Twin Cities drivers would pay $940,000 a day in peak-hour tolls (1996, Table 11). We have adjusted this for inflation up to $1,000,000 a day for our hypothetical example. To reach $250 million annually, we multiplied $1 million × five workdays × 50 weeks per year. Anderson and Mohring proposed a tolling scheme that encompasses 1,200 road miles, which includes some surface roads, but their estimated collections are generated by expressway links, on-ramps and arterial connections to expressways.
19. The Twin Cities ranks 18th in the Texas Transportation Institute's *2005 Urban Mobility Study* (TTI, 2005).
20. We calculated average payment and distribution based on the net distribution of Fiscal Disparities Act transactions.
21. The per capita distribution of toll revenues would be $250 million divided by 1.8 million people living in recipient cities.
22. *New York Daily News*, 'Biz panel rips congestion pricing plan', 2 March 2006.

REFERENCES

Agranoff, R. and M. McGuire (1998), 'The intergovernmental context of local economic development', *State and Local Economic Development Review*, **30**(3): 150–64.

Altshuler, A. and D. Luberoff (2003), *Mega-Projects*, Washington, DC: Brookings Institution.

Anderson, D. and H. Mohring (1996), *Congestion Costs and Congestion Pricing for the Twin Cities*, Report no. 96–32, Minnesota Department of Transportation.

Anderson, D. and H. Mohring (1997), 'Congestion costs and congestion pricing', in D.L. Greene, D.W. Jones and M. Delucchi (eds), *Full Costs and Benefits of Transportation*, Berlin: Springer, pp. 315–36.

Armelius, H. and L. Hultkrantz (2006), 'The politico-economic link between public transport and road pricing: an *ex-ante* study of the Stockholm road-pricing trial', *Transport Policy*, **13**(2), 162–72.

Baum, D. (2001), 'The ultimate jam session', *Wired*, **9**(11), www.wired.com/wired/archive/9.11/singapore_pr.html.

Bertrand, M., D. Karlin, S. Mullainathan, E. Shafir and J. Zinman (2005), 'What's pyschology worth? A field experiment in the consumer credit market', National Bureau of Economic Research, Working Paper 11892, http://ideas.repec.org/p/nbr/nberwo/11892.hml.

Borins, S. (1988), 'Electronic road pricing: an idea whose time may never come', *Transportation Research A*, **22**(1), 37–44.

Calfee, J. and C. Winston (1998), 'The value of automobile travel time: implications for congestion policy', *Journal of Public Economy*, **69**, 83–102.

Cammisa, A.M. (1995), *Governments as Interest Groups*, London: Praeger.

Castelazo, M. and T. Garret (2004), 'Light rail: boon or boondoggle?', *The Regional Economist*, July, 12–13.

Cervero, R. (1998), *Transit Metropolis*, Washington, DC: Island Press.

Deakin, E. and G. Harvey (1996), *Transportation Pricing Strategies for California: An Assessment of Congestion, Emissions, Energy and Equity Impacts: Final Report*, California Air Resource Board, Sacramento, CA.

Drèze, J. (1995), 'Forty years of public economics: a personal perspective', *Journal of Economic Perspectives*, **9**(2), Spring, 111–30.

Eliasson, J. and L. Mattsson (2006), 'Equity effects of congestion pricing: quantitative methodology and a case study for Stockholm', *Transportation Research A*, **40**, 602–20.

Evans, A., M. Gougherty, E. Morris and M. Smirti (2007), 'Politics, public opinion, and project design in California', presented at the 86th Annual Meeting of the Transportation Research Board, Washington, DC.

Fielding, G.J. and D.B. Klein (1997), 'Hot lanes: introducing congestion-pricing one lane at a time', *Access*, **11**, Fall, 10–15.

Garrison, W.L. and J.D. Ward (2000), *Tomorrow's Transportation: Changing Cities, Economies and Lives*, Boston, MA: Artech House.

Giuliano, G. (1992), 'An assessment of the political acceptability of congestion pricing', *Transportation*, **19**(4), 335–58.

Gómez-Ibáñez, J. (1992), 'The political economy of highway tolls and congestion pricing', *Transportation Quarterly*, **46**(3), 343–60.

Goodwin, P. (1989), 'The Rule of Three: a possible solution to the political problem of competing objectives for road pricing', *Traffic Engineering and Control*, **30**(10), 495–7.
Goodwin, P. (1997), 'Solving congestion: Inaugural Lecture of the Professorship of Transport Policy, University College London', http://www.cts.ucl.ac.uk/tsu/pbginau.htm, Accessed 30 April 2006.
Haneman, M. (1991), 'Willingness to pay and willingness to accept: how much can they differ?', *American Economic Review*, **81**(3), 647–53.
Hinze, S. and K. Baker (2005), *Minnesota's Fiscal Disparities Programs: Twin Cities Metropolitan Area and Iron Range*, report, Minnesota State House Research, January.
Kahneman, D., J.L. Knetsch and R. Thaler (1990), 'Experimental tests of the endowment effect and the Coase theorem', *Journal of Political Economy*, **98**(6), 1325–48.
Kahneman, D., J.L. Knetsch and R. Thaler (1991), 'Anomalies: the endowment effect, loss aversion, and the status quo bias', *Journal of Economic Perspectives*, **5**(1), 191–206.
Knetsch, J. (1989), 'The endowment effect and evidence of nonreversible indifference curves', *American Economic Review*, **79**(5), 1277–84.
Kockelman, K.M. and S. Kalmanje (2005), 'Credit-based congestion pricing: a policy proposal and the public's response', *Transportation Research A*, **39**, 671–90.
Marlowe, J. (2003), 'City management and the myth of policy neutrality', Paper presented at the Midwest Political Science Association Annual Meeting, Chicago, IL, 3–6 April.
Olson, M. (1963), *The Logic of Collective Action*, Cambridge, MA: Harvard University Press.
Richmond, J. (2004), *Transport of Delight: The Mythical Conception of Rail Transit in Los Angeles*, Akron, CA: University of Akron Press.
Small, K. (1992), 'Using the revenues from congestion pricing', *Transportation*, **19**(4), 359–82.
Small, K. (2005), 'Road pricing and public transit: unnoticed lessons from London', *Access*, **26**(3), 10–15.
Small, K. and C. Kazimi (2005), 'On the costs of emissions from motor vehicles', *Journal of Transport Economics and Policy*, **29**(1), 7–32.
Smith, Adam (1759), *Theory of Moral Sentiments*, D.D. Raphael and A.L. Macfie (eds), Indianapolis: Liberty Fund.
Texas Transportation Institute (TTI) (2005), *2005 Urban Mobility Study*, http://mobility.tamu.edu/ums/report/.
Transport for London (2003), *Impacts Monitoring: First Annual Report*, http://www.tfl.gov.uk/tfl/cclondon/cc_monitoring-1st-report.shtml.
Tullock, G., A. Seldon and G. Brady (2002), *Government Failure*, Washington, DC: Cato Institute.
US Bureau of Labor Statistics (2004), *National Compensation Survey: Occupational Earnings in the United States*, Washington, DC: US Bureau of Labor Statistics.
US Department of Transportation, Bureau of Transportation Statistics (2003), *Transportation Statistics Annual Report*, BTS03-06, Washington, DC.
Wachs, M. (1994), 'Will congestion pricing ever be adopted?', *Access*, **4**, 15–19.
Wilson, J.Q. (1980), *The Politics of Regulation*, New York: Basic Books.
Winston, C. and C. Shirley (1998), *Alternate Route: Toward Efficient Transportation*, Washington, DC: Brookings Institution.

Index